Tourism Marketing in Bangladesh

Tourism is often a key driver of economic growth in many countries. The recent upward trends of tourism and hospitality education in higher academic institutions in Bangladesh suggests a growing tourism sector. Very little has been written on Bangladesh's tourism industry. This is the first edited volume published from an international publisher which looks at this industry and how it has developed and flourished. The book begins by looking at tourism policy planning and provides a comprehensive overview of topics from tourism products and services in Bangladesh to how they are being marketed. It also discusses how private and public tourism institutions can address future long-term trends.

This book will appeal to those interested to learn more about developing tourism industry in emerging economies and may provide invaluable lessons from Bangladesh's experience and success.

Dr Azizul Hassan is a member of the Tourism Consultants Network of the UK Tourism Society. Hassan's areas of research interest are technology-supported marketing for tourism and hospitality, immersive technology application in the tourism and hospitality industries, technology-influenced marketing suggestions for sustainable tourism and the hospitality industry in developing countries. Hassan has authored over 100 articles and book chapters in leading tourism outlets. He is also part of the editorial team of 15 book projects from Routledge, Springer, CAB International and Emerald Group Publishing Limited. Hassan is a regular reviewer of *Tourism Management, Journal of Hospitality and Tourism Management, Tourism Analysis, The International Journal of Human Resource Management, Journal of Ecotourism, Journal of Business Research, e-Review of Tourism Research (eRTR), International Interdisciplinary Business-Economics Advancement Journal, International Journal of Tourism Cities, Heliyon* and *Technology in Society.*

Routledge Advances in Management and Business Studies

For more information about this series, please visit: www.routledge.com/Routledge-Advances-in-Management-and-Business-Studies/book-series/SE0305

Tourism Marketing in Bangladesh

An Introduction

Edited by Azizul Hassan

Routledge
Taylor & Francis Group

LONDON AND NEW YORK

First published 2021
by Routledge
2 Park Square, Milton Park, Abingdon, Oxon OX14 4RN

and by Routledge
605 Third Avenue, New York, NY 10017

First issued in paperback 2022

Routledge is an imprint of the Taylor & Francis Group, an informa business

Publisher's Note
The publisher has gone to great lengths to ensure the quality of this reprint but points out that some imperfections in the original copies may be apparent.

British Library Cataloguing-in-Publication Data
A catalogue record for this book is available from the British Library

Library of Congress Cataloging-in-Publication Data
A catalog record for this book has been requested

ISBN 13: 978-0-367-55546-7 (pbk)
ISBN 13: 978-0-367-44042-8 (hbk)
ISBN 13: 978-1-003-00724-1 (ebk)

DOI: 10.4324/9781003007241

Typeset in Galliard
by Apex CoVantage, LLC

Contents

viii *Contents*

PART 11
Future trends, implications and challenges   315

18 Potentials of tourism products and services in Bangladesh   317
AZIZUL HASSAN AND HAYWANTEE RAMKISSOON

Index   329

Figures

Tables

Contributors

Md. Nekmahmud is a PhD research fellow in the Doctoral School of Management and Business Administration, Szent Istvan University, Budapest, Hungary, Europe. He was awarded the CAS-Belt and Road master research fellowship programme scholarship in Corporate Management at the University of Chinese Academy of Sciences funded by the world's top research institution Chinese Academy of Sciences, China, and Sino-Danish Centre, Denmark. Moreover, he has completed his MBA in marketing from Begum Rokeya University, Rangpur, Bangladesh. Recently, he has developed a green business value chain process and green two-step manufacturing process, which was published in *Sustainable Production & Consumption*, Elsevier. Nekmahmud published some research works in some prominent scientific journals and attended several international conferences as a paper presenter. He attended the Ninth World Science Forum 2019 conference in Europe. He published book chapters about "Competitiveness in Emerging Markets" in Springer Nature. He involved himself as a responsible reviewer in *Journal of Cleaner Production*, *Journal of Foodservice Business Research*, *SAGE Open*, *International Journal of Organization Theory & Behaviour* and *Future Business Journal*. His research interest areas include green consumer behaviour, environmental marketing, sustainable business, ecotourism, renewable energy and circular economy.

Dr Maria Fekete Farkas is a full professor and head of the Department of Microeconomics at the Faculty of Economics and Social Sciences, Szent Istvan University, Hungary. Her research areas are sustainable development, behaviour economics, new market structures and pricing and the economic and social effects of industry 4.0. She is a member of the organizing committees of several international conferences and serves some international journals as a member of the editorial board, reviewer and author.

Dr Azizul Hassan is a member of the Tourism Consultants Network of the UK Tourism Society. Hassan's areas of research interest are technology-supported marketing for tourism and hospitality, immersive technology application in the tourism and hospitality industry, technology-influenced marketing suggestions for sustainable tourism and hospitality industry in developing countries.

Hassan authored over 100 articles and book chapters in leading tourism outlets. He is also part of the editorial team of 15 book projects from Routledge, Springer, CAB International and Emerald Group Publishing Limited. Hassan is a regular reviewer of *Tourism Management, Journal of Hospitality and Tourism Management, Tourism Analysis, The International Journal of Human Resource Management, Journal of Ecotourism, Journal of Business Research, eReview of Tourism Research (eRTR), International Interdisciplinary Business-Economics Advancement Journal, International Journal of Tourism Cities, Heliyon* and *Technology in Society*.

Jameni Jabed Suchana is Assistant Professor in the Department of Tourism and Hospitality Management at the University of Dhaka. On a broad spectrum, her research interest focuses on tourism development in Bangladesh. Suchana is also keen about researching on tourists' behaviour in the context of various types of tourism such as rural tourism, agri-tourism, sustainable tourism, community-based tourism and so on.

Uchinlayen received a bachelor of business administration and master of business administration in tourism and hospitality management from University of Dhaka and currently teaches at the University of Dhaka as full-time faculty of tourism and hospitality management as assistant professor. His research interests focus on tourism, hospitality, travel agency and tour operation, rural tourism, environment and sustainability and hospitality marketing. He is directly involved in some indigenous students' associations as an advisor on a voluntary basis.

Dr Muhammad Shoeb-Ur-Rahman is Assistant Professor of Tourism and Hospitality Management at University of Dhaka, Bangladesh. He holds a PhD in Tourism Management from Lincoln University, New Zealand and MBA in Tourism and Hospitality Management from University of Dhaka. His research interests include sustainable tourism, tourism management, tourism planning and policies, tourism governance, crisis and resilience in tourism systems, and destination development. Dr Rahman has presented his papers in reputed international tourism conferences and published 13 refereed papers and two book chapters until now. A couple of his notable works concentrating on community governance approach for sustainable tourism development in a developing country context and immediate responses to a global pandemic by the hospitality industry of Bangladesh are under review process in top-tier tourism journals. He is a regular reviewer of a few academic journals including *Tourism Review International*, which is an ABDC ranked journal.

Dr Nor Aida Abdul Rahman is an Associate Professor at Universiti Kuala Lumpur, Malaysian Institute of Aviation Technology (UniKL MIAT) in Subang, Selangor, Malaysia. Currently, she serves as Head of Aviation Management at Universiti Kuala Lumpur. She has worked as internal and external trainer in management, supply chain, halal logistics and postgraduate research. She has written three books on postgraduate research. Her research work has appeared in several reputable academic journals, books and refereed conference proceedings. She earned a PhD degree in supply chain management from Brunel University, London, UK. She is also serving as an external academic advisor in college, a chartered member for Chartered Institute of Logistics and Transport Malaysia (CILTM), HRDF Certified Trainer, Chairman (Academic Committee) for Malaysian Association of Transportation, Logistics and Supply Chain Schools (MyATLAS), Vice President (Research Journal) for Institute for Research in Management and Engineering UK (INRME), JAKIM Halal Certified Trainer, UniKL Halal Professional Board and a member of Academy of Marketing, UK.

Asma Akter Akhy holds BSS and MSS (first position) degrees in sociology from University of Chittagong, Bangladesh. Currently she is working as an assistant professor. She has a teaching experience of eight years. She worked as a guest lecturer in Chittagong College under national university of Bangladesh from 2011 to 2013. Her research interest includes development and environmental issues of sociology and economics. She published a good number of research papers in national and international journals.

Mallika Roy is a PhD candidate and teaching assistant at the Department of Economics and Finance of City University of Hong Kong. She achieved the Prestigious Prime Minister Gold Medal Award Bangladesh for her scholastic academic result. She is working as an assistant professor (on study leave) at the Department of Economics, University of Chittagong, Bangladesh. Her research interests include macroeconomics, economic growth and development. She has a teaching experience of eight years. She worked in Islamia University College and BGC Trust University Bangladesh as a lecturer. After joining University of Chittagong, she worked at University of Professional (BUP) Bangladesh, Premier University and BGC Trust University Bangladesh as an adjunct faculty. She published a good number of her research papers in national and international journals. She also published several newspaper articles.

Ayesha Afrin is a PhD student in Department of Chemistry, University of Chittagong, Bangladesh. Afrin is assistant professor in Department of Applied Chemistry and Chemical Engineering in the same university from where she received her bachelor's degree, master's degree and MPhil degree. Her main research area is analytical chemistry. She has published articles in national and international journals.

Shah Alam Kabir Pramanik completed MBA and BBA (major in marketing) from Rajshahi University, Bangladesh. Now, he is working as a lecturer, Department of Marketing, Islamic University, Kushtia, Bangladesh. He was the senior lecturer and programme coordinator of the Department of Business Administration, Daffodil International University, Bangladesh. He has teaching experience of almost five years at the university level. His teaching and research interests lie in the area of marketing, consumer behaviour, tourism and hospitality marketing, customer relationship marketing (CRM) and strategic marketing. His total seven articles are published widely in international and national journals. His three refereed conference proceedings and five articles are accepted in different journals for publication. He has expertise in structural equation modelling (SEM), AMOS, SPSS and SmartPLS. He also writes in national daily newspapers like *The Financial Express, The Independent* and so on. He is trying to develop a framework for sustainable development of tourism in Bangladesh. He is also focusing on the development of CRM framework to cope up with the digital technology in the financial services industry.

Md. Rakibul Hafiz Khan Rakib completed MBA and BBA in marketing from Rajshahi University, Bangladesh and obtained first position in both the examinations. Now, he is working as a lecturer at Department of Marketing, Begum Rokeya University, Rangpur, Bangladesh. He is the coordinator of MBA (Professional) sixth batch of the Department of Marketing. He has almost three years of teaching experience at the university level. His areas of interest for teaching and research include marketing, consumer behaviour, entrepreneurship development, tourism and hospitality marketing and customer relationship marketing (CRM). His two articles are published in peer-reviewed international and national journals. His one refereed conference proceedings and five articles are accepted in different journals for publication. He has expertise in SPSS, structural equation modelling (SEM) and AMOS. Currently he is trying to develop a framework for sustainable tourism development in Bangladesh.

Dr Md Aslam Mia holds a PhD in development economics from University of Malaya, Malaysia. He is currently a senior lecturer at the School of Management (SOM), Universiti Sains Malaysia (USM). Dr Mia has published around 30 articles related to productivity and efficiency of financial institutions, market structure, microfinance and urban economics in internationally reputed peer-reviewed journals, including the *European Journal of Development Research, Business History, Journal of the Asia Pacific Economy, Singapore Economic Review, Journal of Cleaner Production, Social Indicators Research, International Journal of Social Economics, Cities, Economic Analysis and Policy, Strategic Change* and others. He also presented his research findings over 15 national and international conference/seminar during the last five years. Dr Mia also conducted over 15 workshops/seminar on various research issues/techniques/tools both in and outside of Malaysia. Due to his significant contribution in research and publication, Dr Mia received the PhD Candidate with Highest Impact Publications award by the University of Malaya in 2018.

Dr Célia M. Q. Ramos graduated in computer engineering from the University of Coimbra in Portugal, obtained her master in electrical and computers engineering from the Higher Technical Institute, Lisbon University, and a PhD in econometrics in the University of the Algarve (UALG), Faculty of Economics,

Portugal. She is a professor at the School for Management, Hospitality and Tourism, also in the UALG, where she lectures mainly in information systems. Current research interests include tourism information systems, electronic tourism, business intelligence tools, digital marketing and panel data models. She is a researcher at the Centre for Tourism, Sustainability and Well-being (CinTurs).

Md. Sohel Rana is currently pursuing a PhD in the Faculty of Business and Accountancy, University of Malaya, Malaysia. He holds bachelor and master's degrees in business administration with a major in management studies from the University of Rajshahi, Bangladesh. He has expertise in microfinance and financial inclusion, Islamic microfinance, poverty and development economics, urban economics, finance and banking, productivity and efficiency, tourism, insurance and environmental economics. He has already published a book chapter and eight journal articles in the *Web of Science* and *Scopus* indexed journals. Currently, he is working as a research assistant in the University of Malaya, Malaysia.

Mohammad Fakhrul Islam is an assistant professor in the Department of Business Administration at Stamford University of Bangladesh. Prior to that, he served as a faculty member for Bangladesh University of Business and Technology (BUBT). He has seven years active teaching experience. He has more than ten scholarly research articles in different reputed national and international journals. Mr Islam has also presented eight research papers in different international conferences. Mr Islam has expertise in human resource management, strategic management and organizational behaviour. He holds BBA and MBA degrees with distinctions in management from Rajshahi University, Bangladesh. He also holds ABIA and PGDHRM degrees as well. Mr Islam is actively engaged with leading HR professional bodies in Bangladesh.

Dr Muhammad Khalilur Rahman holds a PhD from University of Malaya (UM). He obtained his master of science in marketing from International Islamic University Malaysia (IIUM) and BBA from International Islamic University Chittagong (IIUC). Dr Rahman has been actively involved in research activities and published more than 40 journal articles in internationally peer-reviewed journals with *ISI*, *Scopus* and non-indexed journals. He has a wide interest in tourism management research which includes medical tourism, eco-tourism, Muslim-friendly/halal tourism, service quality, brand equity, marketing, management, supply chain management and business administration.

Nazmoon Akhter has been a lecturer of Faculty of Business Administration in BGC Trust University Bangladesh for five years. She has completed her BBA and MBA in finance from University of Chittagong. Her interested areas in the research world are corporate finance, economics, marketing, branding, tourism and human resource management. She can operate various software, especially SPSS and STATA. She published a good number of her research papers in national and international journals.

Md Yusuf Hossein Khan is an author, industry expert, researcher and professor. He is currently working as an assistant professor at the College of Tourism and Hospitality Management (CTHM) of International University of Business Agriculture and Technology (IUBAT), Dhaka, Bangladesh. He is also a PhD researcher at the Faculty of Economics at University of Algarve, Portugal. He has a good number of experiences working within the tourism and hospitality industry in several countries. Khan has completed his MSc in international tourism management from the Cardiff Metropolitan University, UK, and MBA International from the Anglia Ruskin University, UK. Besides, being a member of Research Centre for Spatial and Organizational Dynamics (CIEO – Portugal) he is also serving several international journals as an editor and reviewer. Khan is an expert in curriculum design and currently is an active member for CTHM curriculum task force team of IUBAT. His current research interests are in particular, safety and risk in tourism, travellers loyalty, tourist motivation and behaviour and tourism destination image and development. He has good number of publications including books on these areas.

Md Zahid Al Mamun is a member of the Association of Business Executives (ABE). He has earned a postgraduate diploma in business management from The Central College of London and a postgraduate degree in international business (Master of Business Administration, MBA International) from Lord Ashcroft Business School, Anglia Ruskin University, United Kingdom. Besides these, he has also achieved postgraduate and Master of Social Sciences (MSS) degree from the University of Dhaka. His core research areas are human resource management, strategic HRM in organizations, business and marketing research.

FAN Yajing is a PhD student in Department of Marketing, City University of Hong Kong and also a lecturer of finance and economics at Guangxi University. Ms FAN received her bachelor's degree in statistics from Renmin University of China and master's degree in statistics from Hong Kong University. Her main research area is quantitative marketing and empirical modelling. She has published articles in international and Chinese journals covering a wide spectrum of topics in Chinese economy, finance and data mining.

Bablo Biswas achieved his Master in Social Science degree from the Department of Economics, University of Chittagong. After that, he worked at Foyjun Oxygen Plant as a sales executive and ASA NGO as a loan officer, respectively. He has extraordinary achievements in extracurricular activities. He played on the university cricket team and won an award in inter-university cricket tournament for University of Chittagong. He was an active member of Young Economists Society (YES) of the Department of Economics, University of Chittagong. His research interests include macroeconomics, development economics and tourism. He published a good number of research articles in national and international journals.

Yusuf Babatunde Adeneye is a PhD student in finance in the Graduate School of Business at Universiti Sains Malaysia (USM). His research interests lie in the areas of corporate finance, public finance and general management. He has published papers in several journals such as *Journal of Business Studies Quarterly*, *American Journal of Business, Economics and Management* and *Journal of Economics and Public Finance*.

Ei Yet Chu is a senior lecturer in finance in the Graduate School of Business at Universiti Sains Malaysia (USM). His research interests lie in the areas of finance, economics and business in the context of manufacturing and service industries. He has published papers in several journals such as *ASEAN Economic Bulletin, Capital Market Review, Corporate Ownership and Control, International Journal of Business & Society* and *Asian Academy of Management Journal of Accounting & Finance*.

Dr Fathyah Hashim is the Cluster Head of the Graduate School of Business, Universiti Sains Malaysia (USM). She has research interests in voluntary disclosure, intellectual capital, R&D and corporate social responsibility disclosure. She has a good research and publication record.

Shahriar Tanjimul Islam received his bachelor's degree and MSc degree in finance and business from the University of West England, Bristol and University of Ulster, both from United Kingdom. In 2016, he joined as lecturer at the College of Business Administration of International University of Business Agriculture and Technology (IUBAT). His research mainly centred on finance, accounting and business-related fields. Islam teaches several courses on accounting, business and finance. Some names of the courses he specifically teaches are cost accounting, international finance, taxation, business communication and a few more at IUBAT. He has supervised many students during their final year practicum defense. He has attended a number of national and international conferences and presented several papers. He has been invited for talks and panel discussions by different and universities to talk about finance.

Imran Hasnat is a doctoral candidate at the Gaylord College of Journalism and Mass Communication, University of Oklahoma, United States. His research mainly focuses on the use of digital media in public diplomacy. Grassroots media such as community radio is another area of his research focus. He teaches multimedia, data and computational journalism to future journalists. He is a former secretary to the High Commissioner of Canada to Bangladesh and was a UNESCO youth peace ambassador. He has an MA in international relations at Jahangirnagar University, Bangladesh, and an MA in journalism from the University of Oklahoma.

Dr Elanie Steyn is associate professor at the Gaylord College, University of Oklahoma, United States. She teaches and researches in management, leadership and business. She directs a US Department of State/University of Oklahoma grants that involve South Asian students, entrepreneurs and media professionals. She is an editor of *Global Journalism Education in the 21st Century: Challenges and Innovations* (Knight Center for Journalism in the Americas) and co-editor of *Critical Perspectives on Journalists' Beliefs and Actions: Global Experiences* (Routledge). She earned an MA in business communication from Potchefstroom University, South Africa, an MA in communication policy studies from City University, London and a PhD in business management at North-West University, South Africa.

Dr Haywantee Rumi Ramkissoon is a professor at Derby Business School, College of Business, Law and Social Sciences, Derby University, UK; School of Business and Economics, Faculty of Biosciences, Fisheries and Economics, UiT, the Arctic University of Norway, Norway; Monash Business School, Department of Marketing, Monash University, Australia; and College of Business and Economics, University of Johannesburg, South Africa. Professor Ramkissoon's role involves mentoring colleagues in producing and disseminating high-quality research for the benefits of the individuals and society at large. She has published widely. She was commended in the media as one of Australia's five best researchers in business, economics and management (tourism and hospitality) with a career of less than ten years in Australia. She has been honoured with prestigious international awards for research excellence, including the 2017 Global Emerging Scholar of Distinction award from the International Academy for the Study of Tourism. Professor Ramkissoon engages in collaborative multidisciplinary research with academic and industry partners at the local, national and international levels.

Introduction

Azizul Hassan

Since the ancient time, marketing has been featuring as closely associated factors to achieve sales volumes. The appearance of marketing in research is the outcome of pressure infused from the economy, society, technological and business innovation of the generations. Still, the concept of marketing research cannot be granted as very old, rather it appeared in the theoretical research scene in the second half of the 20th century. The extension of marketing research towards the tourism and hospitality industry is the demand of the business and economic growth of the 20th century owing to improved living standards, population growth and the increase of unrestricted time and income. In the shortest time, the tourism and hospitality industry has turned into the leading and essential industry of the global economy. More interests in marketing research actually led to study relevant tourism and hospitality infrastructure construction, ancillary facilities development and creating many other recreational facilities. Over the years, tourism marketing has been staying as an important and conventional area of research for researchers. Tourism marketing research is the study of a coordinated and systematic execution of business policies by both public and private tourism organizations. These organizations generally remain functional at the national, regional and international levels for achieving the maximum possible satisfaction of the identifiable tourist groups' demands and for doing so for achieving proper return.

Interestingly, tourism and hospitality enterprises have recognized the importance of main economic factors as wants, needs and satisfaction to plan and design tourism products and services. Modern tourism and hospitality marketing research has progressed as a reaction of businesses to changes in the socio-economic environment where the most successful tourism bodies or enterprises have shown a keen sense to offer the right of organizational products and structures for the tourists. This is the reason for which a tourist tends to be considered as an exceptional client and any enterprise serving this tourist attitude normally becomes the concern for other competitors.

Tourism and hospitality enterprises of the present world are turning as larger, more automated and more sophisticated in their marketing operations while the tourists as customers also tend to become more experienced, trained, erudite and better quality packages and services demanding. In such instable tourism and hospitality business environment, having adequate knowledge and skill in marketing acquainted through research become the essential elements than the knowledge and enthusiasm on products and services for longer-term growth and development. Tourism and hospitality marketing research thus turns as a recent phenomenon that has both conceptual and application values.

Bangladesh is a country in South Asia having a considerable population. In recent years, the country experienced a steady growth in its economy and socio-cultural developments. With a population of over 170 million the country possesses possibilities in tourism. With the stronghold of a social class having affordability to spend for tourism and leisure activities, the country already placed attention for the development of its tourism industry. The expansion of tourism and hospitality education in the higher academic institutions in Bangladesh is evidence. Arguably, the supportive roles of the government policy are favouring the development of tourism in Bangladesh. In principle, the importance of tourism is on a continuous rise in Bangladesh that in turn deserves attention from the researchers and the academia.

A good number of research studies are conducted outlining the contributions and importance of tourism including the Caribbean and many other developing countries. However, the tourism industry of Bangladesh so far has attracted very few researchers. Some contributions are made but not sufficiently. Considering the ongoing trend of tourism in Bangladesh very few research attempts have been made aiming towards exploring its diverse aspects of tourism. This book focuses on tourism marketing in Bangladesh that becomes one of the leading economic activities in many countries. Even uneven and improperly distributed, the contribution of tourism marketing in the economy of Bangladesh deserves significance in terms of earning opportunities and employment generation.

This book is a contribution towards the very limited knowledge of tourism marketing of Bangladesh. The book is designed to accommodate both conceptual and empirical research studies that link theory with practice. Existing policies, evidences and potential capitalization of tourism marketing are covered as well as some suggestions. The book carefully addresses and unifies the theme or framework of tourism marketing to integrate the theoretical explanations with practices. The book accommodates some critical and rich informative chapters.

In the first chapter, Nekmahmud, Farkas and Hassan argue that the success of tourism marketing is rather a subject to innovation acceptance. This theoretical chapter critically explains the relevant aspects of tourism marketing in Bangladesh, offers examples from existing literature (i.e. marketing mix and marketing strategy model) as a means to support arguments and relates them to the present tourism marketing context of the country. The chapter finds that even though tourism marketing is "interrelatedness", Bangladesh still stays within the conventional side of marketing to completely embrace the latest innovations. This study suggests that marketing with value co-creation, experiences of tourism marketing, technology support, collaboration and alliance, visual media networking, niche marketing and online and social media marketing strategy "interrelatedly" can bring changes. The chapter on such understanding offers some suggestions for managing future challenges and tourism marketing development in Bangladesh.

The second chapter of the book focuses on existing tourism products and services in Bangladesh. Suchana, Uchinlayen and Rahman have summarized some major tourism products and services literature studies with a focus on marketing. Findings of this research explore that Bangladesh possesses a diverse range of tourism products and resources that as tourism-ancillary products and services are expected to be well managed to satisfy tourists. Comprehensively, this chapter makes a positive contribution to tourism marketing research and explores the available tourism products and services in Bangladesh for both local and foreign tourists.

In the third chapter, Rahman, Rahman and Hassan argue that air transport is dominant in the tourism industry. This research explores the role of technology that support air transport and tourism industry in Bangladesh. The aims of this chapter are first, to discuss the importance of technology in the Bangladesh air transport industry; and second, to discuss how technology could improve tourism activity in Bangladesh. This abductive approach with content analysis–based research relies on secondary sources as well as desk research from the Internet. Results of the research present a framework and the implications of technology to air transport and tourism industry. The research proposes that a technology stalemate in air transport is pertinent for helping the sustainability of the air transport and tourism business. The frameworks that are developed in this research can be benefit the practitioners, policy makers and scholars to further develop the area and investigate key challenges. The research stresses on continued research and development of technology in air transport and tourism that is very much required for ensuring tourist safety.

The fourth chapter is contributed by Akhy and Roy. The authors of this chapter define accommodation as a place where tourists can find shelter and food offers in exchange of payment. The chapter believes that in a tourism resourceful country like Bangladesh, accommodation plays a crucial role. The chapter features that accommodation as a tourism product needs to highlight the key elements of any relevant business products. Because many changes have happened in tourist accommodation, this chapter thus focuses on the existing and future potentials of accommodation in Bangladesh. Limiting the research sites on major tourist destinations (i.e. Dhaka, Cox's Bazar, Rangamati, Chittagong city and Foy's Lake), the findings bring out the lack of research in this identified area that acts as a barrier for adequate information supply for the tourists. The chapter identifies the potentials of accommodation in the tourism industry of Bangladesh through SWOT analysis and presents remedial measures of relevant issues.

In the fifth chapter, Afrin and Hassan explain views on tourist transportation in Bangladesh. In the chapter, tourists' opinions stayed as the most important for better exploring the current situation and to discover operators' thoughts on the opportunities for better transit performance. The chapter, on the basis of understanding Bangladesh transport system, finds that the transit services are not providing efficient services to the tourist. The chapter suggests to develop a number of areas as required for promoting tourism in Bangladesh.

Chapter six as contributed by Pramanik and Rakib discusses the competitiveness of tourism products and services of Bangladesh. This chapter makes some theoretical arguments. The chapter then analyses the competitiveness of the tourism industry of Bangladesh by taking two comprehensive models: the Travel Tourism Competitiveness Index and Porters' Five Forces Model. This study considers on the concept of competitiveness and destination competitiveness and other relevant facts and issues while developing and presenting the arguments.

In chapter seven, Rakib and Pramanik make the conceptual analysis of products and services in Bangladesh. This chapter accommodates the concept of promotion and tourism promotion, integrated marketing communication (IMC) for domestic tourists and IMC for foreign tourists with special focus to online and social media promotion. Also, the chapter focuses on the impending challenges to implement promotion tools in the tourism industry of Bangladesh. The authors suggest that tourism product and services marketers need to design effective marketing promotion strategies for stimulating the domestic and foreign tourists towards Bangladeshi destinations.

In chapter eight, Mia argues that even with the necessary government initiatives, the actual arrivals of foreign tourists and earnings are not at par, and rather it is decreasing over the years. Mia then evaluates the present tourism conditions and tourism policies in Bangladesh. This study observed that there is inadequate institutional capacity, coordination problems among various government agencies, lack of infrastructure and a not-so-friendly visa policy that deter the growth of the tourism industry in Bangladesh. Hence, well-planned and well-executed tourism policies will not only increase the international tourist arrivals but also contribute to the development of Bangladesh.

In chapter nine, Hassan and Ramos, on the basis of an innovative technology as augmented reality (AR), analyse conceptual marketing in the tourism industry of Bangladesh. This chapter shows that application of AR as an innovative technology is practically tourism business supportive. Hence, the useful attachment of tourists with this innovative technology can harness productivity and interaction as well as help improve inclusiveness, resource management and overall development. This application of AR as an innovative technology thus can be taken as benefit generating for the overall tourism industry of Bangladesh.

In chapter ten, Rana, Rahman, Islam and Hassan express thoughts in terms of the effects of globalization on tourism marketing in Bangladesh. This conceptual chapter sought for economic opportunities for Bangladesh through potential tourism marketing in the globalization process. Findings of this chapter suggested that the success of tourism marketing of Bangladesh depends on infrastructural development, digital and technological advancement, a traveller-friendly environment, convenient tourism regulations and so on. The chapter finally recommended policies to ensure the presence of Bangladesh in the global tourism market.

In chapter eleven, Akhter and Hassan determine tourists' perceived risks on Bangladesh. The research examines survey data of 320 tourist respondents (6.3% international and 93.7% domestic tourists) staying in Chattogram in Bangladesh. The research with the application of a series of statistical tool (i.e. exploratory factor analysis, principal component analysis, confirmatory factor analysis, binary logistic regression, one way analysis of variance [ANOVA] and ordinal logistic). Results of the research outline six dimensions of tourists' perceived risk on Bangladesh: Financial and Communication Risk, Political Instability and Natural Risk, Health and Time Risk, Physical Risk, Social Risk and Psychological Risk. Results also present that a good number of individual features (i.e. age, type of tourist, purpose of visit and budget of travel) affect perception of travel risk. The research also reports that tourists tend to follow both behavioural modification of consumption and information search as risk reduction strategies where the usage of these strategies relied on the various types of risk that tourists' perceived. Finally, the chapter concludes by stating that tourism planners are required to focus on risks that can cause stress among tourists followed by awareness creation as marketing strategies.

Khan, Mamun and Hassan in chapter twelve elaborately outlined the human resource management (HRM) practices in Bangladesh. By summarizing the findings of 37 research studies, this chapter contributed to the limited knowledge on HRM practices in Bangladesh and compensation and benefits practices, in particular. The authors identified and showed the research gaps and proposed a few areas of compensation and benefits systems in Bangladesh where future studies can be conducted.

In chapter thirteen, Roy, Yajing and Biswas outline the policies for tourism sector development in Bangladesh, the impact of capital investment and revenue on tourism and some implications for the development of the tourism industry. From the theoretical ground of using a vector autoregressive (VAR) model, the chapter also focuses on the total contribution of travel and tourism to GDP, capital investment on travel and tourism and international tourism receipts. Based on findings, this chapter suggests that tourism is a huge contributor to the national economy and without maintaining proper strategies, the tourism industry will not be able to contribute the national economy as expected.

In chapter fourteen, Adeneye, Chu and Hashim investigate the role of capital structure on firms' competitive strategies in Bangladesh. The study aims to test whether high leverage, external equity or retained earnings exert much competitive pressure on competitive strategies of firms in the tourism industry in Bangladesh. A longitudinal panel regression is used to sample all the listed firms in the tourism industry. Three models are analysed: baseline model (nexus between competitive strategy and firm performance), capital structure model (the interaction role of debt financing, equity financing and internal equity financing on the nexus between competitive strategy and firm performance) and robustness test (the use of firm value measured as Tobin's Q and a control of technology, corporate social responsibility and environmental sustainability). This study offers a new model on sustaining competitive strategy through capital structure theoretically and for tourism policymakers.

In chapter fifteen, Khan, Islam and Hassan explain the revenue management techniques and practices in the tourism and hospitality industry of Bangladesh. This literature review–based chapter analyses the key concepts of hotel revenue management (RM), loyalty programmes in RM and its implementation and a case-based scenario assessed. The chapter also looks into the major areas of a hotel for maximizing the return on investment. After addressing the relevant theoretical aspects and limitations, the chapter brings the example of Cox's Bazar, Bangladesh. This chapter suggests to adopt modern technologies for the hoteliers to adapt and in order to gain strategic advantages over their competitors.

In chapter sixteen, Nekmahmud critically analyses the ground of green tourism product in Bangladesh, relates to tourists' purchasing behaviour response and determines the main factors of environmental marketing that are influential to the tourists' purchase intention of green products in Bangladesh. The chapter is conducted based on a mixture of primary and secondary data and proposes an approach to develop the conceptual framework of variables (e.g. green perceived quality, green perceived benefit, green purchase willingness, price consideration, environmental consciousness, safety and health concern, security, accessibility and purchase behaviour intention of green products). A brief summary of the main findings of the study are then presented by using with addressing managerial implications, challenges and suggestions for tourism products and services development in sustainable tourism market in Bangladesh.

The seventeenth chapter of the book is contributed by Hasnat and Steyn and overviews tourism advertising for transforming country image and empowering developing countries. The research considers Bangladesh and Nepal as cases. The chapter starts with an overview of the importance of tourism advertising for any country but specifically developing countries, and then primarily focuses on specific case studies (Bangladesh with its 'Beautiful Bangladesh' campaign and Nepal with its 'Visit Nepal Year 2020' campaign) to illustrate how these countries are implementing tourism marketing. The chapter at the later stage offers strategy recommendations for developing countries interested in tourism marketing.

In chapter eighteen, Hassan and Ramkissoon briefly analyse the potentials of tourism products and services marketing in Bangladesh. The aim of this chapter is to understand the potentials of tourism resources in Bangladesh. Findings of the chapter show that the potentials of tourism products and services in Bangladesh is subject to effective policy planning and implementation. Findings also show that Bangladesh will experience a sharp growth of domestic tourists mostly benefitted from the disposable income and the availability of leisure time. Thus, this research stresses that potentials as a theoretical term needs to be replaced with a more solid and effective set of policy implementations to cater the demands of tourists. Bangladesh as a tourist destination is suggested to redefine the potential of tourism products and services to a concrete base to enable to meet tourist demands.

This book is dominantly contributed by the middle- to early-age researchers either in the middle of their academic career or who recently completed or enrolled in their PhD. Thus, this book is rather an avenue for showcasing their expertise and capacities. Considering contents and scopes, this book is expected to be a reading companion of tourism and hospitality researchers and can be a reference book for the tourism and hospitality students in higher academic institutions in Bangladesh. Apart from them, this book can also be a good knowledge source of the relevant policy planners and industry professionals having interests in tourism marketing in Bangladesh or similar countries. Above all, this book can be appreciated by expatriate and local Bangladeshi researchers who strive for a tourism marketing book on Bangladesh.

Part 1

Tourism marketing overview

1 Tourism marketing in Bangladesh

Md. Nekmahmud, Maria Fekete Farkas and Azizul Hassan

Introduction

Marketing is a discipline that evolves persistently. For this reason, one business organization finds itself very much behind in the competition than the others if they stand still for too long. An instance of this is the evolution that stays as the fundamental changes for the key marketing mix. Once 4Ps tend to explain the mix but this is more commonly accepted that a more advanced 7Ps include a widely demanded added layer of depth to the marketing mix with a number of theorists tending to move further and even further. This research critically explains marketing on the lens of tourism and attaches to the context of Bangladesh.

Tourism marketing research: theoretical underpinnings

Marketing mix

The marketing mix is a well-known and familiar tool of marketing strategy. At the initial stage, this tool was limited to the basic 4Ps as Product, Price, Place and Promotion. The 7Ps were originally developed by E. Jerome McCarthy that was published in 1960 in the book titled *Basic Marketing: A Managerial Approach*.

The 4Ps concept was developed at the time when businesses were more probably selling products than service offers and the customer service role to help brand development was not very well known. As time progressed, Bitner and Booms (1981) added the three extended "service mix P's" (i.e. Participants, Physical evidence, and Processes). At the later stage, Participants were renamed People. In the present business world, this is suggested that the complete 7Ps of the marketing mix are regarded at the time of reviewing competitive strategies. This 7Ps model supports business organizations for reviewing and defining the main issues that influence its products and services marketing. This is very commonly referred to as the 7Ps framework for the digital marketing mix.

The marketing mix 4Ps are as follows:

Product

The product needs to fit the task consumers need it for, this needs to work and this should be what the consumers expect to receive.

Place

The product needs to be available from where the target consumer finds this as the easiest for shopping. This can be mail order, high street or the more recent option of online shopping or via e-commerce.

Price

The product needs to be always viewed as the representation of good value for money. This does not necessarily mean it should be the cheapest pricing option available. One of the key tenets of marketing is that customers usually remain happy for paying a little more for something that performs really well for them.

Promotion

Sales promotion, advertising, PR, personal selling and, in more recent times, social media stay as the basic tools of communication for a business organization. Such tools need to be used for putting across the business organization's message towards the correct audiences in the way that they would most prefer to hear, whether this can be informative or appealing towards their emotions.

The demand to update the marketing mix was widely acknowledged in the late 1970s. It led to the creation of the extended marketing mix in 1981 by Bitner and Booms that resulted the addition of three new elements to the 4Ps principle. This later allowed the extended marketing mix for including products that are services and not simply physical things. However, some of the researchers argued that even though sometimes 4Ps is viewed as dated, this tool is the essential tool for selecting their scope and is specifically beneficial for small business. This tool can generate specific advantages for start-ups to review revenue models and price. Through the application of this tool, business start-ups can produce competitive benefits. Thus, a business enterprise at the very initial stage of its operation can gain a solid position to stay in the market and gradually stronghold their position.

The 7Ps of the service marketing mix are as follows:

Product

The products in the service framework are those services that business organizations offer to the customer. For instance, an information technology (IT) company can offer services in network management, enterprise architecture, software development and more on the basis of customer demand.

Price

The price represents the service costs that the customer mandatorily has to pay. This is serious factor for buyers for their considerations that are based on the other service providers in the industry. Business organizations need to carry out research on their service's optimal pricing on the ground of value to the customer. Business organizations tend to consider their respective industry and factors such as low-cost service providers, competitor's reputation, high-cost service providers and number of players in the industry for determining their pricing strategy. Such factors are a few of the inputs that a business organization has to consider for their service pricing strategy.

Place

Place is the business' physical address where the service professionals interact with customers. For instance, if the business offers IT services to a client, then the place of business will be the office.

Promotion

The promotion of the service is featured as what most of the business organizations concentrate to get more clients. Such business organizations promote their services by applying some specific methods like search engine optimization, public relations, business developers, social media marketing, paid advertising, billboard ads and so on.

People

In the business framework, the people are the employees, consultants and even the freelancers that deliver relevant services to customers. People are one of the most critical factors to provide knowledge-based services. A business owner has to recruit the right people into the organization that would have the capacity to fit in the corporate culture. The business owner needs to find smart people that have the capacity to add value to the relevant business organization. The owner needs to make sure that the taken strategy is good for competing with other innovative business companies and organizations in the industry for acquiring talent. All of the business organizations are commonly reliant on the people who run the organization. They range from the managing director to front-line sales. To place the right people in the right place is essential as they are as much a part of the business and offer as the products/services the business organizations are offering.

Process

The processes are defined as the steps that are required for delivering the service to a customer. One of the basic advantages of service delivery organizations is that they design process maps outlining facts such as activities, function, processes and tasks. Such business organizations can become share these process maps for their employees for making sure that their work is repeatable and successful. The delivery of the products or services is generally done with the customer present so how the service is delivered is once again part of what the consumer pays for.

Physical evidence

The physical evidence is featured as the combination of the branding and environment where the service is offered to a customer by a service representative. The physical evidence capital can be a corporate website, a service brochure, social media accounts or a request for proposal. Most of the services include some physical elements even if the bulk of what the consumer pays for remains intangible.

Is there an 8th P?

In some very specific thought spheres, there are 8Ps in the marketing mix. The final P is productivity (and quality). This appeared from the earlier services marketing mix and stays folded into the extended marketing mix by few marketers.

Productivity and quality

This "P" makes queries whether the business owner offers a good deal to the customer. This stays as less about the owner as a business developing productivity for cost management and more about how the company passes this onto the customers.

How can be the marketing mix model applied?

Tourism business organizations can apply the 7Ps model for setting objectives, conducting a SWOT analysis and for even undertaking competitive analysis. This is a practical framework to evaluate an existing business and to work through appropriate approaches whilst making evaluation. The marketing mix model elements as presented can be asked as (1) products/services: how can the business products or services be developed?; (2) prices/fees: how can the existing pricing model be changed?; (3) place/access: what are the new distribution options available for customers for experiencing a business' products or services (i.e. mobile, online, in-store etc.)?; (4) promotion: how to add or substitute the combination within owned, paid, and earned media channels?; (5) physical evidence: how can the existing customers be reassured (e.g. impressive buildings, well-trained staff, great website)?; (6) people: who are the people and what are their skills gaps?; (7) partners: are the business organizations seeking new partners and managing existing partners well?.

Even after decades, the marketing mix remains very much applicable to a business owner or marketer's daily work. A good marketer can learn to adapt this theory for fitting with not only modern times but their individual business model. Even proposed in the 1980s, the 7Ps still remain widely taught for their basic logic as sound in the marketing environment and the marketers' capacities for adapting the marketing mix. These include changes in communications as social media, updates in the places which a business organization owner can update in the places to sell a product or service or customers' expectations in a repeatedly changing commercial environment.

The "interrelatedness" of tourism marketing

Tourism is an interesting research area creates an "interrelatedness" between some selected areas such as marketing, development, sustainability, innovation and relevancy. This "interrelatedness" stresses identifying and analysing all of the core elements of tourism research. This is one of the reasons why the outcome of the World Conference on Travel and Tourism in Rome in 1963 presents that tourism can devise both positive and negative effects on a country's economy. While for many developing countries, tourism carries economic benefits through the generation of employment opportunities, foreign currency earning and relevancy, the United Nations views tourism as a solid means to contribute to understanding and peace. Thus, the definition of tourism as endorsed by the WTO in 1992 as well as accepted by the United Nations Statistical Commission (UNSC) in 1993 is, "Tourism comprises the activities of persons travelling to and staying in places outside their usual environment for not more than one consecutive year of leisure, business and other purposes". From another perspective, tourism is defined as a travel act for business, services and recreation purposes. This view also accommodated a comprehensive definition: tourism is the service industry with both tangible and intangible components. Tangible items in this regard are transport systems (i.e. road, air, rail, water and the most recent concern of space), hospitality services (i.e. accommodation, foods and beverage, tours, souvenirs) and relevant services (i.e. insurance, banking, safety and security). On the other side, intangible items come with culture, relaxation, escape, new and different experiences, adventure and so on. Thus, tourism is a comprehensively interdependent and interrelated industry as well as tourism is very often applied derogatively implying a thorough interest in the place and society that the tourist visits.

Decoding tourist attractions

The clarification of tourist attractions can sometimes create dilemmas and has never been easy or straightforward. This is a widely popular term having diverse meanings. Still, the common understanding about a tourist attraction is that it is a place that draws attention or attracts general people for visiting a place, attending an event or travelling to a particular location for some key purpose (i.e. enjoyment, recreation, education, information collection or just a normal visit). Thus, in the simplest meaning, tourist attractions are just places that stay as the reason of travel by the people (i.e. man-made tourist attractions as physical structures; natural attractions as physical phenomena as deemed beautiful or unusual). Secondary attractions can also have tourist appeal but cannot be the primary reason for visitation. Negative attraction is rather an area's attributes that tend to cause some market or customers not to visit that particular attraction. The very basic reason for negative attraction can be crime, pollution or terrorism or anything that makes people worry or that feels unfamiliar. Thus, tourist attractions are the places of interest as open for the public for offering education, recreation or historic interest. Tourist attractions can range from leisure complexes, country parks, zoos, museums, historic houses, theme parks and so on.

The impression of tourists

The impressions of tourists towards the tourism services or products are essential for marketers. Impressions that are positive can obviously lead to tourist satisfaction and repeat visit as well as serving as a word-of-mouth advertisement for the relevant destination. Buyers tend to decide whether a service or product is priced fairly or represents value for money before purchasing. It is expected that tourists as buyers will have better value or even equal to their perceived value after making visitation to the tourist destination. Any change in quality or pricing at a certain time can also change the perception of the tourists about the value. Thus, it is essential that reasonability in pricing on key elements of tourism arrangements in Bangladesh are considered.

Marketing strategy (STP) model

Marketing strategy is the combination of segmentation, targeting and positioning (STP) process as the core of marketing strategy which is developed by Philip Kotler (2003) and also, it is part of the strategic business unit (Webster, 2005). In general, the core concept of a marketing strategy consists of the tasks of identifying and selecting the target segments in the place where the actual and potential customers are dwelling and desired to purchase the product. There are three marketing strategies: market segmenting, market targeting and market positioning. In the tourism industry, the marketing STP model is vital for achieving its goals and increasing the potentiality of the tourism sector in Bangladesh. Moreover, STP offers suitable tourism products or services that will serve tourists in distinct interests.

Market segmenting

The term "segmentation" was first introduced by Smith (1956). Market segmentation defines as dividing the complete market into different parts on the basis of several variables (Kotler, 1999). Moreover, it is recognizing that you can't serve all customers with an equal level of satisfaction. In Bangladesh, the market segment plays a vital role in tourism marketing goals and it helps to divide the tourism product and services. Generally, Bangladesh Parjaton Corporation (BPC) articulated the policy in 2010 to regulate the tourism industry. Sarker and Begum (2013) explained there are two types of segmenting the target market in Bangladesh: (1) demographic segmentation and (2) psychographic segmentation. Demographic segmentation includes tourist products that can charm a limited number of groups and ages, and provides the largest range of preferences for defining a large number of groups of people. Nevertheless, psychographic segmentation offers us in-depth information of tourists.

Market targeting

Market targeting includes violation of a market into segments and then focused marketing determinations on one or a few main segments (Kotler, 1999). It tries to fill up the common needs and/or characteristics where the organization decides to serve. Bangladesh Parjaton Corporation (BPC) focus its marketing effort on beaches, forest and jungle, archaeological sites, hills and islands, historical places and so on. Market targeting identifies the actual destination which could help to conduct any research to define where tourists are available and which countries' people are motivated to enjoy the particular tourist place. For example, in the case of less natural tourism, infrastructure, and visitor events an investment may be required.

Market positioning

The position of the market is building your target audience to know exactly how you stand apart from your competitors. According to Kotler (2003), "positioning is the way the product is defined by tourists on important attributes; tourists relatively competitive product that has made this product". Therefore, Bangladesh Parjatan Corporation (BPC) is promoting a bundle of tourist's products and services for the tourists. According to records of MoCAT, BPC recognizes four major tourist's products and services in Bangladesh: beach; forests, hills and islands; historical places; and archaeological sites.

Above these four major tourism products have appeal, market demand and more competitive advantage, even though those products acquired good positioning in the tourist's mind in South Asia, East Asia, Europe and the USA, New Zealand Australia and England. Bangladesh can strengthen its position as a tourist hub of established generating markets at regional and international levels by utilizing different marketing tools such as websites, TVC for satellite and local TV channels and other marketing communication tools (Musa, 2013).

Marketing by the National Tourism Organization

The National Tourism Organization (NTO) of a country generally acts as the public organization responsible for tourism marketing. This organization also performs with both relevant public and private sector organizations engaged in the tourism business of a country. However, the availability of relevant resources required for both marketing and promotional activities needs to be ensured. The gathering and application of experience are also essential. Even with the lack of sufficient funds and experience, NTO creates collaboration with both the public and private agencies with an aim to conduct both marketing and promotional activities for attracting increased number of tourists to particular destinations and thus to leave positive effects on the local economy.

Tourism marketing theoretically includes advertising, publicity, marketing, promotion, personal selling and sales promotion. These tools together can lead to tourism promotion of a particular destination. The dissemination of relevant data and information about a destination are important and social media is also important for attracting more potential tourists. For the purpose, the presentation of relevant information about the particular destination is essential as the tourism business is mostly based on reliable information sharing. NTOs in developing countries such as Bangladesh do not tend to be as innovative to make the latest marketing concepts or elements available for marketing or promotion. The lacking and drawbacks of NTOs mainly suffer from fund shortage and innovative mindsets as well as limited operational capacities.

Tourism marketing research: the past, the present and the future

In a relevant research Dolnicar and Ring (2014) explained the past, present and future of tourism marketing research. The research created a Tourism Marketing Knowledge Grid and applied it as a framework for the review. This grid explored that extant tourism marketing research basically focused on the way service promises are made and kept and widely generated frameworks for improving managerial decision making or offering insights about associations between constructs. Strategic principles as underpinned by the cause-effect relationships understanding are rare. Such results direct to exciting opportunities for further research (i.e. more attention on enabling promises made to tourists with the development of strategic and research principles; more application of experimental, quasi-experimental and longitudinal research designs; unstructured qualitative designs; and increased focus on the study of actual behaviour).

The review on tourism marketing literature suggests some non-conventional approaches to consider, as analysed next.

Cultural context consideration

Using the appraisal theory, Wu (2018) believes that official websites can be medium for tourism marketing. Analysing the discourse of online tourism marketing of two tourist destinations, London and Hangzhou, linguistic resources are found as functioning for promoting destinations. However, cultural contexts impact the marketing and promotion of tourist destinations. Hangzhou, an Asian city, adopts marketing and promotional measures like information sharing and highlighting its history. In this way, this city establishes credibility among the visitors who have interest in culture, marketing and promotion. London as a city of Europe promotes tourism by foregrounding its present attractions and inviting the visitors to learn its history and culture.

Tourist behaviour consideration

Koc and Boz (2014) offered a unique approach to consumer/tourist behaviour called psychoneurobiochemistry and explored the likely and potential influences of psychoneurobiochemical factors on tourism marketing. Having a multidisciplinary approach, the researchers analysed and synthesized the neurological, psychological, biological and chemical research findings on the ground of their implications for tourism marketing. The researchers specifically searched at neurotransmitters such as melatonin hormone; serotonin and dopamine; circadian and photoperiod rhythm; and emotions.

Relationship marketing and value co-creation

Ensuring tourism innovation through relationship marketing and value co-creation can help tourism marketing. Casais et al. (2020) discuss tourism innovation as applied for sharing accommodation by the hosts on the basis of guests' value co-creation outcome. This process stands relationship marketing as the central feature of peer-to-peer business models that is analysed as an innovation catalyst. A close relationship with guests during their stay is established that is critical for the co-creation of the tourism experience and innovation increment. User-generated online review contents and constant interpersonal contact established between the guests and hosts lead to incremental and connected innovation followed by amenities, facilities and partnerships with relevant businesses.

Collaboration and alliance

Khalilzadeh and Wang (2018) argued that the base of destination collaboration is the interdependency of the organizations involved in producing destination products. The demand for conflict studies is underscored by the higher rate of destination collaboration failure. This research was somehow different from earlier studies in a way that suggests a new approach for defining its utility functions on the basis of motivational and attitudinal values. The researchers applied the network theory for defining the utility function of four key players and the game theory for examining three distribution solutions of coalitional activities' values. Findings of the research supported the notion of "free riders" stated in collaboration studies and explained the reasons for free riding in tourism destinations' marketing activity natural phenomenon. The research also accepted that individual entities and hospitality are the two basic players having the highest admission fee and the least contribution and thus, concepts like fairness and stability need to be regarded in incentive policies for encouraging collaboration within the higher admission players.

Reid et al. (2008) argue that strategic alliances can turn into a general tourism marketing strategy. Such alliances can take diverse forms and become operational with diverse objectives. Still very often, alliances are formed without accurate expectations, pure operating procedures or objectives criteria for success evaluation. In this case study, the researchers review a tourism marketing alliance, the Atlantic Canada Tourism Partnership, that was created successfully and operated for many years. Such partnership joins together tourism ministers of the four Canadian provinces, four industry associates and the federal government. The primary role of this type partnership is to promote Atlantic Canada in selected overseas markets and the USA. This case analyses the strategic approach and partnership results and thus concludes lessons learnt from the case study and identifies areas for further developments in the partnership.

Technology support

In their research, Ying and Peters (2011) suggested a newer of system engineering for marketing decision support titled Tourism Marketing Information System (Tour MIS). This system supported the tourism industry, educational and research institutions to collect, store, process and disseminate information. Austria has already developed and promoted a Tour MIS that can meet the demand of tourism enterprise and tourism destinations' decision making. Considering the fast development of tourism in China, the authors demanded the necessity to set up a Tour MIS that was already implemented in Australia.

Visual media elaboration

Visual media can play crucial role in marketing and promoting tourism services in the present connected marketplace. On the theoretical ground of the elaboration likelihood model (Petty and Cacioppo, 1986), John and De'Villiers (2020) examine the way through which visual media influences potential audiences' perception towards a particular educational tourism destination. The researchers made comparison between the central and peripheral individual persuasion routes through social media that support marketers' to understand the pattern within which visual media affects consumer's purchase decision. An affirmative relationship was found between perceived destination images of international students as tourists with audience engagement, audience involvement, argument quality and source credibility.

SoCoMo marketing

Advanced technology can make users able to amalgamate information from many sources on their mobile devices. Users can personalize their profile through applications and social networks. Also, they can interact dynamically with their context. Buhalis and Foerste (2015) connected the different concepts of context-based marketing, social media and personalization, as well as mobile devices and suggested social context mobile (SoCoMo) marketing as a new framework. This framework enables marketers to increase value for all stakeholders at the destination. The research found that contextual information becomes widely relevant when big data collected by a wide range of sensors in a smart destination can offer real-time information. This can influence the tourist experience and thus SoCoMo marketing has the capacity to introduce a new paradigm for travel and tourism. This marketing enables tourism organizations and destinations to revolutionize their offering and co-create products and services dynamically with their consumers. The suggested SoCoMo conceptual model outlines the emerging opportunities and challenges for all stakeholders.

Bassanoa et al. (2019) believed that storytelling about places is a recognized tool for enhancing regional reputation as regions compete for tourism and economic development spending in this digital age. With the support of digital media, people can be inspired to tell their stories of tourism and to share their experiences. For enhancing brand competitiveness, storytelling is manageable in a local service system. The system within which cultural organizations and local governments encourage and understand storytelling about places can leave a significant effect on the region's competition for tourism and development spending. "Place storytelling" means local stakeholders can tell their personal stories about their beloved places. The analysis and cases of this research show that place storytelling allows strategic communication that cares for sustainable competitive benefits. A place storytelling model is developed in this research allowing for cultural organization and local government's use for encouraging and managing stakeholder engagement in a multilevel process to improve regional service system marketing and communications in the digital age.

Cross-country experiences of tourism marketing

Gulbahar and Yildirim (2015) opined that a good number of the companies both follow and adapt technological advancements in communication supported by higher web page usage ratio percentage and mobile application based service offers and linkage with social media channels. The usage and acceptance of the Internet and social media appear as an essential channel in almost all areas of tourism. Tourism is one of the leading sectors in adopting technological developments, social media channels and tools of technological communications. Almost all of the Turkish tourism enterprises are getting updated with these developments when the Internet framework coverage of the mobile communication stays as powerful and this relates to other emerging economies. There is a lack of considerable data that examine the social media effect on tourism marketing in Turkey but in reality, this is a necessity for analysing social media's position in tourism marketing and its effects. There is also a question in terms of social media channel preference for tourism marketing in Turkey as well as the relevant company's use of channels for marketing and CRM. The research described a framework for electronic communication usage and social media usage for marketing in Turkey that allows tourism enterprises to benchmark and road map for future efforts.

Ely (2013) informed that the Mexico Tourism Board, by positioning Mexico as a cultural/historical destination as well as a sun-and-sea destination, sought to diversify its tourism industry and increase the number of visitors to archaeological sites. This has raised a debate between experts about the positive and negative effects of tourism on the people of Mexico and the conservation of its archaeological sites. The research found that effective marketing can serve to promote specific tourist values that enhance the positive and decrease negative effects. The key result of this research presented that in the past, archaeological and touristic goals were thought to remain at odds. However, at the time these two groups connect forces for defining a specific site's financial and non-financial objectives, opportunities tend to exist for the collaborative creation of marketing materials that promote those objectives and offer benefits for both groups.

Xiao (2013) addressed a shift of focus in China tourism by discussing the contexts for, and issues of, domestic tourism development. The research stressed that policy was oriented towards domestic tourism as quality of life and highlighted the agenda for destination marketing and management.

Kotoua and Ilkan (2017), in the context of Ghana, developed a model for investigating the relationships between intention to visit and tourists' satisfaction as a source of mediation for travellers through information search and e-word of mouth. Results indicated simple websites no longer had an impact on destination marketing because of technological advancement. The research suggested that websites were required to offer diverse tools and marketing channels for facilitating the surfing and information demands of tourists. This research through using the instruments of online word of mouth and information search by modifying the theory of planned behaviour to consider the context of intention to visit thus suggested that that the dimensions of tourists' satisfaction as a mediator affect the overall tourists' intention to visit.

Bangladesh

The geographical location

Bangladesh, a developing country, is situated in South Asia. The country has an area of 147,570 square kilometres and borders with India to the north, east and west. Myanmar stays on the southeast and the Bay of Bengal to the south. The exclusive economic zone of Bangladesh is 200 nautical miles while the territorial waters of Bangladesh extend 12 nautical miles. Bangladesh has a large coastline with large marshy jungle on the Bay of Bengal that is well known as the "Sundarbans", the world's largest mangrove forest and the home of the royal Bengal tiger.

Geographically, the country is located in the Ganges Delta. This is the largest delta in the world and has densely vegetated land areas. This land area is often called the Green Delta. The confluence of the Brahmaputra (Jamuna), Ganges (Padma) and Meghna Rivers with their numerous tributaries have formed this delta. These rivers flow down from the Himalayas which are adjacent to the north-western frontier of the country.

Tourism attractions

The traditional image of Bangladesh before the global tourists is as a poverty-stricken, flood-ravaged and disaster-prone tourism zone. The country has a rich tradition, history and natural settings, as well as man-made architectural excellences. The beauties of Bangladesh include vast greenery, mighty rivers, sea and river beaches and tribal life with its cultures and celebrations. The country offers tourism resources that can meet the demands of both domestic and foreign tourists with the availability of sea and river beaches, lakes, sanctuaries, forests, wild-lives, hills, archaeological attractions with monuments, handicrafts, religious festivals, cultural heritage, folklore, customs and so on. The beauties of Bangladesh have attracted foreign tourists since the remote past. The French traveller Francois Bernie experienced such beauties and stated

> Egypt has been represented in every age as the finest and most fruitful country in the world, and even our modern writers deny that there is any other land of peculiarly favoured by nature; but the knowledge I have acquired of Bengal during two visits paid to that kingdom inclines me to believe that pre-eminence ascribed to Egypt is rather due to Bengal.
>
> (François, 1826 p. 181)

Tourism marketing in Bangladesh

The past

Previously, the tourism industry in Bangladesh was not in good condition because the Bangladesh government focused on the agricultural sector and textile industry and the government may be unwilling to invest in the development of tourism. At that time, transportation and accommodation was also not in good condition for a tourism destination. Only local and domestic tourists were the target customers for tourism. A decade ago, there were only a few tourism agencies where their marketing strategy was only traditional methods, for example person to person marketing and word-of-mouth. Even investments of both public and private sectors in tourism were limited.

The present

At present, Bangladesh has improved and overcome a few of the basic limitations in relation to present tourism attractions to the tourists. Many relevant facilities related to tourism as visa and immigration, transportation, accommodation, catering and other ancillary services are modernized both for domestic and foreign tourists. These activities in reality can generate benefits for the entire tourism industry.

The future

Bangladesh is holding high potentiality for tourism and it can play a key role to contribute to the national economy of the country. In this country, there are lots of natural beautiful hills, vales, deep and mangrove forests, rivers and the longest beach in the world. Moreover, the scope of nature-based tourism, culture-based tourism, historical tourism, eco-tourism and research-based tourism is quite evident. In the future, the tourism sector will be one of the economic earning sources regarding foreign exchange earnings as well as the creation of employment. For developing the tourism industry, both the private sector and the government should increase investment and create international base facilities for foreign tourists. An online and social media marketing strategy and niche marketing strategy can help inform and attract foreign tourists. It should use innovative technology to spread positive news to the tourists and use digitized information systems for historical tourism products.

Critical explanations: drawback and prospect analysis

In Bangladesh, there are some disadvantages of the tourism sector such as problems of safety, security and hygiene, lack of entertainment facilities, low-quality services, new investment, non-professional tour operators, unskilled and profitable tourist agencies, political instability and finally lack of infrastructural development. The Bangladesh tourism sector has a lack of online marketing facilities. The policy of the tourism industry is not adapting to the development of tourism growth because there is no policy-related sustainable tourism. Green marketing strategies such as green marketing mix, eco-friendly services, online marketing and customized services are not available in the tourism industry in Bangladesh. Hossain (1999) demonstrated that the tourism industry failed to grow properly due to a lack of sustainable and effective tourism marketing strategies and the reluctant attitude of different governments. Several foreigners have a negative perception of Bangladesh. Foreign tourists feel Bangladesh is a country of poverty, floods, beggars and political conflict. Besides, the ministry of civil aviation and tourism, BPC and other private tour operators did not apply proper marketing tools and concepts to reach their tourist products and service to the target consumers.

On the other hand, the Bangladesh tourism sector has a huge potential scope for expansion of world tourism such as the low price of products and services, rich historical-cultural heritage, diversification of the tourism product portfolio, prominence on development of tourism and hospitality skilled manpower. Moreover, Bangladesh has many rivers, the longest sea beach in the world, a historical tourism place and ancient architecture.

Bangladesh tourism is facing many problems including marketing strategy such as marketing mix, promotional tools, pricing strategy and green marketing and green value chain (Nekmahmud and Rahman, 2018; Hasan, 2019). By applying niche market strategies, Bangladesh can earn more foreign currency as

Bangladesh has four main tourism products: beaches; hills and islands; forests; and historical and archaeological sites that are most attractive to many tourists (Sarker and Begum, 2013). Now the government focuses not only on tourism but also on hospitality management such as hotels, motels, transportation, restaurants, security, environment and entertainment.

Conclusion and Recommendations

Tourism marketing is one of the parts of marketing which contend with the tourism industry. The main purpose of tourism marketing is to attract visitors to the particular tourism place which can be a country, a city, heritage site or tourist places and others. Nowadays people around the world like to travel for exploration and entertainment. As a result, skilled tourism marketing can attract visitors and make a lot of money. Bangladesh is famous for its natural beauty, greenery, archaeological and historical sites and hospitality. Therefore, tourism marketing is essential to attract potential foreign tourists in proper ways and it plays a significant role in socio-economic development.

Tourist spots should be more attractive and a promotional strategy of tourism marketing can develop the tourism culture to attract tourists. However, the government should focus on not only tourism but also hospitality management like hotels, motels, restaurants, transportation, security and entertainment and continued research and development are needed for the policymakers.

This chapter explained that tourism marketing requires interrelatedness while destination marketing in tourism requires innovativeness. Innovations and information technology are vital for Bangladesh to develop its tourism growth to attract foreigners. A solution to information technologies could increase the effectiveness of the tourism business by introducing a model that consists of eight components (information, users, information resources, suppliers, information systems, booking and sales systems, information processes and ensuring of information systems and technology). Then, the model is tested in the hospitality and tourism industry companies (Saifullin and Lomovtseva, 2019). If Bangladesh wants to establish generating markets at international and county levels, it has to use different types of marketing tools for example online marketing, website development, TVC for satellite and local TV channel, social media marketing, green marketing and other marketing communication tools, e.g. advertising, newsletters, catalogues, direct mail or email campaigns, public relations and sales promotions and so on. Mobile advertising can play a vital role in attracting the tourists and informing about tourism destinations, products and services (Huq et al., 2015). In addition, it is time to update and use Airbnb app, which is a home-sharing mobile platform that enables tourists and local hosts to connect. Airbnb mobile app allows hosts to the location available for rent and helps travellers to attract and find accommodation, phone reservations or bookings by travel agents (Nathan et al., 2020). Bangladesh can achieve tourism goals and create potential markets by observing the changing design and requirements for different types of markets such as China, Japan, Eastern Europe, and South America and South Asian countries. The government of Bangladesh can promote international tourism such as Sundarbans and Cox's Bazar, Saint Martin's Island and some regional heritage sites such as Fort William, and House of Ahsan Manzil and Sonargaon as the main reason for leading Bangladesh in international tourism. Moreover, BPC pays interest in terms of promotion and product diversification strategies, in particular travellers who are arrivals from South Asia, East Asia, USA, Europe, Gulf Cooperation Council (GCC) countries, Australia and New Zealand and others. Developing packages of points of interest and services is a good way of catching the interest of tourists. The government should ensure easy access transportation, international standard accommodation, entertainment facilities, sports and cultural amenities.

Moreover, visitors are now more conscious of the environment and health issues. Therefore, a green marketing strategy and an environmentally friendly product strategy could help to develop the tourism industry by fascinating foreign tourists. Marketing with value co-creation, e.g. cross-country experiences of tourism marketing, technology support, collaboration and alliance and visual media networking also helps to expand the tourism market in Bangladesh. Moreover, proper marketing strategy, marketing mix, communication mix, investment and research and innovations could support to develop the tourism industry in Bangladesh successfully by offering attractive and appealing tourism products and services.

References

Bassanoa, C., Barileb, S., Piciocchic, P., Spohrerd, J. C. Iandolob, F. and Fisk, R. (2019). Storytelling about places: Tourism marketing in the digital age. *Cities*, 87, pp. 10–20.

Bernier, F. (1826). *Travels in the Mogul empire* (Vol. 1). London: W. Pickering, p. 181.

Bitner, M. J. and Booms, H. (1981). Marketing strategies and organization: Structure for service firms. In J. H. Donnelly and W. R. George (eds.), *Marketing of services, conference proceedings*. Chicago, IL: American Marketing Association, pp. 47–52.

Buhalis, D. and Foerste, M. (2015). SoCoMo marketing for travel and tourism: Empowering co-creation of value. *Journal of Destination Marketing & Management*, 4(3), pp. 151–161.

Casais, B., Fernandes, J. and Sarmento, M. (2020). Tourism innovation through relationship marketing and value co-creation: A study on peer-to-peer online platforms for sharing accommodation. *Journal of Hospitality and Tourism Management*, 42, pp. 51–57.

Dolnicar, S. and Ring, A. (2014). Tourism marketing research: Past, present and future. *Annals of Tourism Research*, 47, pp. 31–47.

Ely, P. A. (2013). Selling Mexico: Marketing and tourism values. *Tourism Management Perspectives*, 8, pp. 80–89.

Gulbahar, M. O. and Yildirim, F. (2015). Marketing efforts related to social media channels and mobile application usage in tourism: Case study in Istanbul. *Procedia – Social and Behavioural Sciences*, 195, pp. 453–462.

Hasan, M. M., Nekmahmud, M., Yajuan, L. and Patwary, M. A. (2019). Green business value chain: A systematic review. *Sustainable Production and Consumption*, 20, pp. 326–339.

Hossain, M. A. (1999). Marketing of tourism industry in Bangladesh: An empirical study of performance and strategies. *Unpublished PhD Thesis*. Maharashtra: University of Pune.

Huq, S. M., Alam, S. S., Nekmahmud, M., Aktar, M. S. and Alam, S. S. (2015). Customer's attitude towards mobile advertising in Bangladesh. *International Journal of Business and Economics Research*, 4(6), pp. 281–292.

John, S. P. and De'Villiers, R. (2020). Elaboration of marketing communication through visual media: An empirical analysis. *Journal of Retailing and Consumer Services*, 54, p. 102052.

Khalilzadeh, J. and Wang, Y. (2018). The economics of attitudes: A different approach to utility functions of players in tourism marketing coalitional networks. *Tourism Management*, 65, pp. 14–28.

Koc, E. and Boz, H. (2014). Psychoneurobiochemistry of tourism marketing. *Tourism Management*, 44, pp. 140–148.

Kotler, P. (1999). *Marketing management: The millennium edition*. Upper Saddle River, NJ: Prentice Hall.

Kotler, P. (2003). *Marketing for hospitality and tourism*. New Delhi: Pearson Education.

Kotoua, S. and Ilkan, M. (2017). Tourism destination marketing and information technology in Ghana. *Journal of Destination Marketing & Management*, 6(2), pp. 127–135.

McCarthy, E. J. (1964). *Basic marketing*. Homewood, IL: Richard D. Irwin.

Musa, M. (2013). "Marketing" a tool to develop Bangladesh's tourism sector. *International Journal of Business, Economics and Law*, 2(1), pp. 6–9.

Nathan, R. J., Victor, V., Tan, M. and Fekete Farkas, M. (2020). Tourists' use of Airbnb app for visiting a historical city. *Information Technology & Tourism*, 22, pp. 217–242.

Nekmahmud, M. and Rahman, S. (2018). Measuring the competitiveness factors in telecommunication markets. In K. Datis, F. Mike and M. Wilfried (eds.), *Competitiveness in emerging markets*. Cham: Springer, pp. 339–372.

Petty, R. E. and Cacioppo, J. T. (1986). The elaboration likelihood model of persuasion. *Communication and persuasion*. New York, NY: Springer, pp. 1–24.

Reid, L. J., Smith, S. L. J. and McCloskey, R. (2008). The effectiveness of regional marketing alliances: A case study of the Atlantic Canada Tourism Partnership 2000–2006. *Tourism Management*, 29(3), pp. 581–593.

Saifullin, T. and Lomovtseva, M. (2019). Information services industry in tourism. In V. Vasile (ed.), *Caring and sharing: The cultural heritage environment as an agent for change*. Cham: Springer, pp. 63–72.

Sarker, M. A. H. and Begum, S. (2013). Marketing strategies for tourism industry in Bangladesh: Emphasize on niche market strategy for attracting foreign tourists. *Journal of Arts, Science & Commerce*, 4(1), pp. 103–106.

Smith, W. R. (1956). Product differentiation and market segmentation as alternative marketing strategies. *Journal of Marketing*, 21(1), pp. 3–8.

Webster, F. E. (2005). Back to the future: Integrating marketing as tactics, strategy, and organizational culture. *Journal of Marketing*, 69(4), pp. 4–6.

Wu, G. (2018). Official websites as a tourism marketing medium: A contrastive analysis from the perspective of appraisal theory. *Journal of Destination Marketing & Management*, 10, pp. 164–171.

Xiao, H. (2013). Dynamics of China tourism and challenges for destination marketing and management. *Journal of Destination Marketing & Management*, 2(1), pp. 1–3.

Ying, L. and Peters, M. (2011). Setting up the tourism engineering marketing information system of China. *Systems Engineering Procedia*, 1, pp. 301–308.

Part 2

Services and products of tourism

2 Existing tourism products and service offers in Bangladesh

*Jameni Jabed Suchana, Uchinlayen
and Muhammad Shoeb-Ur-Rahman*

Introduction

Tourism products serve various needs of a tourist. These products can be considered as the objects of the transactions between tourists and businesses (Koutoulas, 2004). Representing a significant part of tourism industry, tourism products and services gain attention from both tourism academics as well as practitioners. On this note, gaining a deep understanding of tourism products and services including their complex nature and relationship remains a challenging task (Mrnjavac, 1992). Tourism products help to extend tourist seasons, foster market awareness, position or re-position destinations, support investment, increase revenue, and improve the local economy. Accordingly, contribution of different tourism products towards tourism development is inevitable.

Bangladesh, a potential destination in the South Asian region attracts increasingly both domestic and foreign tourists by using an array of tourism products and services. However, the literature reveals little effort has been given on categorizing these tourism products and services. Identification of different categories can be helpful to target desirable tourism market segments and positions thereon upon the mind of the visitors. Considering such a limitation, this chapter attempts to provide an overview of existing tourism products and services in Bangladesh. In this regard a theoretical perspective has been developed and within the theoretical frame various tourism products and services have been categorized thereafter.

Tourism products and services

Tourism products and services are invariably identified in association with tourism experiences and satisfying various tourists' needs. Tourists in the postmodern era have been exhibiting varying interests (Rahman and Shahid, 2012). These interests raise the needs of tourists, which are satisfied through travelling (outside their usual place of residence) and experimenting with different tourism products. Seemingly, tourists enjoy several benefits from tourism products. Understanding the concept of tourism products is necessary both from supply and demand viewpoints. On the supply side, destinations can offer, develop, and manage profitable packages to tourists once the concept is clear. Correspondingly, an increased number of satisfied tourists creates a balance at the demand side.

Armstrong and Kotler (2016, p. 260) define products as a means to satisfy consumer demands which can be found in the form of "physical objects, services, persons, places, organizations, and ideas". The theoretical underpinning of "tourism products" is equally all encompassing and entails tourism services within an existing frame (Cirikovic, 2014; Lewis and Chambers, 1989; Smith, 1994). The notion of tourism products has been elaborated by World Tourism Organization (UNWTO) (2019) as "a combination of tangible and intangible elements, such as natural, cultural and man-made resources, attractions, facilities, services and activities around a specific centre of interest which represents the core of the destination marketing mix and creates an overall visitor experience including emotional aspects for the potential customers". Such a definition puts forward a generic and element-based view of tourism products.

Tourism products are mainly service-based products or services that have several characteristics including intangibility, psychological involvement, perishability, heterogeneity, ownership complexity, and so on. For instance, to promote a particular destination in the off-season, events like fairs and festivals, which are offered to the tourists, are perishable and variable in nature. Again, targeting the specific market segments such as business tourism, different hotels and convention centres means offering different services like conference planning, organizing and management services. Some benefits pre-exist in different components of tourist' products signifying the "physical plant" of "generic tourism product" (Smith, 1994). On the other hand, producers (industry suppliers) incorporate consciously other ancillary benefits (i.e. services to ensure complete and satisfying experiences to tourists). From this perspective, tourism product can also be defined as a service that can be enjoyed by tourists from the place of origin, in a tourist destination, until returning to the residence.

Smith (1994) identifies five important elements to define tourism products: physical plant, service, hospitality, freedom of choice, and involvement. At the very basic level, physical plant covers natural resources, conditions in the physical environment, and the design and layout of various physical components at a destination. The provision of services add utility to core products, i.e. physical plants, whereas hospitality indicates the extent of warmth in delivering services. Physical plants incorporate a destination-specific focus while both services and hospitality generate an employee-oriented focus in which developing human skills (capital) is essential to shape overall tourism products. Finally, freedom of choice and involvement elements take the stance from tourists' or visitors' side. Freedom of choice ensures that tourists must be provided with alternatives and involvement indicates the physical and psychological engagement of tourists into the experience-generating process. Taking all these elements into consideration, this study has adapted tourism production function (as shown in Figure 2.1) to theorize the concept of tourism products in general.

Given the element-based discussions, tourism products entail all the four components (resources, facilities, services, and outputs) in the tourism production

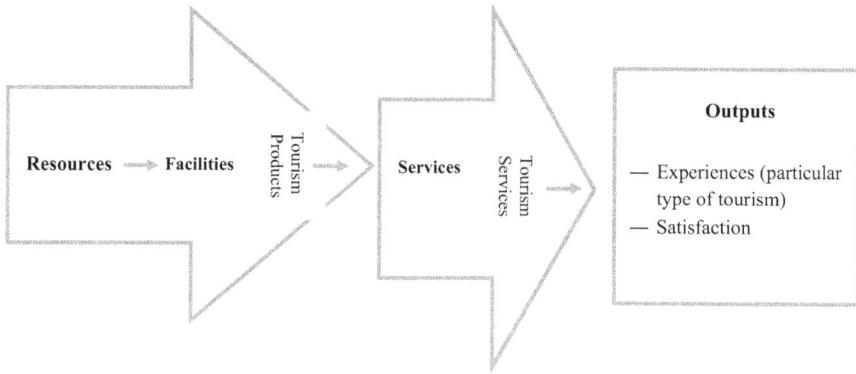

Figure 2.1 Tourism products and services along the line of tourism production function
Source: Adapted from Smith (1994, p. 591)

function. However, resources and facilities are mostly destination-centric and signify the physical plant element of tourism products. Services inherently include the hospitality essence whereas desired outputs necessitate freedom of choice and involvement of tourists.

Methods

This study has been anchored with a qualitative research focus. The authors focus on a desk-based literature survey. The desk-based literature review made reference to both peer-reviewed publications as well as "grey literature" like secondary data sources. The literature search was framed by the keywords from the title of this paper such as tourism products, tourism services, Bangladesh tourism and so on. The thematic screening of literature has been further scoped and discussed within the conceptual framework as developed in the previous section (see Figure 2.1).

Brief profiling of tourism products and services in Bangladesh

Bangladesh is a country in the South Asian region being blessed with a variety of attractive tourism products. However, lack of proper identification and categorization of these products left the country behind in utilizing such resources to attract appropriate tourists both from inbound and domestic tourism considerations. This chapter attempts to address the gap by concentrating on the identification, categorization, and discussion of various existing tourism products in Bangladesh.

Tourism products are a combination of tangible and intangible elements in the tourism production processes. These products can vary from events, things,

Table 2.1 Categories of existing tourism products in Bangladesh

Generic Types	Identification	Elements/Components
Resources	Core or physical plant	Climate (Topography), Beaches, Islands, Countryside, Scenic Landscape, Flora and Fauna, Forests, Waterfall, Wildlife, etc.
Facilities	Facilitate service offerings and broaden freedom of choice	Places of Historical Significance, Celebrations, Museums and Art Galleries, Parks and Zoos, Religious Institutions, Traditions, Fairs and Festivals, Native Life and Customs, Arts and Handicrafts, Shopping Malls, Cinema Halls and Theatres, Theme Parks and Amusement Parks, etc.
Services (Secondary or Peripheral)	Creating a firsthand impact on outputs	Accommodation, Transportation, Travel Agent and Tour Operators, Tourist Police, Convention Centres and Conferences, etc.
Outputs	Focus on tourism experiences	Sites Tourism, Eco-Tourism, Rural Tourism, Cultural Tourism, Ethnic Tourism, Spiritual/ Religious Tourism, Medical Tourism, Dark Tourism, etc.

or places to people and even organizations that motivate tourists to visit a destination. Following the conceptual framework, existing tourism products can be categorized as shown in Table 2.1.

The following sub-sections discuss briefly the various categories of tourism products, which are summarized in Table 2.1.

Resources

Resources indicate the core of a tourism product and these usually represent natural resources and environmental components that attract visitors at a particular destination. Good weather and climate play a significant role in making an enjoyable holiday. The recent growth trend of resorts (in remote places) in Bangladesh to some extent corresponds with the favourable weather and climate. This is particularly meaningful in a situation where tourists from countries with extreme weather usually prefer destinations with favourable weather. Bangladesh has diverse natural resources which can be marketed to attract tourists. For example, the country has the longest unbroken sandy sea beach in the world (i.e. Cox's Bazar). Besides this beach, there are also numerous beaches that can attract more and more tourists such as Inani Beach, Patenga Sea Beach, and Kuakata Sea Beach. Inani Beach in real terms can be counted as a part of Cox's Bazar. Patenga Sea Beach is located in Chittagong whereas Kuakata Sea Beach is located in the southeastern part of the country in Patuakhali District. In recent times, it has been observed that in all these beach areas tourism activities are increasing including parasailing, motorboat rides, surfing, swimming facilities and so on.

Topographical features of Bangladesh have exhibited six different seasons each of which beautifies with unique pattern. Such features also minimize the vulnerabilities associated with seasonality in tourism demand (Rahman et al., 2020). For nature and adventure lovers, islands can also be an enthusing choice. Islands of Bangladesh offer natural scenic beauty with exotic flora and fauna; places of worship like churches and temples; local lifestyles, local foods, fairs and festivals and so on. Popular island destinations in Bangladesh include St. Martin's, Moheskhali, and Nijhum Dip. However, more island destinations particularly in the Barisal division are approaching such as Bhola Island and Monpura Island.

Bangladesh is mostly a low-lying country except the Chittagong Hill Tracts (CHTs) in the southeast and Sylhet's northeast region. The formation of hills in these regions is soil-based unlike many other places in the world where rock-based surfaces are quite visible. Such kind of formation impacted upon the vegetation and featured with tropical semi-deciduous forests. Altogether the regions provide scenic landscapes as core tourism products. Alongside these, a few protected areas are increasingly becoming an attraction for visitors. Bangladesh has 19 national parks and 20 wildlife sanctuaries along with five noteworthy conservation sites and one World Heritage Site (i.e. Sundarban Mangrove Forest) (Forest Department, 2019). Beside forests, there are also some noteworthy waterfalls that attract tourists including Madhabkunda, Ham Ham, Jadipai, Nafa Khum, Shuvolong, and Shitakundo. Another tourist attraction is the valleys, also known as dale. In recent years, Sajek Valley in the Rangamati Hill District is found as the most prominent one in this segment. The valley is approximately 1,800 feet above sea level and located near the Indian border city of Mizoram.

In terms of water-based tourism, Bangladesh has immense potential since the country represents the biggest river delta in the world. The numbers, sizes, and networks of rivers in Bangladesh raise the prospect of "riverine tourism". However, river cruises in the country focused mostly on the Sundarban area with a typical three- to four-night stay (Siddiqi, 2018). In addition to rivers, a few lakes are also contributing in tourism such as Foy's Lake, Meghla Lake, Kaptai Lake, Mahamaya Lake, Madhobpur Lake, and Bagakain Lake.

Facilities

Facilities are human-created elements that facilitate service offerings and broaden freedom of choice for visitors. Industry stakeholders generate a combination of tour packages in concentration of various facilities at destinations. The culture of a host community perhaps nowadays becomes a great source of inspiration for visitors, which is created by humans and transformed from one generation to another. There are numerous cultural elements that exist in society that help tourism businesses to broaden the range of offerings to visitors. One of the most important elements can be found as places of historical significance. Bangladesh has a number of archaeological sites being criticized on the ground of lack of facilities to market as tourism products. A few mentionable names are Dharmarajika Buddha Vihara, Mahasthangarh, Paharpur Mahavihara, Golden Temple,

Kantaji Temple, Shaat Gambuj Mosque, Ahsan Manjil, National Martyr's Memorial, Central Shahid Minar, Lalbag Fort and so on. Some of these sites are also regarded as holy places for religious tourists such as Shaat Gambuj Mosque for Muslims. Among these sites, UNESCO has declared Paharpur Mahavihara and Shaat Gambuj Mosque as World Heritage Sites.

The qualities and varieties of museums and art galleries at a destination are good indicators in terms of destination heritage and cultural richness as well as edification from an evolutionary perspective. These are the places to acquire knowledge and information about a destination (resources). There are varieties of museums around Bangladesh to be marketed as tourism products. In this vein, poor management of tourism facilities has been observed in most of the settings. A few notable museums and galleries include Bangladesh National Museum, Bangabandhu Memorial Museum, National Museum of Science and Technology, Cox's Bazar Fisheries Museum, Fish Museum and Biodiversity Centre, Bangladesh Maritime Museum, Liberation War Museum, Tribal Museum, National Art Gallery and so on. Besides museums and art galleries, the country has many attractive zoos and parks to provide entertainment to tourists. Some significant zoos of the country are Dhaka National Zoo, Gazipur Borendra Park, Nijhum Dhip Park, Dulhazra Safari Park, Bangabandhu Sheikh Mujib Safari Park and so on.

Bangladesh has different religious events and festivals that involve great gatherings to share spiritual thoughts. Religious places are destinations where tourists of different religions can visit for holy purposes. People from home and abroad travel to these sacred places and experience events with joy and religious spirits (Alauddin et al., 2014). Bishwa (World) Ijtema, a religious congregation of Muslims after the *hajj* can be cited in this connection. Similarly, some other events can be identified for other religions, for example, Durga Puja and Krishna Janmashtami for Hindus, Buddha's birthday and Modhu Purnima for Buddhists, and Christmas for Christians. These events are promising tourism products to promote VFR (Visiting Friends and Relatives) tourism. Apart from the individual religious perspective, there are events in Bangladesh that offer a cosmopolitan view. For instance, Bengali New Year or *Pahela Baishakh* (in Bengali), International Mother Language Day, Lalon Fair and so on. In summary, Bangladesh has immense prospects in terms of event- and festival-focused tourism (Rahman et al., 2019).

The ethnic lifestyles of different communities at different parts of the country are considered as distinct tourism products (facilities). A few ethnic communities in Bangladesh can be noted as Chakma, Marma, Tripura, Tanchangya, Mro, Santals, Khasi, Rakhain, Garos, and Manipuri. Most of these communities formed their ancestral root to the "mongoloid" group. They possess unique culture, religion, beliefs, and cultural resources which appeal to tourism facilities. Protection and preservation of ethnic lifestyles remain key considerations for a pro-tourism government (who develops tourism-friendly policies). Additionally, native life and customs of ethnic communities can be a good subject to research and generate knowledge.

Bangladesh has a strong cultural background for dance and music. The most remarkable dances in Bangladesh are folk dances, Baul dances, Chhau dance, Dak dance, Dhali dance, Fakir dance, Gambhira dance, Ghatu dance, Jari dance, Kali dance and so on. Some ethnic dances may include Chakma bamboo dance, Marma prodip (lamp) dance, Santal dance, Manipuri dance, and so on. Lack of preservation and practice efforts lead to the extinction of some ethnic dances. Along with dance, the music of Bangladesh has a rich heritage whether it is classical or folk. In order to promote the music of Bangladesh, different music festivals are organized on regular periods such as the Dhaka World Music Festival, Dhaka International FolkFest and so on. Finally, arts and handicrafts have always remained important for tourists to recreate memories after years. Tourists travel to a destination and buy handicrafts or art works as souvenirs. In Bangladesh, destination-centric (local) souvenir shop facilities are not available while most of the shops are located at city or urban level. This situation impacts negatively on local supply chain effectiveness and economic leakage.

Beside cultural and historical elements, commercial theme parks or amusement parks are recreational facilities where visitors can enjoy their leisure time. In Bangladesh, these kinds of recreational facilities are located at different destinations. For example, Fantasy Kingdom Amusement Park (theme park), Shishu (Children) Park, Heritage Park Concord, Water Kingdom, Foy's Lake Amusement World (theme park), Bangladesh Butterfly Park, Dream Holiday Park and so on.

Good shopping facilities are an essential requirement to satisfy tourists. This indicates a place where tourists can spend their time shopping for necessary products as well as souvenir items. To provide shopping facilities, large malls and markets are chiefly available in city-based destinations in Bangladesh such as Jamuna Future Park and Bashundhara City Shopping Complex at Dhaka, "new markets" at district levels, duty-free shopping facilities at different airports, and so on. These places also come up with sources of entertainment and recreation for visitors. For instance, different movie theatres (e.g. STAR Cineplex, Blockbuster Cinemas) provide leisure and recreation opportunities by watching or enjoying movies.

Services

The tourism industry requires a third-party intervention to ensure services provision. This industry is complex in nature and services are provided to optimize tourists' satisfaction. The varied tourism demands and needs of the tourists make it difficult to distinguish tourism products and services and often researchers identify the term as "tourism products" in general. Tourism services are usually catered by the intervention of tourism and hospitality industry (Goeldner and Ritchie, 2009).

To provide accommodation services with modern facilities and aesthetic appeals, many world-class and mid-range or economy-class hotels and resorts are established in Bangladesh. Some popular hotels and resorts in Bangladesh are

Intercontinental Hotel, Pan Pacific Sonargaon Hotel, Dhaka Regency Hotel and Resorts, Radisson Blu Water Garden Hotel, Dusai Resort and Spa, Elenga Resort (in Tangail), Padma Resort (in Munshigonj), Arunima Resort (in Norail), Royal Tulip Sea Pearl Beach Resort and Spa (at Inani Beach), Mermaid Eco Resort, Grand Sultan Tea Resort and Golf (in Sylhet), the Palace (in Sylhet), and so on. These hotels and resorts provide core services of lodging or augmented services of gym, spa, conference, convention, food and beverage, and many others.

Accessibility is one of the most important elements for tourism development in any country. A wide variety of transportation systems are provided in Bangladesh to ensure greater accessibility. The surface transports in Bangladesh include road vehicles (bus, rickshaw, auto-rickshaw, tuk-tuk, cars, taxis, etc.), railway, and waterway vehicles (ships, launches, ferries, boats). For air transport, many local airlines in Bangladesh are performing both domestic and international flights such as Biman Bangladesh Airlines (the national carrier), US Bangla Airlines, Regent Airways, Novo Air and so on. Both the accommodation and accessibility services are made available to visitors through wholesale and retail services, in which tour operators provide the wholesale services and travel agents deliver retail services. In Bangladesh, there are around 586 agencies operating tours at home and abroad (Tour Operators Association of Bangladesh, 2019). Being the key industry stakeholders, travel agents and tour operators handle all the tour-related activities including hotel room reservation and cancellation, package design, ticket reservation, networking with local and international tourism ministry/industry. On this note, the industry requires highly trained professionals with necessary expertise to keep up destination developments initiatives.

Different people travel for different purposes. Of them, business travellers are a rewarding market segment. These types of travellers travel with a purpose to arrange or attend seminars, workshops, conferences, and the like. A few notable conference and convention centres in Bangladesh are Bangabandhu International Conference Centre (BICC), BRAC Centre, Krishibid Institution Bangladesh, Spectra Convention Centre (SCC), Bangladesh China Friendship Conference Centre (BCFCC) and so on. Various types of training, conferences, conventions, seminars, meetings, and exhibitions are hosted at these venues. These venues aim to provide standard services, luxurious banquet halls, business conference facilities, car parking facilities, and so on. The most recent modern and well-furnished convention hall and conference centre located in Bangladesh are the place for weddings, business conferences, seminars, and socio-cultural gatherings.

Outputs

Based on the quality and quantity of resources, facilities and services provision diverse types of tourism experiences can lead to satisfaction or dissatisfaction of visitors. Bangladesh as a potential tourism market in the South Asian region has the latent features to produce the following types of tourism experiences.

Sites tourism

Visiting scenic beauty is one of the major reasons of tourism. In Bangladesh, there are many natural and manmade attractions. Both domestic and foreign tourists can visit Bangladesh to enjoy such attractions at various destinations within the country (Das and Chakraborty, 2012). At a specific level, river-based tourism can be a good example.

Eco-tourism

Given the diverse natural resource base, Bangladesh has potential to explore ecotourism destinations (Tuhin and Majumder, 2011). However, lack of proper planning and unstructured involvement of the private sectors raise the ultimate challenges for ecotourism success in Bangladesh (Rahman and Shahid, 2012).

Rural tourism

Rural tourism is tourism that basically takes place in a rural setting by observing and/or experiencing local lifestyles and resources. The prospect of rural or village tourism is immense in Bangladesh provided the number of villages across Bangladesh. In a broader spectrum, rural tourism focuses three core elements: people, space, and products (Bran et al., 1997). The product elements of rural tourism emphasize various attractions and activities to engage visitors in the most satisfying state of affairs (Ahmed and Jahan, 2013; Rahman et al., 2018).

Cultural tourism

With a diverse range of cultural elements, Bangladesh has exhibited opportunities for cultural tourism such as ethnic tourism, religious tourism, and gastronomy. For example, the country holds a secular philosophy and visitors from the major religions (e.g. Buddhism, Islam, Hindu, Christianity) pay visit to numerous religious monuments and sacred places to satisfy their spiritual souls (Al-Masud, 2015).

Medical tourism

Medical tourism is a recent dimension in Bangladesh that is based on providing health care facilities and services to tourists (both domestic and international). The country fails to develop this niche market of tourism in comparison to neighbouring country India (Hassan et al., 2015). The rapid growth of numerous medical centres, hospitals, and pharmaceutical sectors bear a clear indication of improvement in terms of medical facilities. However, to get benefit from this particular segment, alternative forms of treatment can give a good result such as ayurveda or yoga.

Dark tourism

Dark tourism means travelling to places associated with tragedy, grief, and similar themes. Bangladesh is a destination exhibiting a number of historical sites with the appeal of dark tourism. For example, the Central Shaheed Minar, the Bangabandhu Memorial Museum (where the father of the nation was assassinated), War Cemetery and the Liberation War Museum are a few distinguished dark tourism sites. To conceptualize, the Central Shaheed Minar is a monument inside the University of Dhaka campus that signifies the sacrifice made by the Bengali people for their mother language.

Conclusion

This study explores that Bangladesh has a wide range of tourism products to attract both inbound and domestic tourists and thereby enable different forms of tourism. Although tourism products and services are used sometimes as separate terms, literature indicates "tourism product" is a wider concept encapsulating "tourism services" within the definitional frame of the former term. Accordingly, tourism products emphasize the core products or resources whereas services highlight ancillary activities that shape ultimately the core products and enhance marketability of a destination. The categories of tourism products include resources, facilities, services, and outputs. The basis of categorization of tourism products has followed the generic tourism production process. The current study identifies an array of existing tourism products and services, which can be used for tourism promotion in Bangladesh. In addition, proper balancing of demands and supply needs to be considered that can also be supplemented by effective policy planning and ensuring efficient management of these products and services.

The implication is quite obvious to policymakers who can develop appropriate plans and strategies based on the categorization of existing tourism products. This study however is exclusively based on secondary sources of information. Future research can add different primary sources such as site visits, industry expert interviews and so on to ensure rigorousness of the findings. Besides the methodological issue, an operational focus can add value in terms of effective marketing and management of a destination at national scale i.e. Bangladesh or at some local scales such as regional level (e.g. the Chittagong Hill Tracts region).

References

Ahmed, I. and Jahan, N. (2013). Rural tourism-prospects in rustic Bengal. *European Journal of Business and Management*, 5(16), pp. 163–172.

Alauddin, M., Shah, M. G. H. and Ullah, H. (2014). Tourism in Bangladesh: A prospects analysis. *Information and Knowledge Management*, 4(5), pp. 67–73.

Al-Masud, T. (2015). Tourism marketing in Bangladesh: What, why and how. *Asian Business Review*, 5(1), pp. 13–19.

Armstrong, G. and Kotler, P. (2016). *Marketing: An introduction*. New York, NY: Pearson.

Bran, F., Marin, D. and Simon, T. (1997). *Rural tourism: The European model.* Bucharest: Economic Publishing House.

Cirikovic, E. (2014). Marketing mix in tourism. *Academic Journal of Interdisciplinary Studies*, 3(2), pp. 111–115.

Das, R. K. and Chakraborty, J. (2012). An evaluative study on tourism in Bangladesh. *Research Journal of Finance and Accounting*, 3(1), pp. 84–95.

Forest Department. (2019). *Management and conservation: Protected areas.* Retrieved from: http://bforest.gov.bd/index.php/protected-areas (accessed: the 30th October 2019).

Goeldner, C. R. and Ritchie, J. B. (2009). *Tourism principles, practices, philosophies.* Hoboken, NJ: John Wiley and Sons.

Hassan, A., Ahamed, M. U. and Rahman, M. S. -U. (2015). The development, nature, and impact of medical tourism in Bangladesh. In M. Cooper, K. Vafadari and M. Hieda (eds.), *Current issues and emerging trends in medical tourism.* Hershey, PA: IGI Global, pp. 294–309.

Koutoulas, D. (2004). *Understanding the tourist product.* Retrieved from: www.researchgate.net/publication/280317594 (accessed: the 15th March 2019).

Lewis, C. and Chamber, R. (1989). *Marketing leadership in hospitality.* New York, NY: Van Nostrand Reinhold.

Mrnjavac, Z. (1992). Defining tourist product. *Acta Turistica*, 4(2), pp. 114–124.

Rahman, M. S.-U., Kabir, K. and Hassan, A. (2019). Tourism events in South Asia: Brief profiling of cultural celebrations. In A. Hassan and A. Sharma (eds.), *Tourism events in Asia.* Oxon: Routledge, pp. 20–35.

Rahman, M. S.-U., Muneem, A. A., Avi, M. A. R. and Sobhan, S. (2018). Can rural tourism promote sustainable development goals? Scoping rural tourism prospects in rustic Bangladesh. *Rajshahi University Journal of Business Studies*, 11(1), pp. 131–144.

Rahman, M. S.-U. and Shahid, R. B. (2012). A growing dilemma of tourism diffusion and sustainability: Wows and woes for Bangladesh eco-tourism! *UTMS Journal of Economics*, 3(1), pp. 57–69.

Rahman, M. S.-U., Simmons, D. G., Shone, M. and Ratna, N. (2020). Co-management of capitals for community wellbeing and sustainable tourism development: A conceptual framework. *Tourism Planning & Development*, 17(2), pp. 225–236.

Siddiqi, R. (2018). *Prospects of riverine tourism in BD.* Retrieved from: www.observerbd.com/details.php?id=129918 (accessed: the 25th June 2019).

Smith, S. (1994). The tourism product. *Annals of Tourism Research*, 21(3), pp. 582–595.

Tour Operators Association of Bangladesh (TOAB). (2019). *Member directory of tour operators association of Bangladesh.* Retrieved from: www.toab.org/index.php?action=search_name (accessed: the 5th November 2019).

Tuhin, M. K. W. and Majumder, M. T. H. (2011). An appraisal of tourism industry development in Bangladesh. *European Journal of Business and Management*, 3(3), pp. 287–299.

World Tourism Organization (UNWTO) (2019). *Tourism products.* Retrieved from: http://marketintelligence.unwto.org/content/tourism-products (accessed: the 20th May 2019).

3 Tourism and air transport sustainability in Bangladesh

The role of technology

Nor Aida Abdul Rahman,
Muhammad Shoeb-Ur-Rahman
and Azizul Hassan

Introduction

Bangladesh is located in the northeastern part of the Indian subcontinent. From the recent update by United Nations data published in Worldometer (2020), the current population of Bangladesh is 164,040,611 which is equivalent to 2.1% of the total world population. The development of the tourism industry in Bangladesh began about 20 years ago. As reported in CEIC data, Bangladesh receives tourists from many countries such as Australia, Canada, India, France, Germany, Italy, Japan, Netherlands, New Zealand, Norway and many more. There is a significant growth of tourists visiting Bangladesh. For example, Bangladesh visitor arrival grew from 23.6% in December 2017 to 29.1% in 2018. Among Bangladesh's tourist attractions are include beaches, resorts, forest, wildlife species, historical monuments and many more. In Bangladesh, both the Ministry of Tourism and the Bangladesh Civil Air Transport Ministry are working together in designing the policy for air transport and promoting the tourism industry.

Currently, we are living in the era of technology. Our economy is driven by the technology advancement. In relation to tourism, technology has changed the way we travel. For instance, we can recognize the development of technology in the air transport industry. We can see rapid development of the airline industry especially on the innovation activity in aircraft manufacturing. Martin (2018) highlights the four key developments in air transport technology worldwide including Bangladesh. They are maintenance repair and overhaul (MRO), cloud computing, drones and safety and security systems. Table 3.1 sheds light on the role of each development and how it affects air travel and the tourism sector.

Tourism in Bangladesh – popular tourist attractions

The global air transport and tourism industry is growing, including in Bangladesh. Even though Bangladesh is small in terms of size, however the country has its own attractions and natural splendour. Bangladesh has a strong prospect in tourism activity from its natural beauty forest and its culture. As highlighted by

Table 3.1 Technology and four key developments in air transport technology

Key Development on Technology in Air Transport	Description	Affect to Air Transport and Tourism Activity
Maintenance, repair and overhaul (MRO)	Involve inspection activities of the aircraft including scheduled and preventive maintenance to ensure its airworthiness. All air transport players (airline) should comply with their local and international regulator and standards. Has to ensure safety and airworthiness of all aircrafts by international standards.	Support airline fleet and reduce possibilities of the aircraft to be delayed. To support aircraft to fly as scheduled.
Cloud computing	Refer to data storing and analysing data. Cloud computing helps the air transport player to improve their customer service activity. Help the air transport player to store their huge data at one safe place. Data stored and analysed here consist of data from maintenance, flying schedule, passenger's data etc.	Since consoles connect the data from the cloud computing with regards to passenger information and flight schedule, this technology helps tourists or air passengers to print their own labels and boarding passes. By using cloud computing, all data are also accessible to the authorized airline staff to access and update the data.
Drones	Drones or also known as UAV aircraft (unmanned aerial vehicle) refer to aircrafts with no pilot. Normally they are used to enhance visual checks made by engineers, for example fuselage engineers.	By using this technology, it could help the airlines to ensure the flight schedule is followed as planned. No delay leads to increased passenger/tourist satisfaction.
Safety and security system	Refer to guidelines or standards established by local and international authority. This safety and security system alerts the responsibilities of the airlines, airport operator, ground handler and all air transport players to implement safety procedures and security.	Safety ensures the aircraft are safe to fly and would not lead to the injury or loss of passengers or tourists. Security systems help to gather passenger intelligence or information, which follow pre-boarding and boarding procedure, which ensure the airlines to monitor their passenger activity. Only authorized passengers allowed onboard.

Source: Developed by the authors, 2020

Table 3.2 Hills as key popular attractions in Bangladesh

Hill's Name	Location in Bangladesh	Elevation
Chittagong Hills	Chattogram	1052m
Saka Haphong Hills	Bangladesh-Burma border	1063m
Mowdok Mual	Bangladesh-Myanmar border	1022m
Keokradong	Bandarban	986m
Zow Tiang	Bangladesh-Myanmar border	1021m
Dumlong Peak	Belaichori, Rangamati	1010m
Chimbuk Hills	Bandarban	762m

Source: Developed by the authors, 2020

Mondal (2017), Bangladesh is a popular country with its forest and biodiversity. Its cultural product has a strong attraction for local and international tourists worldwide to come and visit Bangladesh. An earlier study from Islam and Nath (2014) emphasized that the strong attraction point of Bangladesh is the hills. Bangladesh is a popular country and recognized by its hills. For instance, Chittagong Hills, Saka Haphong Hills, Mowdok Mual, Keokradong, Zow Tiang, Dumlong Peak, Chimbuk Hill and many more (see Table 3.2 for details). These attractions and popular places promise the tourist exciting and interactive experiences in Bangladesh.

It is acknowledged that the tourism industry supports many other businesses such as food, medical, as well as transportation including air transport. Technological advancement has changed the way tourists travel and this technology development also promises exciting experience to the tourist and provides guidance to the tourist in searching for the best food, facilities, places, transport and many more. With regards to technology, tourism and policy in Bangladesh, the first policy related to tourism and air transport was introduced in Bangladesh in 1992. Among the key organizations on the air transport and tourism policy development in Bangladesh are Bangladesh Parjatan Corporation, Ministry of Civil Air Transport and Tourism and Bangladesh Tourism Board (Hassan and Burns, 2014).

Tourism, air transport and technology

The global air transport and tourism industry is growing. It is reported that the current global airline revenues already exceed £600 billion per annum (IATA). While the tourism industry is also expected to reach USD 11,382 billion by year 2025. According to International Air Transport Association (IATA) (2019), it is expected that air transport will carry approximately 7.8 billion passengers by year 2036. Technology plays a significant role for the sustainability of any business. It is undeniable that at present, technology plays a critical role to many industry including air transport and tourism and it change the business landscape to become more effective and efficient (Rahman et al., 2019a).

Given the significance of the air transport and tourism industry on social and economic spheres, the focus of this paper is on the technology innovation and adoption of technology in air transport and tourism industry in Bangladesh. It is vital to highlight this issue since technology is documented as a key factor for business sustainability in the 21st century. Hence, if it is left unaddressed, air transport and the tourism industry may fail to grow and become insufficient to compete with other countries. With that, there is a strong reason to look at the technology issue in relation to air transport and air transport business sustainability (Kim et al., 2019). In fact, there is a critical need to further look into how technology play roles across air transport or the air transport industry and tourism industry in Bangladesh. The following research questions are developed in this study: First, why is technology important for the Bangladesh air transport industry? Second, how could technology improve tourism activity in Bangladesh?

From the review of literature, there are many studies that look into air transport technology in mobile application. For instance, Lee et al. (2012) explore technology readiness as a means to predict passenger adoption of check-in kiosks. Taylor (2016) also focused on technology adoption and adoption behaviours specifically relating to mobile applications among the passengers at retail. More recently, Martin-Domingo and Martín (2016) have also explored the use of mobile apps among the tourists at European airports. Recently, Sivarajah et al. (2019) investigate technology usage and digital transformation in helping the industry to gather competitive intelligence in facilitating business sustainability.

Since we are living in the 21st century era, known as the industry 4.0 era, there is a need to explore technology components in more detail. In this technology connectivity era, there is a need to shape the discussion on nine pillars of technology as highlighted by Rahman et al. (2019b). The nine pillars of technology discussed are autonomous robot, simulation, horizontal and vertical integration, internet of things, cybersecurity and block chain, cloud computing, additive manufacturing, augmented reality and big data analytics.

The next section will briefly discuss the nine pillars of technology and air transport in supporting the tourism industry in Bangladesh.

The importance of technology for the Bangladesh air transport industry

Technology development in Bangladesh is vital for economic growth. Historically, the cultivation of technology or modern science was started many years ago since British ruling in Bangladesh via formal education institution establishment in 1921 in the field of science. The internet was introduced a bit late in Bangladesh in 1996. As mentioned by Islam and Rahman (2006), compared to other developing countries, the use of information technology is not at par. This happened because of the lack of awareness about technology and the importance of information technology for industry developments. The first industry that used the internet and information technology was banking, followed by other industries including air transport and tourism.

From the air transport point of view, technology development is crucial in air travel as technology is the main platform for effective communication (Abdul Rahman, 2012). In fact, technology also enhances communication between air traffic controller and pilot, airline with their ground handler and airport authority, and also with their passengers or travellers. From the operation point of view, technology also serves as a tool for improvement for better airline operation, improved ordering catering system for in-flight meals, improved communication with passenger and many more. Universally, the air transport industry is recognized as a highly regulated industry and known as one of the crucial industries that boosts tourism activity and economic growth of any country. The main players in the air transport industry are coming from airline, airport and cargo sectors. Recent technology advancements in the service industry such as aviation have led to the conversion of service delivery from face-to-face to self-service technology. As published in the *International Airport Review* (2017), there are six key technologies that will revolutionize the aviation and airport industry worldwide including in Bangladesh. They are block chain technology, drone technology, augmented reality, artificial intelligence or also called as AI, airline new distribution capability and beacon technology, which is related to indoor positioning systems.

The value of technology is highly emphasized in the air transport industry including the use of social media (Rahman et al., 2017). For instance, by using social media platforms, the airport or the airline could easily communicate and update any information to the travellers via Facebook, Instagram, Twitter or any other social media platform. Industry sectors in air transport particularly are natural adopters of cutting-edge technologies apart from transportation, automotive, telecommunication, electrical and electronics (Karaman et al., 2018). Technology has a ubiquitous presence in day-to-day lives for both consumer and business organizations. While for air transport and tourism scholars, these new and emergent technologies present exciting opportunities to manage these exchanges through the ability to collect and access large volume of data from the tourist or travellers. Yet, seemingly the only way that air transport industry players can remain relevant and competitive is via adopting technology and getting their business up to date with current technology developments.

How the nine pillars of technology could improve Bangladesh air transport and the tourism industry

Air transport or the air transport industry in Bangladesh would have more value in achieving economic development by concentrating and embedding the elements of technology in their supply chain activity. Technology usage not only helps the tourism and air transport players to improve communication systems, operations and business activities, but also help their supply chain to be more efficient according to the market trends and customer demand in term of strategies, products and services offered (Rahman et al., 2019b). Figure 3.1 below shows the nine pillars of technology discussed in this study and Table 3.3 discusses the notion of the technology and why is it important for the Bangladesh air transport industry.

| Augmented reality |
| Internet of Things |
| Horizontal and Vertical integration |
| Artifical Intelligence |
| Block Chain |
| Cloud Computing |
| Big Data |
| Autonomous Robot |
| Cybersecurity |

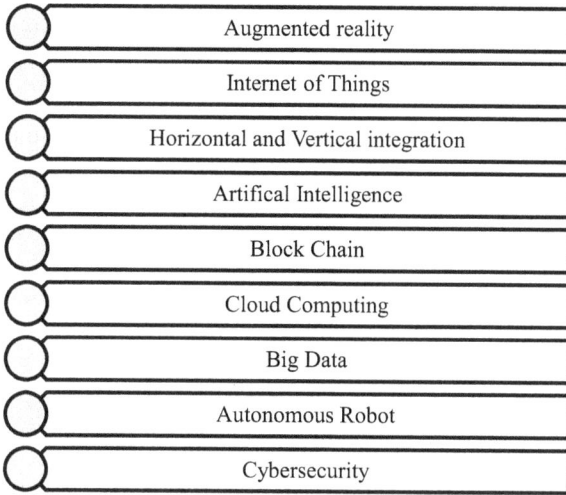

Figure 3.1 Nine pillars of technology in Bangladesh air transport

Source: Developed by the authors, 2020

Table 3.3 The notion of nine pillars of technology and why it is significant for the air transport industry or air transport in Bangladesh

Technology Pillars and References	Description	Importance to Bangladesh's Air Transport and Tourism Industry
Augmented reality (Safi et al., 2019)	Augmented reality refers to superimposed technology that is able to generate a real image on a user's view. It is a computer-generated image technology which is mostly used in training and simulation activity in many industries including air transport.	This technology allows travel agency to show the physical location of tourist attraction to their travellers. At the same time, for air transport, this augmented technology could improve air traffic control. It helps the air traffic controller (ATC) at the airport control tower to guide the aircraft to maintain safe distance of the aircraft. This is significant to ensure the travellers safety onboard.
Internet of things (IOT) (Xia et al., 2012)	Refers to connectivity or interaction of human to human or human to computer. This activity emerges from interrelated computing system or devices. Today, we live in internet of things in our daily life, for example in sending and receiving data, posting the image etc.	Internet of things (IOT) could streamline the operation among travel agency and the airlines by using the same system, process or internet devices.

(Continued)

Table 3.3 (Continued)

Technology Pillars and References	Description	Importance to Bangladesh's Air Transport and Tourism Industry
Horizontal and vertical integration (Cruijssen et al., 2007)	Refers to a communication system that integrates vertical and horizontal network in any organization.	The similar communication system used may help the travel agency and airline to improve their communication performance.
Artificial intelligence (Min, 2010)	Is a technology that helps to improve the decision-making process via data analysis and the trend of information.	This technology helps the travel agency or air transport player to provide assistance online to the passenger or traveller.
Block chain (Yang et al., 2019)	Refers to a time-stamped technology that records the transaction data using block.	Since the data in block chain is decentralized and information can be shared to peer-to-peer network, the data is more transparent, traceable and secured
Cloud computing (Qing et al., 2013)	Refers to the centre of data that gathers and stores all the information. The user could open and access their data from many different places.	This technology could help to streamline the resources from production, marketing, sales, hospitality, e-commerce, finance etc. In the air transport sector, it may help the airline for example to estimate travel time, aircraft identification traffic and emission.
Big data (Li et al., 2019)	Refers to a system with an extremely large size of data that consists of various information.	Could help the travel agency to identify their tourist target market, new clients and create opportunities for new business activity. In fact, by using this technology, the travel agency could also analyse the tourist trends by collecting all information from different air transport or consumer centres and develop a new strategy to market their travel product.
Autonomous Robot/ Drones (Ingrand and Ghallab, 2017)	Refers to unmanned aerial vehicles without the existence of a pilot and made of light composite materials to reduce weight and increase manoeuvrability.	Provide a bird's eye view and record footage for tourist activity for safety reasons. Could also monitor passenger or tourist cargo at the warehouse.

Source: Developed by the authors, 2020

Conclusion

To conclude, the recent wave of technology development has brought some limitations and contributions which open up new opportunities for both air transport and tourism sector to explore and improve their technological advancements in Bangladesh. This is vital to get updated with the technology to make both industries possible to sustain for the long run. From the academic point of view, there are many opportunities for scholars to further explore the technology development and its impact in the air transport and tourism perspectives. There is still a dearth of studies that look into usage of technology in Bangladesh from the perspective of air transport and tourism. In fact, scholars may also look at how technology affects the sustainability of the aviation business such as from airline, travel agent, airport, flight kitchen, air cargo provider and other perspectives. Another interesting topic to explore is to see how policymakers support the technology development in Bangladesh from the Ministry of Aviation and tourism perspectives. Since this study is still lacking, there are a lot of opportunities for future scholars to explore the issue of technology in aviation and the tourism sector. This study is limited to discussion on technology in relation to the nine pillars of technology only. Future scholars may also further develop these nine pillars of technology and empirically test how they affect the operational and performance of air transport and the tourism sector. This study contributes to the scholars, practitioners and policymakers in three different ways. Scholars could further discuss on the issue of technology more specifically from different perspectives of technology components in air transport and the tourism industry. Future research is encouraged to be more specific, for instance into type of data and how that data could be analysed and used in decision making and planning for every type of air transport organization including airline, ground handler, caterer, airport and MROs, as well as travel agencies.

References

Abdul Rahman, N. A. (2012). The car manufacturer (CM) and third party logistics provider (TPLP) relationship in the outbound delivery channel: A qualitative study of the Malaysian automotive industry. *PhD Thesis*. London: Brunel University Library.

CEIC (2020). *Bangladesh visitors arrivals: Annual*. Retrieved from: www.ceicdata. com/en/bangladesh/visitors-arrivals-annual (accessed: the 8th January 2020).

Cruijssen, F., Cools, M. and Dullaert, W. (2007). Horizontal cooperation in logistics: Opportunities and impediments. *Transportation Research Part E: Logistics and Transportation Review*, 43(2), pp. 129–142.

Hassan, A. and Burns, P. (2014). Tourism policies of Bangladesh – a contextual analysis. *Tourism Planning and Development*, 11(4), pp. 463–466.

IATA (2019). *Annual review*. Retrieved from: www.iata.org/publications/Documents/ iata-annual-review-2019.pdf (assessed: the 29th December 2019).

Ingrand, F. and Ghallab, M. (2017). Deliberation for autonomous robots: A survey. *Artificial Intelligence*, 247, pp. 10–44.

International Airport Review (2017). *6 technologies that will revolutionize the aviation and airport industry in 2017*. Retrieved from: www.internationalairportreview. com/article/26374/technology-revolutionise-aviation-2017/ (accessed: the 25th December 2019).

Islam, A. and Rahman, A. (2006). Growth and development of information and communication technologies in Bangladesh. *The Electronic Library*, 24(2), pp. 135–146.

Islam, M. J. and Nath, T. K. (2014). Forest-based betel leaf and betel nut farming of the Khasia indigenous People in Bangladesh: Approach to biodiversity conservation in Lawachara National Park (LNP). *Journal of Forestry Research*, 25(2), pp. 419–427.

Karaman, A., Kilic, M. and Uyar, A. (2018). Sustainability reporting in the aviation industry: Worldwide evidence. *Sustainability Accounting, Management and Policy Journal*, 9(4), pp. 362–391.

Kim, Y., Lee, J. and Ahn, J. (2019). Innovation towards sustainable technologies: A socio-technical perspective on accelerating transition to air transport biofuel. *Technological Forecasting and Social Change*, 145, pp. 317–329.

Lee, W., Castellanos, C. and Choi, H. S. C. (2012). The effect of technology readiness on customers' attitudes toward self service technology and its adoption; the empirical study of US airline self service check in Kiosk. *Journal of Travel and Tourism Marketing*, 29(8), pp. 731–743.

Li, M. Z., Ryerson, M. S. and Balakrishnan, H. (2019). Topological data analysis for air transport applications. *Transportation Research Part E: Logistics and Transportation Review*, 128, pp. 149–174.

Martin, M. (2018). 2018's four major air transport technology developments. Retrieved from: www.aircraftinteriorsinternational.com/industry-opinion/2018s-four-major-air transport-technology-developments.html (accessed: the 29th December 2019).

Martin-Domingo, L. and Martín, J. C. (2016). Airport mobile internet an innovation. *Journal of Air Transport Management*, 55, pp. 102–112.

Min, H. (2010). Artificial intelligence in supply chain management: Theory and applications. *International Journal of Logistics Research and Applications*, 13(1), pp. 13–39.

Mondal, M. S. H. (2017). SWOT analysis and strategies to develop sustainable tourism in Bangladesh. *UTMS Journal of Economics*, 8(2), pp. 159–167.

Qing, Li., Ze-yuan, Wang., Wei-hua, Li., Jun, Li., Cheng, Wang and Rui-yang, Du. (2013). Applications integration in a hybrid cloud computing environment: Modelling and platform. *Enterprise Information Systems*, 7(3), pp. 237–271.

Rahman, N. A. A., Mohammad, M. F., Hassan, R., Ahmad, M. F. and Kadir, S. A. (2017). Shipper's perceptions of aviation logistics service quality (ALSQ) of air freight provider. *Journal of Engineering and Applied Science*, 12(3), pp. 699–704.

Rahman, N. A. A., Muda, J., Mohammad, M. F., Ahmad, M. F., Rahim, S. A. and Mayor-Vitoria, F. (2019a). Digitalization and leap frogging strategy among the supply chain member: Facing GIG economy and why should logistics players care? *International Journal of Supply Chain Management*, 8(2), pp. 1042–1048.

Rahman, N. A. A., Rahman, N. A. A., Mohammad, M. F., Ahmad, M. F., Rahim, S. A. and Mayor-Vitoria, F. (2019b). Technology connectivity for air transport supply chain sustainability: A conceptual model. *Test Engineering and Management*, 81, pp. 5791–5798.

Safi, M., Chung, J. and Pradhan, P. (2019). Review of augmented reality in aerospace industry. *Aircraft Engineering and Aerospace Technology*, 91(9), pp. 1187–1194.

Sivarajah, U., Irani, Z., Gupta, S. and Mahroof, K. (2019). Role of big data and social media analytics for business to business sustainability: A participatory web context. *Industrial Marketing Management*, 4.

Taylor, E. (2016). Mobile payment technologies in retail: A review of potential benefits and risks. *International Journal of Retail & Distribution Management*, 44(2), pp. 159–177.

Worldometer (2020). *Bangladesh population*. Retrieved from: www.worldometers. info/world-population/bangladesh-population/ (accessed: the 1st February 2020).

Xia, F., Yang, L. T. and Vinel, W. A. (2012). Internet of things. *International Journal of Communication System*, 25, pp. 1101–1102.

Yang, A., Li., C., Liu, C. Li, J. Zhang, Y. and Wang, J. (2019). Research on logistics and supply chain of iron and steel enterprise based on block chain technology. *Future Generation Computer System*, 101, pp. 635–645.

4 Socio-economic impacts of accommodation on tourism development

Bangladesh perspective

Asma Akter Akhy and Mallika Roy

Introduction

Tourism can be characterized as those who travel for the reason of entertainment and relaxation beyond their normal environment for maximum one year so that they can have the option to get the services for mental and physical fulfilment. Accommodation is one of the basic needs of any tourism operation. These are establishments that provide the tourist with a place to stay i.e. hotel facilities that are paid by the tourist for the duration of the stay. There are different types of accommodation that are commonly used by visitors.

The Bangladesh government announced that 2016 would be the "Tourism Year" targeting one million tourists. The World Travel & Tourism Council (WTTC) (2016), the global authority responsible for travel and tourism's economic and social contribution, said travel and tourism's direct contribution to Bangladesh's GDP in 2015 was BDT407.6bn (2.4 percent of GDP).

The hotel industry is concerned solely with offering guest accommodation services. Conversely, in a more general sense, the hospitality industry is concerned with leisure. It therefore requires housing, restaurants, bars, cafes, nightlife and a variety of tourism and travel facilities. According to Rahman (2010), the contribution of the tourism sector was around 4.4% to GDP, 3.8% to employment generation and 1.5% to investment in 2013. In spite of the fact that the nation has enormous potential for economic growth utilizing the tourism industry, in correlation with neighbour nations, Bangladesh is the most minimal beneficiary of international tourist and the size of both international and domestic tourism industry is very insignificant. Due to some limitations, Bangladesh has failed to introduce itself as a tourist destination in the world.

Bangladesh has recently begun realizing that the nation has a massive potential in the tourism industry development in future. Although Bangladesh has not advanced much as far as planning and implementing physical plans of its urban communities and towns and different territories of monetary and resource potentials, current efforts in preparing the plan may assist the development of tourism in different sectors.

Literature review

Ali (2004) found that from the ancient period this area is famous for beauty. Foreign tourists praised this country for its natural beauty, rich cultural heritage and hospitality of the people.

Alegre and Garau (2009) examined dissatisfaction of sun and destination, namely the island of Majorca in the Balearic Islands. The following attributes were rated in terms of satisfaction: climate, cleanliness and hygienic scenery, peace and family, interaction with other tourists, night life, sports activities, tourist attractions, prior visits to the destination, ease of access, facilities for children, easy access to information, local cuisine, local lifestyle and affordable facilities.

Amin (2007) argued that tourism is now one of the largest industries in the world contributing over 10% to global GDP. Economically, travel and tourism creates jobs and contributes to a country's GDP as well as bringing in capital investment and exports.

World Economic Forum (2013) said that over the past several decades, travel and tourism has become a key sector in the world economy.

Baisakalova (2009) argued that the tourism development is a win-win strategy as it leads to creating jobs, developing infrastructure and improving sustainability of the country. Improving the business environment, eliminating barriers such visa obtaining procedures, providing incentives for innovations and knowledge development can result in growth of the inbound and domestic tourism.

Deegan and Moloney (2007) argued that in the context of the overall macro-economic activity there is good reason to believe that tourism can be a significant contributor to economic development in the year ahead.

Elena et al. (2012) described that tourism is one of the profitable sectors in Bangladesh.

The basic from of supplementary accommodation are youth hostels, motels, camping sites, bed and breakfasts, tourist holiday village, inns, guest houses, farmhouses, time share and hostels (Raju, 2009).

According to Ghosh (2001), there are two basic elements in tourism: the journey to the destination and the stay. In short tourism means the business of providing information, transportation, accommodation and other services to travellers.

Hossain and Nazmin (2006) found that according to the foreign tourists of Bangladesh, scenic beauty ranked first, cost of services second, attitude of the people third and so on down to facilities ranking tenth.

Islam and Islam (2006) argued that Bangladesh is a country of the Asian region holding high potentiality for tourism. Since long past, Bangladesh was an attractive destination to the tourists. But at present her position is not significant in terms of international tourism.

Johannesburg Summit (2002) stated that tourism helps a country directly in building necessary infrastructures which not only facilitate the tourist but the local community. In a destination country, tourism initiates the development of different relevant physical facilities and infrastructure.

Nath (2007) observed that the level of satisfaction with the overall tourism facilities and services is at a level of 51%. These tourists judge very poorly the level of night entertainment, tour information, advertisement and travel agencies.

Okaka (2007) described that the media can spur the current prospects for the African countries to actively collaborate in a wide range of eco-tourism enhancement activities which include: joint product or service development, human resources development and management, exchange of tourism expertise and tourism information within the existing economic and regional blocks.

Energy production, transportation and industrial process are the main factors in carbon dioxide emission. One big contributor to carbon dioxide emission is air transportation, which is closely related to the accommodation business. Airplanes cause by far the biggest carbon dioxide emission compared to train and buses (Planet Green, 2013).

Roy and Roy (2015) discussed that tourism can bring many economic, social and environmental benefits, particularly in rural areas and developing countries, although mass tourism is also associated with negative effects. She has shown that, due to the variety of tourist spots, tourism policy can contribute in the economy.

Sandip (2014) mentioned that the development of service industry will accelerate our economic growth. The study developed some competitive strategies in the light of vision 2021 and therefore economic growth.

Tuhin and Majumder (2011) opined that in recent times, tourism is the most significant and up-to-date business all over the world. International tourism ranks fourth after fuels, chemicals and automotive products in term of generation of export income.

According to UNWTO (2008), in Bangladesh, the tourism sector got recognition as an industry in 1999 (Dhaka Mirror, 2012) but it could not receive much attention to date from the government as a vibrant industry, where globally it is regarded as the fastest rising industry.

Again, according to UNWTO (2002), tourism can provide material benefits to the poor in many forms directly and indirectly. It can also bring long-term earning opportunity, cultural pride and sense of ownership and reduce vulnerability through diversification and development of personal skills of the poor.

Wang and Qu (2006) investigated tourist satisfaction using 12 variables: accommodation, shopping facilities, restaurant facilities, quality of accommodation, personal safety, tourist information, beach cleanness, the state of the roads, beach promenades, drinkable water, traffic flow and parking facilities.

Yilmez (2008) commented that as one of the largest and fastest growing sectors of the global economy, tourism consists of many small- and medium-size enterprises which try to be successful in an extremely competitive and rapidly changing business environment.

Zulfikar (1998) pointed out that tourism may be broadly divided into domestic and international tourism. In domestic tourism, people move within their own country whereas in international tourism, the barriers exist in travelling destinations beyond national boundaries.

It has been well described by different authors that tourism marketing is essential for economic development of a country. This study intends to do research work considering both primary and secondary sources. Most of the authors do either conceptual frameworks or secondary sources. To mitigate the gap of the research work this study used both quantitative and qualitative methods.

Thus, the objectives of this study are first, to assess the present accommodation conditions and opportunities of tourism sector in Bangladesh. Second, to express overall conditions by SWOT analysis on accommodation of tourism sector. Third, to sort out the limitations of the development of tourism in Bangladesh. And finally, to offer policy instruments that can be used to utilize tourism accommodation facilities properly.

Background of Bangladesh

Geography

Bangladesh is honoured with profound, rich and prolific soil, a blessing from the three significant streams (the Ganges, Brahmaputra and Meghna Rivers) that structure the deltaic plain whereupon it sits. This extravagance comes at a substantial expense, be that as it may. Bangladesh is on the whole level, and aside from certain slopes along the Burmese fringe, it as a rule adrift level. Accordingly, the nation is consistently overwhelmed by the waterways, by tropical cyclones off the Bay of Bengal, and by tidal bores. Bangladesh is bordered by India all around it, except a short border with Burma (Myanmar) in the southeast.

Climate of Bangladesh

The atmosphere in Bangladesh is tropical and monsoonal. In the dry season, from October to March, temperatures are gentle and lovely. The climate turns hot and damp from March to June, anticipating the rainstorm downpours. From June to October, the skies open and drop the greater part of the nation's complete yearly precipitation, as much as 224 inches for each year (6,950 mm). As referenced, Bangladesh regularly experiences flooding and cyclone strikes – a normal of 16 typhoons hit for each decade.

Economy

Bangladesh is a developing nation, with per capita GDP of about USD$4,200 every year starting in 2017. In any case, the economy is developing quickly, with around a 6% annual growth rate from 2005 to 2017. In spite of the fact that manufacturing and services are expanding in significance, practically 50% of the Bangladeshi workers are engaged in agriculture.

Major tourism attractions

Although there are opportunities for developing tourism attractions and facilities across the country, at present only a handful tourism spots are available for visits by tourists. These limited facilities are concentrated in places like Dhaka, Chittagong and Sylhet divisions. In Chittagong division, the sightseeing facilities are mainly located at Cox's Bazar, Rangamati, Khagrachari and Bandarban districts. Mainamati at Cumilla is an important archaeological site and Lalbagh Fort in Dhaka is a great historic site. In Sylhet division, the major tourist attractions are located mainly at Jaflong, Madhabkunda, the tea gardens and the shrines of Harzat Shah Jalal and Shah Poran. Some renowned attractions in Rajshahi division are Kantaji's Temple, Swapnapuri, Ramsagor and Rajbari at Dinajpur, Paharpur at Noagaon and Mahasthangarh at Bogra. In Khulna division, some attractions are Shatgombuj Mosque at Bagerhat and Sundarban, the largest mangrove forest in the world, is located in the southern part of Khulna division along the Bay of Bengal. This mangrove forest has been declared as a World Heritage Site (Ministry of Civil Aviation and Tourism, 2020). Some major tourism attractions are listed in Table 4.1.

Table 4.1 Major tourism attractions

Serial Number	Category	Site Name
1	Archaeological sites	Lalbag Fort, Mughal Eidgah, Ahsan Manjil, Sonargoan, Wari Bateshawar, Mainamati, Paherpur, Mahasthangar, Kantajew Temple, Sixty Dome Mosque
2	Beaches	Patenga, Parki, Cox's Bazar, Teknaf, Kuakata, Kotka
3	Religious places	Mosques, Hindu temples, churches, Buddhist temples
4	Hills and islands	Rangamati (The Lake District), Kaptai (The Lake Town), Bandarban (The Roof of Bangladesh), Khagrachari (The Hilltop Town), Mymensingh, Sylhet, Moheskhali Island, Sonadia Island, St. Martin's Island
5	Historical places	Museum of Father of the Nation, National Memorial, Central Shahid Minar, Martyred Intellectual Minar, National Poet's Grave, Curzon Hall, Baldha Garden, Sohrawardi Park, Old High Court Building, Bahdur Shah Park, Dighapatiya Palace, World War II Cemetery, Shilaidaha Kuthibri, Sagordari-Jessor, Mujibnagar Memorial, Trishal, Gandhi Asram
6	Forests and swamp forests	Sundarbans, Ratargul Swamp Forest
7	Other attractions	National Assembly Building, Bangabhaban, Shankhari Bazar, National Park, National Botanical Garden, National Zoological Garden, Batali Hill, DC Hill, Rajshahi, Jamuna Bridge, Kirtankhola, Madhabkundu, Jaflong

Source: Bangladesh Parjatan Corporation, 2020

Types of accommodation in Bangladesh

Camping and caravan sites can be in very basic field with few utilities provided compared to very sophisticated resort locations, including a high range of comfort services with lots of leisure, food services and retail choices (Cooper et al. 2008; Saxena 2008).

There are many different types of accommodation on offer, with each providing various facilities and different experiences. Types of accommodations available in Bangladesh are hostels, hotels, motels, resorts, capsule hotels, floating hotels, guest houses, cottages, camping and vacation rentals.

Hostels

A place while travelling, people, especially young people, can stay cheaply with shared rooms and sometimes private rooms.

Hotels

A hotel offers paying tourist accommodation and can also have various other facilities or amenities, such as a bar, swimming pool and/or spa. Most hotels organize conferences and seminars, making them appropriate places for meetings. Generally a one- to five-star rating exists as a way of measuring the quality of the hotels and their facilities.

Motels

Motel is a word derived from the words *motorist hotel*. They are situated on highways and open places. Tourists who are on a transitory and cost-conscious mode prefer to stay in such accommodations. These motels offer not only parking space but also some amenities such as television, swimming pools or restaurants.

Resort

A spot where a lot of people go for relaxation, sport or other stated purpose.

Cottage

In today's tourism sector, cottage is used to describe a small vacation house typically in a rural area.

Theoretical background

Figure 4.1 depends on the prevailing definitional components of the tourist experience found in the literature. The phasic nature is spoken to utilizing Clawson and Knetsch's (1966) Five Stages Model, yet in addition consolidates

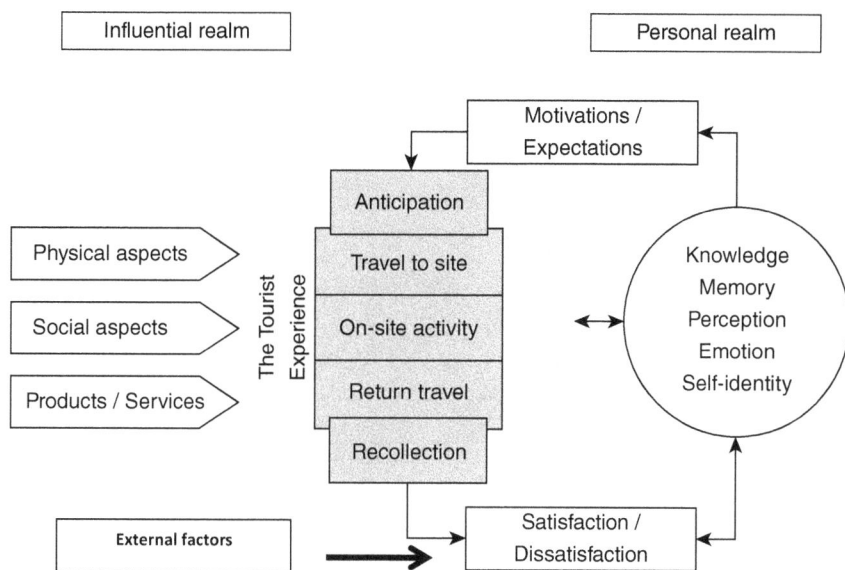

Figure 4.1 Tourist experience conceptual model of influences and outcomes
Source: Developed by the authors, 2020

impacts and individual results. Right now, tourists' experience is all that occurs during a visitor occasion (travel to site, nearby action and travel back). That being expressed, the expectant stage and memory period of the vacationer experience are still displayed, exhibiting how the traveller experience is arranged and foreseen before an outing takes place and recalled long after an excursion has completed. The expectation and memory stages likewise spill into the experience itself. This depends on the possibility that during movement to a site, the traveller could in any case be creating and refining desires for the goal similarly as return travel could include reflection on the excursion which has quite recently occurred.

During the experience, three classes of impacts are exhibited, including those components which are outside the person. The physical viewpoints include spatial and place-based components of the destination, while social perspectives incorporate the different social effects on experience. The impact of goods and services focuses on the elements, for example, service quality, relaxation activities accessible and the sort of traveller-related goods available.

The model is a combination of the influential realm and personal realm. Based on the tourist experience conceptual model of influences and outcomes, we tried to test the model to identify the variables of tourists' satisfaction and dissatisfaction.

Methodology, data analysis and results

Methodology

The study is based on primary and secondary data sources. Primary data were collected through a questionnaire responded by randomly selected domestic and foreign tourists of the country during the period from 14 October to 28 November 2019. As the primary data is concentrated on the tourists only, from the primary source the study purports to see the demand side views of the tourism sector. For the primary sources one questionnaire has been prepared. The questionnaire was kept succinct and consisted of both close-ended and open-ended questions, with a mixture of formats to keep the participants attentive and less likely to induce quick completion, and thus vague and absent-minded answering, which helped us for in-depth analysis. Structured and unstructured questionnaires had been used to collect data. In addition to this, some selected respondents were exclusively interviewed for in-depth analyses to strengthen the dimension of the study. An unstructured questionnaire was used to take interview. Interviews were mainly taken when tourists were visiting spots. However, respondents were so free and frank that collection of information was not so hard. Structured questionnaires were supplied to hotel counters with a request letter. Then after seven days those were collected from there. The major sources of secondary sources include publications of various authors, research reports, journals, websites and so on that are indicated in the references.

Sample size

The study is an empirical effort based on primary and secondary data. For collecting data, 20 foreign and 30 domestics tourists were selected randomly from five famous tourist spots, for a total of ten tourists from each spot: Dhaka, Cox's Bazar, Rangamati, Chittagong and Foy's Lake.

Respondents' profile

The survey group consisted of 50 respondents. The majority of them, that is, 30 respondents, were from Bangladesh. Among the selected 20 foreign tourists, 3 were from Nepal, 5 from Sri Lanka, 5 from Thailand and 7 from Korea.

Classification

The questionnaire has some classification questions, as the demographics of an individual has a large impact on that person's pre-environmental knowledge, perceptions, beliefs, interpretation and outcome of such a visit, as well as prejudice in filling out the questionnaire. The classification questions believed to be most relevant to the study are:

Question – Age
Question – Gender

Question – Nationality
Question – Marital status
Question – Employment
Question – Income

The age range for participants is 18+ for more appropriate completion and also adults are capable of answering the questions properly. Descriptive statistics for income and age are given in Table 4.2.

Married and unmarried respondents are quite similar in quantity in our survey. On the other hand, we collected information from various occupation holders such as students, service holders, businessmen, the unemployed and also those who have been retired from their job for an in-depth analysis of demographic behaviour of the respondents which will help us to get a good explanation.

Results

Pre-knowledge

Visitors can discover the travel industry data on web journals, discussions with friends or known ones, sites of focal points and so forth. It should be able to provide tourism information based on the user's preferences and current location. A rich website can provide better information than the people. In our result, 62% of visitors collected their required information from websites, whereas only 38% of people visited the tourist places recommended by their friends or known persons.

Table 4.2 Descriptive statistics for income and age

Income		Age	
Mean	52000	Mean	36.38
Standard Error	5952.19	Standard Error	1.929765
Median	50000	Median	35
Mode	100000	Mode	35
Standard Deviation	42088.34	Standard Deviation	13.6455
Sample Variance	1.77E+09	Sample Variance	186.1996
Kurtosis	−1.66035	Kurtosis	−0.27556
Skewness	−0.0089	Skewness	0.599866
Range	100000	Range	53
Minimum	0	Minimum	17
Maximum	100000	Maximum	70
Sum	2600000	Sum	1819
Count	50	Count	50
Confidence Level (95.0%)	11961.37	Confidence Level (95.0%)	3.878007

Source: Developed by the authors, 2020

Satisfaction/dissatisfaction based on motivation/expectation

Table 4.3a, 4.3b, 4.3c Summary statistics of the variables of accommodation of a tourist place

Natural Settings		Room Quality		Spaciousness	
Mean	4.72	Mean	4.62	Mean	4.52
Standard Error	0.0641427	Standard Error	0.06934092	Standard Error	0.071371406
Median	5	Median	5	Median	5
Mode	5	Mode	5	Mode	5
Standard Deviation	0.45355737	Standard Deviation	0.49031435	Standard Deviation	0.50467205
Sample Variance	0.20571429	Sample Variance	0.24040816	Sample Variance	0.254693878
Kurtosis	-1.0213505	Kurtosis	-1.8142904	Kurtosis	-2.07800566
Skewness	-1.01053	Skewness	-0.509877	Skewness	-0.082561867
Range	1	Range	1	Range	1
Minimum	4	Minimum	4	Minimum	4
Maximum	5	Maximum	5	Maximum	5
Sum	236	Sum	231	Sum	226
Count	50	Count	50	Count	50
Confidence Level (95.0%)	0.12889958	Confidence Level (95.0%)	0.1393458	Confidence Level (95.0%)	0.14342621

Food quality		Food cost		Transport cost	
Mean	4.9	Mean	4.8	Mean	4.12
Standard Error	0.042857	Standard Error	0.057143	Standard Error	0.101579
Median	5	Median	5	Median	4
Mode	5	Mode	5	Mode	4
Standard Deviation	0.303046	Standard Deviation	0.404061	Standard Deviation	0.718275
Sample Variance	0.091837	Sample Variance	0.163265	Sample Variance	0.515918
Kurtosis	5.791962	Kurtosis	0.407247	Kurtosis	-0.99714

Skewness	−2.74986
Range	1
Minimum	4
Maximum	5
Sum	245
Count	50
Confidence Level (95.0%)	0.086125

Skewness	−1.5468
Range	1
Minimum	4
Maximum	5
Sum	240
Count	50
Confidence Level (95.0%)	0.114833

Skewness	−0.18284
Range	2
Minimum	3
Maximum	5
Sum	206
Count	50
Confidence Level (95.0%)	0.204131

Location

Mean	4.92
Standard Error	0.0387562
Median	5
Mode	5
Standard Deviation	0.2740475
Sample Variance	0.075102
Kurtosis	8.5344877
Skewness	−3.192877
Range	1
Minimum	4
Maximum	5
Sum	246
Count	50
Confidence Level (95.0%)	0.0778834

Safe and secured

Mean	3.44
Standard Error	0.0709124
Median	3
Mode	3
Standard Deviation	0.5014265
Sample Variance	0.2514286
Kurtosis	−2.02037
Skewness	0.2492888
Range	1
Minimum	3
Maximum	4
Sum	172
Count	50
Confidence Level (95.0%)	0.1425038

Behaviour of hotel staff

Mean	4.04
Standard Error	0.056856
Median	4
Mode	4
Standard Deviation	0.402036
Sample Variance	0.161633
Kurtosis	3.655877
Skewness	0.342324
Range	2
Minimum	3
Maximum	5
Sum	202
Count	50
Confidence Level (95.0%)	0.114257

Cleanliness

Mean	4
Standard Error	0.02857143
Median	4
Mode	4
Standard Deviation	0.20203051
Sample Variance	0.04081633
Kurtosis	24.5
Skewness	0
Range	2
Minimum	3
Maximum	5
Sum	200
Count	50
Confidence Level (95.0%)	0.05741644

Source: Developed by the authors, 2020

These summary statistics shed light on the overall facilities of an accommodation place. We can see that according to tourist satisfaction mean of location, food quality, reasonable food cost, natural setting, room quality and spaciousness are 4.92, 4.9, 4.8, 4.72, 4.62 and 4.52, respectively, which indicates that these variables are in excellent condition. On the other hand, transport cost, behaviour of hotel staff and cleanliness scored means of 4.12, 4.04 and 4, respectively, which implies very good qualities of these variables. Only safety and security scored less, but obviously in quite good condition. Thus we can realize that the tourists are very satisfied.

Indoor and outdoor facilities

Accommodation is one of the essential requirements for any tourism activity. Explorers and vacationers need accommodation for rest while they are on a visit. Accommodation from low-budget lodgings to world-class lavish inns is accessible at all the significant traveller destinations to give the vacationer a home away from home. These are foundations that give a spot to the visitor to remain, and are paid for the span of the stay by the vacationer. There are different kinds of accommodation which are being utilized by sightseers consistently. Based on our research we have drawn a clear picture of indoor and outdoor facilities which are provided in various tourist spots in Bangladesh. All accommodation places have separate toilets and also balcony facilities (Figure 4.2a). So in this case, customers are quite satisfied.

From our study, we can see that room facilities are in excellent condition (Figure 4.2b). Most of the hotels, motels or cottages have TV (96%), AC (94%), Wi-Fi (96%) and land phone (96%). But 64% of them have refrigerators (fridge), which can be improved. Satisfaction level of overall indoor facilities is very high.

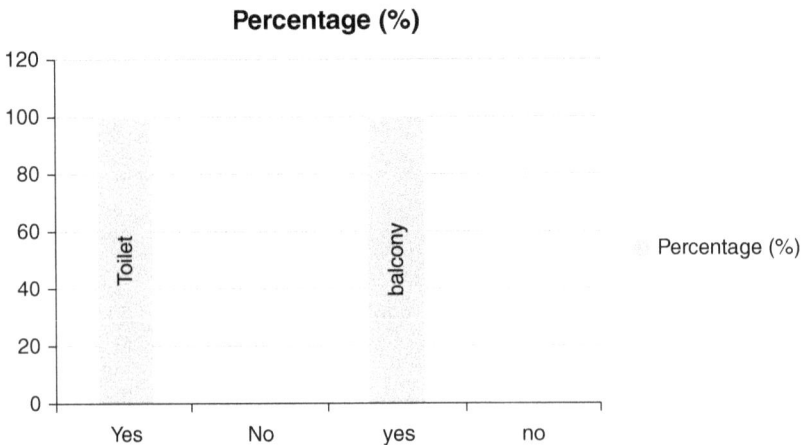

Figure 4.2a Respondents' facilities of toilet and balcony

Source: Developed by the authors, 2020

Room facilities (%)

yes ▪ no

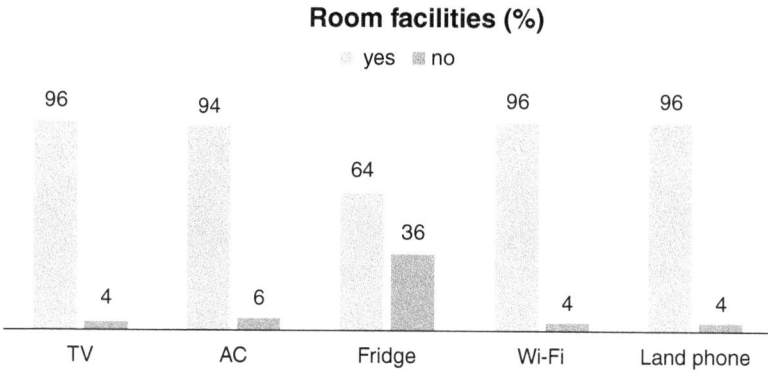

Figure 4.2b Respondents' facilities of TV, AC, fridge, Wi-Fi and land phone
Source: Developed by the authors, 2020

Some amenities (%)

▨ Some amenities (%)

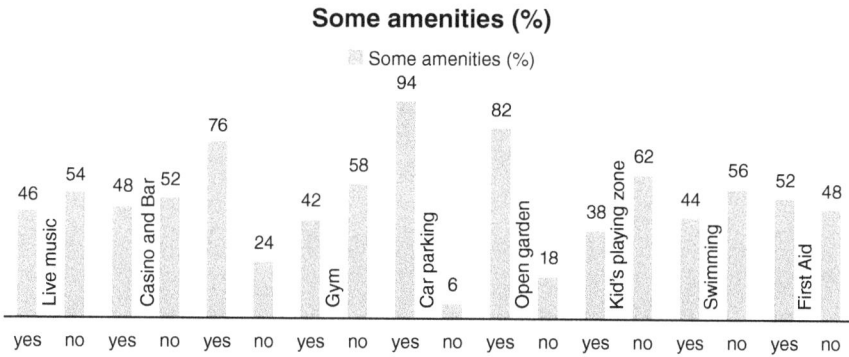

Figure 4.3 Respondents' outdoor facilities
Source: Developed by the authors, 2020

Some questions believed to be most relevant to the study are:

Question – Hospital far or near
Question – Food options available or not
Question – Pre-booking
Question – Pay in advance
Question – Breakfast facility
Question – Own transport

Hospital

A hospital is an institution that is constructed, staffed and prepared for the diagnosis of disease; for the treatment, both clinical and surgical, of the ill and the injured; and for their lodging during this procedure. In life, people can become

Hospital (%)

Hospital

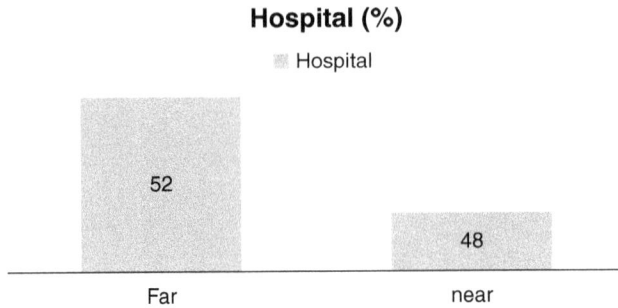

Figure 4.4 Respondents' hospital facilities

Source: Developed by the authors, 2020

sick anytime. Again, as we all know just by life experience, anyone can have an accident, and it can happen anytime or anywhere. On top of that, there are countless different types of accidents. So tourists should aware of this also. Hospitals near tourist sites are very helpful. In our study, we found that 48% of hospitals are near whereas 52% are far away. In our opinion, policy makers should implement policies for establishing more hospitals near the tourist places to provide medical services if needed.

Food options

Food is a significant segment in the tourism industry. Food as a noteworthy fascination while individuals travel (Cohen and Avieli, 2004). Food has been perceived as a viable promotional and positioning tool of a tourist place (Hjalager and Richards, 2002).

For example, France, Italy, Thailand, Mexico and India have been known for their cuisine. The importance of the connection between food and tourism cannot be ignored. Authentic and interesting food can attract visitors to a destination. The destination (tourist place) will utilize food as the fundamental fascination and will create promoting methodologies that will concentrate on the food.

For the geographic location, although Bangladesh is rich in foods, various restaurants in tourist places offer different types of foods such as Bangladeshi traditional foods, Indian food, Italian-Mexican food, Chinese cuisine, Thai cuisine, mixed items and so on. In our study, we found that 20% offer only local food whereas 76% offer mixed (different country's cuisine including local). And only 4% offer Italian-Mexican food. Tourists are highly satisfied as there are many food options. Bangladeshi foods are globally famous for authentic quality that attracts people from all over the world to come to Bangladesh to taste them. That's why some restaurants in tourist places should focus on Bangladeshi traditional foods more.

Food options (%)

Food options

Italian-Mexican 4

Local 20

mixed 76

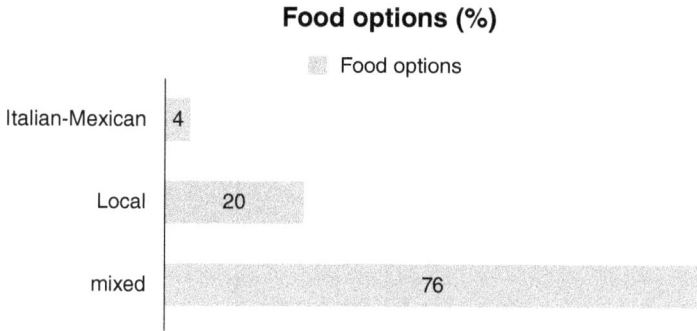

Figure 4.5a Respondents' food options
Source: Developed by the authors, 2020

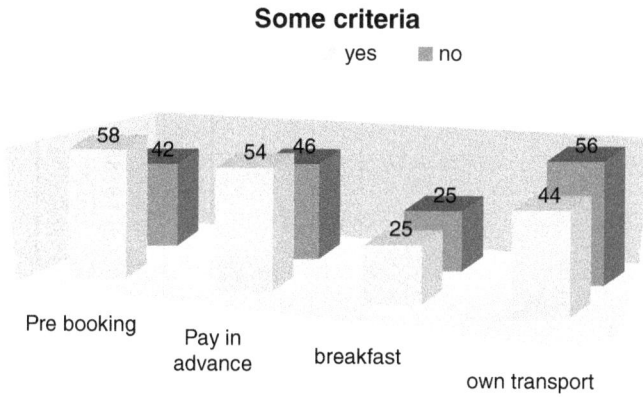

Some criteria

yes   no

58 42 54 46 56

25 25 44

Pre booking

Pay in advance

breakfast

own transport

Figure 4.5b Respondents' food options
Source: Developed by the authors, 2020

Regression analysis result

Room service, environment and safety-security level on overall satisfaction

We hypothesized that there is a very close connection with room service, environment and security level with overall satisfaction of accommodation. Our regression model considered overall satisfaction as dependent variable. It depends on room service, environment and safety-security level. The following variables have been chosen to estimate regression equation:

Accommodation= f (room service, environment, safety-security)

Table 4.4 Regression analysis result

Summary Output

Regression Statistics

Multiple R	0.9760187
R Square	0.9526126
Adjusted R Square	0.9494534
Standard Error	0.1298032

ANOVA

	df	SS	MS	F	Significance F
Regression	3	15.2418013	5.0806	301.53972	8.473E-30
Residual	45	0.758198698	0.016849		
Total	48	16			

	Coefficients	Standard Error	t Stat	P-value	Lower 95%	Upper 95%	Lower 95.0%	Upper 95.0%
Intercept	0.0202556	0.302208542	0.067025	0.9468585	-0.5884236	0.6289349	-0.588423643	0.62893486
5	-0.1058597	0.0787916	-1.34354	0.1858333	-0.2645541	0.0528348	-0.264554086	0.05283477
5	0.9526766	0.037911107	25.12922	4.016E-28	0.8763197	1.0290335	0.876319744	1.02903352
4	0.1880275	0.052552436	3.577902	0.0008431	0.0821815	0.2938735	0.082181451	0.29387353

Source: Developed by the authors, 2020

The a priori relationship is as follows: Room services are positively related to tourism, i.e. when room services can be provided on the basis of the different demand criteria of the tourist then it will have positive impact on increasing tourism. Environment is one of the determining factors for accommodation, which is positively related. And lastly, when security measures are sufficient, tourists will feel safe to tour.

In our analysis, R square and adjusted R squares are 0.952 and 0.959, respectively; 95.2% indicates that the model explains almost all the variability of the response data around its mean. Also the result of adjusted R square fits well. Moreover, significance F (8.473E-30) indicates that there is almost no chance to affect by random change. P-values (4.016E-28 and 0.0008431) showed significant results.

SWOT analysis

Based on the survey, we have made a SWOT table for the tourism sector of Bangladesh (Table 4.5).

Recommendations to promote accommodation in the tourism industry of Bangladesh

Tending to the major and multi-faceted difficulties looked at by the travel industry requires an incorporated way to deal with strategy improvement across numerous government departments. Rationality and consistency are basic in the structure

Table 4.5 SWOT analysis result

Strength	Weakness
• Friendliness and hospitality of Bangladeshi people • Hotels, motels, resorts and cottages are available • Natural beauties • Rich arts and culture • Wildlife and hill tracts • Archaeological sites • Low- and high-range accommodations are available • Reasonable food costs • Various food options	• Lack of awareness • Lack of investment • Lack of innovation • Low-quality service • Communication problems • Lack of safety and security
Opportunities • Promote tourism • Promote tourism by media • New tourism products and innovation • Need to make tourist sites more attractive • Government should take more action • Properly utilize the natural resources • Public-private partnership (PPP) could be encouraged	**Threats** • • Natural disasters • Corruption • Threats of security • Communication gaps

Source: Developed by the authors, 2020

and application of strategies between all degrees of government to guarantee that travel industry approaches are viable.

Governments and other stakeholders can assist goals and the travel industry in improving their intensity on the world tourism market through proper policy making, planning, arrangements and directed projects including those outlined next.

Policies and an institutional structure that add to a business domain are helpful for tourism growth and development

These policies ought to incorporate stable macroeconomic and well-planned basic arrangements in territories that encroach on sustainable tourism. They however are not restricted to employment and education or training, small and medium-sized enterprises (SMEs) and businesses, sustainable development, transport and infrastructure, local improvement, culture and creative industries, trade and investment and safety and security.

To take full advantage of the tourism industry's advancements, the government needs to step forward with necessary policies. First, build up an extensive policy framework for accommodation; second, advance a reasonable policy framework for accommodation through an "entire of government" approach; third, empower a culture of co-activity among tourism industry accommodation characters; and finally, actualize assessment and execution appraisal of government approaches and programmes influencing tourism industry accommodation improvement.

Target programmes

Appropriate projects may include the stimulation of advanced accommodation components, efficiency-based development and quality by first, advancing the sharing of innovative practices as far as association, enterprise or process improvement for a superior utilization of existing higher labour productivity and a revival of the tourism industry accommodation components supply (for example new products). Second, urging SMEs to adjust their ideas to the consumer's desire by advancing the accommodation product quality in solace and services through travel industry accreditations. Third, encouraging co-activity, the formation of networks and working groups in the accommodation sector of the tourism industry to accomplish economies of scale and scope. Fourth, improving and advancing the allure of work in the accommodation sector by impacting upgrades in the travel industry labour market to make appealing working places; expanding the pioneering/management capacity of travel industry SMEs through support and training programmes; advancing training and aptitude advancement for workers of the accommodation sector through education and vocational training; and urging organizations among enterprises to offer better vocational points of view.

Conclusion

The discoveries of this research demonstrate that tourist evaluation of destination accommodation is still the most significant marker of visitor fulfilment. Tourists at different places had divergent opinions on various service indicators selected for this study. As we selected five specific locations for this study, we discovered different findings based on place accommodation. Tourists were overwhelmed by the natural beauty, archaeology, beaches and other attractions. However, the lowest level of satisfaction was observed with respect to attributes such as safety and security, communication, information services about the destination, distance between the places and hospitals. On the other hand, the prevalence of high satisfaction of tourists staying at the resorts/hotels/motels indicated that staying cost was reasonable and affordable. Amenities also were sufficient. Based on our regression analysis it is clear that environment and safety-security are the important indicators of accommodation facilities. That's why Bangladesh needs to concentrate on these indicators to compete with other tourist destinations.

Despite the fact that improvement of different multi-storied lodging projects has been seen in the region, it appears that development of tourism does not only mean developing hotels or tourist accommodation, but rather requires total development of the region as well as enhanced competitiveness of those components that are directly or indirectly involved in tourism. In coming years, global competitiveness will increase. To achieve a good rank in the tourism sector, Bangladesh should improve the components of tourism.

References

Alegre, J. and Garau, J. (2009). Tourist satisfaction indices: A critical approach. *Investigations Regional*, 14, pp. 5–26.

Ali, M. (2004). *Bangladesher Sandhanay*. Dhaka: Student Ways (in Bengali).

Amin, S. b. (2007). Tourism and economic development: An analytical framework. Paper Presented at the *XVI Biennial Conference on Participatory Development: External and Internal Challenges*. Dhaka: Bangladesh Economic Association. The 12th -15th December 2007.

Baisakalova, A. (2009). Economy diversification: Tourism cluster competitiveness and sustainability. *The Second International Conference organized by the Global Business and Management Forum on "World Financial Crisis and Global Business Challenges"*. Dhaka: University of Dhaka.

Bangladesh Parjatan Corporation (2020). *Home*. Retrieved from: www.parjatan.gov.bd/ (accessed: the 18th March 2020).

Clawson, M. and Knetsch, J. L. (1966). *Economics of outdoor recreation*. Baltimore: Johns Hopkins Press.

Cohen, E. and Avieli, N. (2004). Food in tourism: Attraction and impediment. *Annals of Tourism Research*, 31, pp. 755–778.

Cooper, C., Fletcher, J., Fyall, A., Gilbert, D. and Wanhill, S. (2008). *Tourism principle and practice*. Harlow: Prentice Hall.

Deegan, J. and Moloney, R. (2007). *Understanding the economic contribution of tourism: The case of the West of Ireland*. Worcester, MA: Global Business & Economics Anthology, Business & Economics Society International.

Dhaka Mirror (2012). *Tourism in Bangladesh: Problems and prospects.* Retrieved from: www.dhakamirror.com/feature/tourism-in-bangladesh-problems-and-prospects/ (accessed: the 18th March 2020).

Elena, M., Lee, M. H., Suhartono, H., Hossein, I., Rahman, N. H. A. and Bazilah, N. A. (2012). Fuzzy series and Saima model forecasting tourist arrivals to Bali. *Jurnal Teknologi*, 57(1), pp. 69–81.

Ghosh, B. 2001. *Tourism and travel management.* New Delhi: Vikas Publishing House Pvt. Ltd.

Hjalager, A-M. and Richards, G. (2002). Still undigested: Research issues in tourism and gastronomy. In A-M. Hjalager and G. Richards (eds.), *Tourism and gastronomy.* Oxon: Routledge, pp. 224–234.

Hossain, M. A. and Nazmin, S. (2006). Building an image of Bangladesh as a tourist destination: Some strategic guideline. *Dhaka University Journal of Marketing*, 9.

Islam, F. and Islam, N. (2006). *Tourism in Bangladesh: An analysis of foreign tourist arrivals.* Retrieved from: http://stad.adu.edu.tr/TURKCE/makaleler/stadbah2004/makale040103 (accessed: the 18th March 2020).

Johannesburg Summit (2002). *Bangladesh country profile.* New York, NY: The United Nations.

Ministry of Civil Aviation and Tourism (2020). *Home.* Retrieved from: https://mocat.gov.bd/ (accessed: the 18th March 2020).

Nath, N. C. (2007). Tourism sector in Bangladesh: Insight from a micro level survey. Paper Presented at the *XVI Biennial Conference on Participatory Development: External and Internal Challenges.* Dhaka: Bangladesh Economic Association. The 12th – 15th December 2007.

Okaka, W. (2007). *The role of media communications in developing tourism policy and cross-cultural communication for peace, security for sustainable tourism industry in Africa.* Retrieved from: https://pdfs.semanticscholar.org/6ef4/ed9f8d1affb560ac59a5a315c27a65a570e9.pdf (accessed: the 18th March 2020).

Planet Green (2013). *Technology & transport, planes, trains, automobiles (and Buses): Which is the greenest way to travel long distance in the US?* Retrieved from: http://tlc.howstuffworks.com/family/plane-train-automobile-travel.htm (accessed: the 18th March 2020).

Rahman, M. L., Hossain, S. M. N., Miti, S. S. and Kalam, A. K. M. A. (2010). An overview of present status and future prospects of the tourism sector in Bangladesh. *Journal of Bangladesh Institute of Planners*, 3, pp. 65–75.

Raju, G. P. (2009). *Tourism marketing and management.* Retrieved from: www.abebooks.com/Tourism-Marketing-Management-G-P-Raju/1267247003/bd (accessed: the 18th March 2020).

Roy, S. C. and Roy, M. (2015). Tourism in Bangladesh: Present status and future prospects. *International Journal of Management Science and Business Administration*, 1(8), pp. 53–61.

Sandip, S. (2014). Competitive marketing strategies for tourism industry in the light of "Vision 2021 of Bangladesh". *European Journal of Business and Management*, 6(4), pp. 210–220.

Saxena, A. (2008). *New trends in tourism and hotel industry.* Retrieved from: http://site.ebrary.com/lib/cop/docDetail.action?docID=10416183&page=198 (accessed: the 18th March 2020).

Tuhin, M. K. W. and Majumder, M. T. H. (2011). An appraisal of tourism industry development in Bangladesh. *European Journal of Business and Management*, 3(3), pp. 287–298.

Wang, S. and Qu, H. (2006). A study of tourist satisfaction determinants in the context of the pearl river delta sub regional destinations. *Journal of Hospitality and Leisure Marketing*, 14(3), pp. 49–63.

World Economic Forum (2013). *The travel and tourism competitiveness report.* Retrieved from: http://www3.weforum.org/docs/WEF_TT_Competitiveness_Report_2013.pdf (accessed: the 18th March 2020).

World Tourism Organization (UNWTO) (2002). *Tourism and poverty alleviation.* Madrid: UNWTO.

World Tourism Organization (UNWTO) (2008). *Tourism highlights 2015.* Madrid: UNWTO.

World Travel and Tourism Council (WTTC) (2016). *Travel and tourism economic impact report (2015).* London: WTTC.

Yilmez, B. S. (2008). Competitive advantage strategies for SME'S in tourism sector. *Burcu Selin Yilmaz*, 3(1), pp. 106–117.

Zulfikar, M. (1998). *Tourism and hotel industry.* New Delhi: Vikas Publishing House Pvt. Ltd.

5 Tourist transportation in Bangladesh

Ayesha Afrin and Azizul Hassan

Introduction

As an ancillary service, transportation in Bangladesh is a crucial element for tourist experience delivery. Tourist transportation in this country can hardly be claimed as the most updated but there are scopes for further development. The transportation system in Bangladesh is multimodal consisting of air, rail, road and water. Tourists in Bangladesh can experience certain forms of transportation which are less available in other parts of the world, such as rickshaws (Sultana, 2013). Different innovations in transportation are the responsible factors of the rapid spread of tourism. The transportation system of a country is meant to have a vital place in the global network system and is one of the most important components of the tourism infrastructure. The tourism industry needs a supportive transportation system to carry tourists from one place to another when one of the key purposes of tourism is experience sharing through mobility. Bangladesh is turning into a considerable destination for both regional and international tourism business. This paper briefly explains the transportation system in Bangladesh to cover the specific demands of both domestic and international tourists.

Tourism trade in Bangladesh

Tourism is one of the world's fastest growing industries. This is a major foreign currency earning source and employment generation industry for many countries. The trade of tourism is centuries old: Guyer and Feuler first offered an acceptable definition of tourism in 1905 (Ugurlu, 2010). Later, in order to prevent confusion around the definition of "tourism", the World Tourism Organization (UNWTO) (2011, p. 1) defined it as: "Tourism comprises the activities of persons travelling and staying in places outside their usual environment for not more than one consecutive year for leisure, business and other purposes". According to the World Travel and Tourism Council (WTTC) report (2019), the direct contribution of travel and tourism to the GDP of Bangladesh in 2018 was US$2,750.7bn (3.2% of GDP). This is forecasted to rise by 3.6% to US$ 2849.2 billion in 2019 and by 6.5% per annum to 4.7% of GDP by 2024. All of these figures present that the tourism trade in Bangladesh will have a significant rise that should have a positive influence on her transportation system and network.

```
                    Tourist Transportation


Air Transport          Land Transport          Water Transport


                    Roads      Railways
```

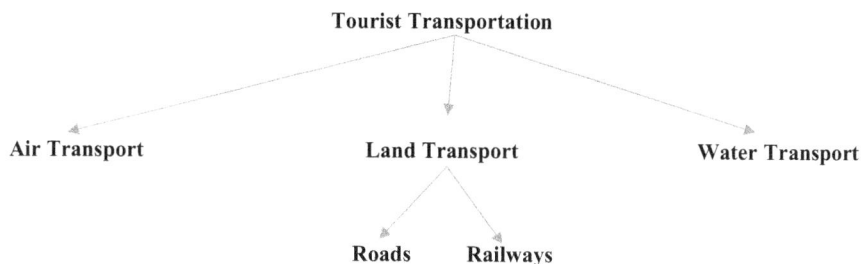

Figure 5.1 Tourism transportation
Source: TourismNotes, 2019

Tourist transportation

A tourist is generally offered to travel by any means of transportation in accordance to the demand and convenience (World Bank, 2009). It is important for a tourist to become well aware about the various modes of transport available in a country like Bangladesh. From a wider context, the various modes of transport can be classified as:

Tourist transportation in Bangladesh

Transportation is an integral part of the tourism industry and the national economy. Bangladesh witnessed a rapid growth in transportation since independence. The overall annual growth rate is nearly 8.2% for freight transport and 8.4% for passenger transport (Mahmud et al., 2006). Still, the transport intensity of the Bangladesh economy is considerably lower than that of many other developing countries. The modes of tourist transportation in Bangladesh are discussed next.

First, the rickshaw is the most conventional feature of tourist transportation in Bangladesh. Characteristically, this is the traditional mode of transportation in Bangladesh and is environmentally friendly due to pedal operation. Dhaka, the capital city of Bangladesh, is known as the "Rickshaw Capital of the World". In all major cities in Bangladesh, tourists will find the rickshaw the most inexpensive and convenient mode of transportation (Ahmed et al., 2017).

Second, air transport is by far the most effective transport mode mainly for foreign tourists. According to Rodrigue (2017), because of prices, only 12.5% of the tourists travel by plane but for international travel this share is around 40%. Air transport has revolutionized the geographical aspect of distances. The most remote areas of the world can now be reached from any part of the world in less travelling time. Business travellers are among the largest users of airline facilities but low-cost airlines have also been able to attract significant market segment mainly for tourist transportation. Bangladesh has three international airports: Hazrat Shahjalal International Airport in Dhaka, Shah Amanat International Airport in Chittagong and Osmani Airport in Sylhet. Visa on arrival is available at the international airports for nationals of certain countries. There are also airports to cater domestic and private airlines. All of the 11 operational airports in the country (i.e. Dhaka, Barisal, Chattogram, Cumilla, Cox's Bazar, Ishwardi, Jessore, Rajshahi, Syedpur, Sylhet and Thakurgaon) serve domestic flights according to the Civil Aviation Authority of Bangladesh (2019). Low-cost domestic flights to all major destinations in Bangladesh are provided by local airlines such as United Airways, Regent Airways, Novo Air, US-Bangla Airlines and the national flag carrier Biman Bangladesh Airlines (Discovery Bangladesh, 2019). In Bangladesh, most of the airports have closer proximity to the city centre that makes the tourist transportation experience more convenient and useful. Figure 5.1 below shows the distance between the specific airports to the relevant city centre with the available mode of transportation.

Third, railway transport can be both enjoyable and time saving for tourists. Nationally operated the Bangladesh Railway provides service throughout Bangladesh. About 32% of the total area of Bangladesh is covered by the railway network (Discovery Bangladesh, 2019). The Bangladesh Railway services places of interest such as Chattogram, Sylhet, Khulna, Mymensingh, Bogra, Rajshahi and Dinajpur starting from Dhaka. The Inter-city Express Service is available to and from important cities at a reasonable fare. There are also local trains that serve passengers at cheaper rates but can take longer time to reach a destination. During the festive seasons (i.e. the Eid, the Puja, the winter season from November to January and others) the demand for train tickets to travel between Dhaka, Chattogram, Cox's Bazar and the rest of Bangladesh remains at the peak. Tickets can be purchased from the official website of the Bangladesh Railway and train schedules can also be viewed online. Since the process is not straightforward, it

Airport	Distance	Taxi and Ride Sharing	Bus	Train	Car Rental
Hazrat Shahjalal International Airport, Dhaka	20 km	Yes	Yes	Yes	Yes
Shah Amanat International Airport, Chattogram	9 km	Yes	Yes	No	Yes
Osmani International Airport, Sylhet	7.8 km	Yes	Yes	No	Yes
Cox's Bazar Airport, Cox's Bazar	4.7 km	Yes	Yes	No	Yes

Figure 5.2 Distance to the city centre from the airport

Source: Nordea, 2019

is highly advisable for foreign tourists to read the English instructions before attempting to purchase train tickets online.

Fourth, waterway transportation: Bangladesh is believed to have one of the largest inland waterway networks in the world. The landscape of Bangladesh is dominated by about 250 rivers flowing from the north to the south towards the Bay of Bengal. Almost two-thirds of the geographical area of Bangladesh is wetland laced with a dense network of creeks, canals and rivers. Water transport is the only means of transportation available in almost 10% of the total area of the country. The navigable waterways in Bangladesh vary from 8,372 kilometre during the monsoon to 5,200 kilometre during the dry season (Discovery Bangladesh, 2019). Bangladesh Inland Water Transport Authority (BIWTA) is established by the government of Bangladesh to maintain the navigability of ports and channels while the state-owned Bangladesh Inland Water Transport Corporation or BIWTC (BIWTC) provides passenger and cargo services in inland waterways and coastal areas of the country. As a country crisscrossed by hundreds of rivers, Bangladesh boasts of many different conventional locally made boats. The sampan is one of them that has hundreds of years of traditions of Chattogram in Bangladesh. In Bangladesh, these country-made boats are most commonly used as carriers and can be seen in the rivers, canals and creeks. They do not only carry passengers but also merchandise, both small- and large-scale shipments. Along with these traditional boats, mechanized waterway transportation is also available. Speedboats having the latest technologies are generally used as a transportation mode through the Padma River and Maowa Ferry Ghat of Dhaka, Bangladesh. A journey by Rocket Steamer Service from Dhaka (Sadarghat) to Khulna, the gateway to Sundarbans can possibly be a rewarding experience for tourists. As Bangladesh is a riverine country, ship, boats and ferries play important roles in the transportation system of Bangladesh. A ferry in Bangladesh is sometimes called a "launch," by which tourists can travel to Khulna, Barisal, Patuakhali, Chandpur and most other destinations of the country. However, this is strongly advisable to purchase first-class tickets when travelling by steamer or launch for better comfort and lavishness when the price difference between first and other class tickets is mostly negligible (Sultana, 2013). Tourists can enjoy the natural beauty of Bangladesh with water transport that is rather inexpensive and enjoyable.

Fifth, land transportation by bus/coach services is an unavoidable feature of tourist transportation in Bangladesh. In the country, road transport operation in domestic routes is mostly dominated by private entrepreneurs. Bus or coach fair in Bangladesh is most likely among the cheapest in the world. Dhaka as the capital city along with other major divisional cities like Chattogram, Rajshahi, Sylhet, Barisal, Rangpur and others have created the road transportation network. Luxury express and non-stop bus/coach services are made available to principal cities from the three major bus terminals of Dhaka (i.e. Gabtoli, Saidabad and Mohakhai). Bangladesh Road Transport Corporation (BRTC) as the government agency also maintains a countrywide network of bus and truck services. In the most recent time, Dhaka–Kolkata–Dhaka direct daily bus services via

Benapole of Jessore is introduced. Tourists can easily avail air-conditioned bus/coach services to almost all major tourist destinations like Cox's Bazar.

Sixth, car rental, that is private car hire service, is available in Dhaka and most of the major cities. Car rental companies are usually less interested to provide a vehicle without a driver. However, the service that is offered is relatively comfortable for tourists having very poor knowledge about Bangladesh. Drivers with a rented car can act as a tourist guide to turn the experience as more enjoyable. In Bangladesh, renting a car is relatively inexpensive while in every city, taxi cabs and green-coloured auto-rickshaws are used for travelling short distances. From the institutional context, Bangladesh Parjatan Corporation (BPC), a government organization, has a fleet of air-conditioned and non-air-conditioned cars, microbuses and jeeps. BPC also offers transfer service for tourists between Dhaka airport and main city points or hotels.

Seventh, mobile app–based taxi service and share riding (i.e. Uber, Pathao and OBHAI) are becoming popular. Uber Technologies, Inc., widely known as Uber, is a US-based multinational ride-hailing company. This company offers transportation services (i.e. ride service hailing, peer-to-peer ridesharing, a micromobility system with electric bikes and scooters and food delivery). Operational areas of Uber are over 785 metropolitan areas across the world. Uber platforms can be accessed through its mobile apps and websites (Uber, 2019). Pathao offers almost all of its services through one app that is rather a super app. Pathao regularly updates its app for making the overall app navigation better, more natural and straightforward for users. Both the passengers and users of Pathao are required to have access to an IOS- or Android-based smartphone with GPS enabled and Internet. This company uses a location-based system for matching the passenger with a nearby driver heading towards the desired destination. Pathao at present offers ride-sharing, food delivery, parcel, and on-demand transport sharing services in three major cities of Bangladesh (i.e. Dhaka, Sylhet and Chattogram), and Kathmandu of Nepal. At present, Pathao has its food delivery services in Chattogram and Dhaka metropolitan areas. Pathao offers on-demand ride-sharing services through cars and motorcycles. Currently, Pathao has a registered fleet of over 200,000 vehicles (Pathao, 2019). OBHAI is a ride-sharing app that enables passengers to select a microbus, car, CNG or autorickshaw for specific destinations. The mission of OBHAI is to offer reliable, safe and convenient transportation for everyone across Bangladesh (OBHAI, 2019). Ride-sharing businesses have become popular and a few other companies are either in operation or will appear in the transportation industry of Bangladesh.

Tourist transportation administration in Bangladesh

Since tourists very often face urban transport issues (i.e. mobility, congestion, safety and environmental aspects), they become vulnerable to any harmful effect. The rapid urbanization process, high vehicular population growth and mobility followed by inadequate transport facilities and policies, varied traffic issues with non-concentration on non-motorized vehicles, the absence of reliable public transport system, inadequate traffic management practices and parking facilities are challenges for the responsible government agencies and ministries to tackle (Mahmud et al., 2006). Primarily, the Ministry of Communication is the highest authority responsible for transit, construction, development, expansion and maintenance of roads and railway transportation that help tourists compare to other countries. The Bangladesh Road Transport Corporation (BRTC) is the only public transport service provider in the country, established under the Bangladesh Road Transport Corporation Ordinance of 1961. It provides bus services that are international, inter-city and intra-city. Bangladesh has five ministries responsible for transportation within country with specific responsibilities both directly and indirectly. These are first, the Ministry of Road Transport and Bridges that is responsible for ensuring road safety; second, the Ministry of Civil Aviation and Tourism as responsible for civil aviation; third, the Ministry of Shipping to look after ensuring maritime transport; fourth, the Ministry of Railways for railway transportation; and fifth, the Ministry of Road Transport and Bridges with two divisions, the Road Transport and Highways Division and the Bridges Division (Zulfiker, 2017). This is the general tourist transportation administration in Bangladesh having sufficient concentration to develop tourist transportation facilities and offerings.

Comparative advantages of modes of tourist transportation in Bangladesh

Tourists have a wide variety of transport options available today. There are several advantages and disadvantages of all the modes of transport. Air transport has direct routes, approximately high speed, social and political significance and luxurious travels as its advantages. However, issues like high cost, jet lag, unsuitable for heavy bulk cargo, accidents always fatal and international rule are the disadvantages of air transportation. Second, road transport has flexibility, reliability, door-to-door service, affordability, supplements other modes of transport and quick transit for short distances as its advantages. Still, slow speed, limited carrying capacity, accidents, non-air-conditioned coaches and bad conditions of roads are the disadvantages of road transport that are required to be considered. Third, railway transport has the advantages of cheap long-distance travel, large carrying capacity, dependable service, quicker than road transportation and the ability to view scenery en route. However, inflexibility, unfit to hilly regions, difficulties in rural areas and dining car facilities not always available are the disadvantages of railways that have to be considered. Fourth, water transport is rather economical with a large carrying capacity. Still, weakness of the system or insecurity can become disadvantages of water transport for tourists that require attention to resolve.

Transport as a promoting element of tourist destination

Whether transportation plays an important role in enriching the travel experience of a tourist depends on the mode of transportation and the frequency of use. Transportation can be turned into a separate tourism product (i.e. cruising,

Orient Express trains, boat trips, etc.). The effective factors in choosing the transportation mode in tourism can be (Schiller et al., 2010): time limit, distance, status, comfort, security, benefit, price, geographical position and competition. Bangladesh as a vacation land has many facets. Her tourist attractions include archaeological sites, historic mosques and monuments, resorts, beaches, picnic spots, forest and wildlife that brings one in close touch with mother nature. Among many others, in Rajshahi division that is the northern part of the country, there are archaeological sites. In Chattogram division that is the south-eastern part, there are natural and hilly areas like along with sandy sea beaches. In Khulna division, that is the south-western part, there is the Sundarbans, the largest mangrove forest of the world with royal Bengal tiger and spotted deer. In Sylhet, that is the north-eastern part, there is a green carpet of tea plants on small hillocks. Expenses in other areas can be much lower. Lonely Planet ranked Bangladesh as the best value destination for the year 2011 (Lonely Planet, 2019). In comparison to most other countries of the world, tourists may find Bangladesh as one of the best tourist destinations. However, smooth tourist transportation is essential to let the tourists enjoy and experience these beauties. In Dhaka city, it is usually recommended that tourists rent a car, with a driver, for personal use during long stays in the city. In recent years, construction of a number of bridges such as the Bangabandhu Jamuna Bridge, Meghna Bridge and Meghna-Gumti Bridge, Bangladesh-China Friendship Bridge, Shambhuganj Bridge and Mahananda Bridge have been completed. It has established a strategic link between the east and the west of Bangladesh that has integrated the country and is generating multifaceted benefits to the tourist apart from quick movement of tourist/passenger traffic.

To attract tourists, destination developers may use many forms of transport to move people around. These novel modes of transport ensure that major exhibits are viewed in a certain sequence and ensure that the crowd moves through at a reliable pace. Overcrowding should be avoided at all costs to prevent untoward incidents and to maintain the beauty of the place. Tourists can cover the entire park in a shorter duration with the help of these modes of transport. Transportation is the most crucial component of the tourism infrastructure. It is required not only for reaching the destination but also visiting the site and moving about at the destination. Variety in modes of transportation adds colour to the overall tourism experience. Unusual forms of transportation are also an attraction such as the cable cars in hilly terrain, the funicular railway, or jet boating. The choice of mode of transport is vast and tourists can choose a mode to suit their budget. They can opt for scheduled or non-scheduled transport such as the hiring of vehicles, boats, coaches or trains so that they can travel with their group.

Tourism in Bangladesh is a slowly developing as a foreign currency earner. The country has much to attract international and domestic tourists. The transport sector in Bangladesh faces the challenge of providing equitable services and opportunities in an extremely densely populated country and an environment prone to disasters, potentially aggravated in the future by the impact of climate change. Transport is, thus, one of the major components of the tourism industry.

To develop any place of tourist attraction there has to be proper, efficient and safe modes of transportation. While the government of Bangladesh has achieved fast expansion of the road network, thus providing considerable benefits to a population lacking accessibility, this created a tradeoff with other modes of transport such as waterways and railways, safety of transport users and the environment. Facing these challenges requires us to reexamine priorities and to build sustainable, safe and quality road infrastructure and an integrated modern mass transport system for achieving desired socio-economic development or to develop a strategy which allows for a sustainable contribution of the transport sector to the national development objectives.

Tourist mobility impacts the environment

Many studies have shown that tourist mobility is related to accidents, traffic jams and air pollution and has a negative impact upon local residents (Banister, 2008). Public transportation is also a significant environmental concern because it can bring a lot of people together in one space, such as on a train or bus, as well as reduce traffic jams and the number of private car accidents. Tourism-related public transportation is important. It is really important for policy makers to implement appropriate transportation policies and facilities when they set up their tourism policies and plans (Black, 2010). In addition, in order to develop transportation to support tourism and local residents in Bangladesh, the municipality should host an initiative and bring all stakeholders together to solve the problem. However, research has found that the facilitation of transportation services may threaten tourism and decrease the number of visitors due to concerns about fairness. Also, the national government and provincial government have to make rules to control and facilitate transportation together, because the topic is so important to the image of the province and the tourism industry. The topic of government management of transportation services requires a great deal of management, such as registering and training transportation personnel and facilitating tourism. Providing guidelines for solving the transportation problems for visitors or activities related to tourism is also important, as the issues must be managed and need to be improved and developed sustainably (Goldman and Gorham, 2006). So, the guidelines to improve the tourist transportation system or major steps that believe would improve and strengthen relationships include: first, to develop more railway routes that connect to multiple provinces; second, to improve the transportation network, especially the major and minor roads and support transportation channels to cover increases in traffic lanes, and manage transportation planning systems for the province; third, better communication and regular meetings between top agency leaders; fourth, more formal written guidelines; and fifth, strong involvement by private sector tourism and transportation interests with government. There needs to be more investment in transportation resources to support tourism; all respondents believe that the tourism industry is not engaged enough in transportation decision making.

Tourist transportation in Bangladesh: some challenges

Because the cities are crowded and there is competition for public transit, tourists may find it difficult to use the transportation system. They may face difficulties to get public vehicles like buses, trains, and taxis especially in Dhaka city. Outside Dhaka hiring an auto (widely known is CNG) or easy bike is a piece of cake.

An increase in traffic due to world tourism growth puts pressure on transportation facilities, and this can have adverse effects. Those negative effects are (Ritchie, 2012): first, congestion that means delays which leads to waste of time and energy. Serious congestions may have a negative effect on transportation modes, especially on airports and roads during peak times. Second, safety and security that make sure that the transportation mode is safe and secure is a basic and important requirement for tourism. Third, environment that is an increase in traffic may have disastrous effect on the environment if that area does not have the carrying capacity for additional tourists. Last, seasonality, that is the seasonal pattern of travel demand creates overcrowding at certain times. Adversely low occupancies and load factors will occur at other periods.

With continued economic development, Dhaka (the capital city of Bangladesh) is beginning to experience severe traffic congestion. With more than 250,000 motorized vehicles in this city alone, traffic congestion wastes fuel and time and makes tourist travel difficult. This also turns the existing public transport inefficient, adding unsafe levels of noise and air pollution. This is impacting the quality of life for inhabitants of the metropolitan area, the nation's largest. Traffic congestion varies during the day, necessitating planning and longer trips; this impacts productivity, cutting across social and economic status. Many government and public-transport agencies drafted policies, undertook projects and implemented programmes to solve the problem.

Passenger and pedestrian safety in the roads is currently a burning issue in Bangladesh. Death counts in the highways are rising every day at an alarming rate. Although the government is undertaking a number of significant steps addressing the issue, situations in the highways do not seem to get much positive change too soon. Public unrest and riots demanding safe roads tend to occur in quite an unpredictable manner as both the authority and the highway section of the police prove to fail in bringing discipline in the road transportation system and the source of all problems. The High Court of Bangladesh recently directed the government to constitute a national independent inquiry committee, including at least 15 experts, to conduct a survey on the fitness of public motor vehicles in order to avoid road accidents across the country (The Daily Star, 2019).

Tourist transportation in Bangladesh: suggestions

Transportation is not just about moving an object from point A to point B. Rather, it is a process of value delivery: sending things all over the country, carrying customers to upper floors or building a warehouse for cargos, all of which require products in transportation (Tolley, 2003). From the basic understanding of Bangladesh transport system, it appears that efficient services to the tourists are hardly offered. In spite of all difficulties recently we won in tourism the South Asian Travel Award (STA) 2019, by the choice of tourist which motivates us to do better work for the tourist in the future (South Asian Travel Awards, 2019).

There are always opportunities to improve the tourist transportation in Bangladesh. All the transit stakeholders need to point out critical improvement opportunities according to their views. There can be some contrasting views in their response on questions related to transit challenges. However, this is normal that regarding the improvement options, majority of them can share a common view. This can mostly validate the arguments that the tourist transportation in Bangladesh is yet to be developed to meet those challenges as previously mentioned.

For evaluating transit performance, it is important to identify significant indicators to measure overall performance of the transit operators. Tourist opinion is the most important part of this study because it allows one to explore the current situation and to discover operators' thoughts on the opportunities for better transit performance. For this study, common and widely used indicators are selected to measure transit performance through data envelopment analysis (DEA). With a view to identify the relationship between the input indicators (i.e. labour, fuel, capital) and output indicators (i.e. vehicle-km, technology) within the transit (Tripadvisor, 2016).

Safety standard of the vehicles used for tourist transportation need to be ensured. For this, training facilities for drivers, mechanics and transport management need to be provided; strategic interventional roles at the time of emergency need to be played; the private sector in transport service and introduction of new routes need to be introduced; and road transport services for tourists need to be operated.

Apart from these, major administrative authorities as the Dhaka Metropolitan Police Traffic have to take actions against traffic violations, ensure smooth traffic flow every day, take measures to reduce road accidents. There are no specific studies about the efficiency of transit services in Bangladesh cities. Most studies have looked at the overall scenario of transport in cities like Dhaka and the reasons behind the poor quality transport there. A significant number of studies have been carried out to explore the potential for mass transit in the city (World Bank, 2009). Modern and innovative technologies (i.e. Satellite Navigation System) need to be applied including Wi-Fi and GPRS system in the vehicles for better and easier transportation for tourists.

Tourism is all about travelling. To sum up, in order to develop and increase the role of transportation in tourism countries should pay attention to the following points: first, transportation modes specific to the regions must be developed; second, the transportation costs must always be kept competitive; third, the passengers must be attracted to sea and railway modes of transportation; fourth, new embarkation ports must be established in order to develop cruise travel; fifth, new fast train lines must be established and new fast trains must be bought and foreign experience must be applied; sixth, the distance from stations must be kept less; seventh, new coaches must be brought to the country; eighth, new technologically advanced aircraft must be put into airlines; ninth, the personnel must be trained; finally, the governments should allocate financial support for the development of transportation. It is required not only for reaching the destination but also visiting the tourism sites and moving about at the destination. Variety in modes of transportation adds colour to the overall tourism experience. Tourist transportation in Bangladesh in coming years can possibly implement more state-of-the-art technologies being implemented for the best of tourism.

Conclusion

Travel and tourism is a huge and diverse industry. To be a tourist is to be concerned about driving time, driving safety, driving costs, and driving frustration. Travellers and economic growth are clearly best served by cooperative and collaborative relationships between those who formulate and implement public policies pertaining to highways and travel and tourism. Following the lead of earlier research, the current study finds out first, there is not enough cooperation between the top transportation management and tourism agencies; second, the perceptions of the tourism agencies are relatively weak; third, in most areas, the private tourism/recreation operators seem to play non-active roles in the development of transportation projects affecting their interests. The recommendation focuses on traveller information services, including a listing of 13 information services media available to tourists (displays at transportation terminals, billboards, tourist-oriented road signage, information "logo" signs, variable message signs, historical markers, highway welcome centres, interactive video kiosks, tourist-oriented road maps, promotional-informational brochures, in-vehicle tourist information, tourist-oriented radio channels and special road condition maps/advisories). Some areas were identified where priority attention should be given to maximize improvements to traveller information services. Future research studies can focus on explaining the collaboration between the Bangladesh government with local administration organizations (LAO) in all areas of Bangladesh to rethink the national development plan in all transportation networks, as well as implement a comprehensive transportation network that will reach all areas. The number of people in this area is increasing and many foreigners are deciding to make their holidays in Bangladesh, and public transportation still cannot support a good quality of life. Lack of public transportation is an obstacle to development in all social aspects, such as economy, tourism, education and social development.

References

Ahmed, I., Noor -E -Alam and Warda, F. (2017). *A sustainable urban transport initiative in Dhaka: Introducing bus rapid transit system.* Retrieved from: https://bit.ly/37tM8oV (accessed: the 1st December 2019).

Banister, D. (2008). The sustainable mobility paradigm. *Transport Policy*, 15(2), pp. 73–80.

Black, W. R. (2010). *Sustainable transportation: Problems and solutions.* New York, NY: Guilford Press.

Civil Aviation Authority of Bangladesh (2019). *Home.* Retrieved from: www.caab.gov.bd (accessed: the 1st December 2019).

Discovery Bangladesh (2019). *Transportation & communication of Bangladesh.* Retrieved from: www.discoverybangladesh.com/transportation.html (accessed: the 1st December 2019).

Goldman, T. and Gorham, R. (2006). Sustainable urban transport: Four innovative directions. *Technology in Society*, 28, pp. 261–273.

Lonely Planet (2019). *Getting around Bangladesh on local transport.* Retrieved from: www.lonelyplanet.com/bangladesh/transport/getting-around/local-transport (accessed: the 1st December 2019).

Mahmud, S. M. S., Rahmanm, M. W. and Rabbi, H. S. (2006). *Transport system in Bangladesh: Issues and options for sustainable development.* Dhaka: Accident Research Institute.

Nordea (2019). *Maaprofiili Bangladesh.* Retrieved from: www.nordeatrade.com/fi/explore-new-market/bangladesh/travel-moving-in) (accessed: the 1st December 2019).

OBHAI (2019). *Home.* Retrieved from: obhai.com (accessed: the 5th September 2019).

Pathao.com (2019). *Home.* Retrieved from: https://pathao.com/ (accessed: the 1st March 2019).

Ritchie, R. C. (2012). *The Duke's province: A study of New York politics and society, 1664–1691.* Chapel Hill, NC: University of North Carolina Press.

Rodrigue, J-P (2017). Transportation and the internet. In B. Warf (ed.), *The Sage encyclopedia of the internet.* London: SAGE.

Schiller, P. L., Bruun, E. C. and Kenworthy, J. R. (2010). *An introduction to sustainable transportation.* London: Earthscan.

South Asian Travel Awards (2019). *Home.* Retrieved from: www.southasiantravelawards.com/en/home (accessed: the 1st December 2019).

Sultana, N. (2013). *Efficiency analysis of public transit systems in Bangladesh: A case study of Dhaka city.* Retrieved from: https://bit.ly/319K2qS (accessed: the 30th July 2019).

The Daily Star (2019). *Bangladesh public transport.* Retrieved from: www.thedailystar.net/tags/bangladesh-public-transport (accessed: the 1st December 2019).

Tolley, R. (ed.) (2003). *Sustainable transport: Planning for walking and cycling in urban environments.* New York, NY: CRC Press.

TourismNotes (2019). *Tourism transportation.* Retrieved from: https://tourismnotes.com/tourism-transportation (accessed: the 30th December 2019).

Tripadvisor (2016). *Bangladesh: Transportation in Bangladesh.* Retrieved from: www.tripadvisor.com/Travelg293935c157253/Bangladesh:Transportation.In.Bangladesh.html (accessed: the 1st December 2019).

Ugurlu, T. (2010). *Definition of tourism (UNWTO-Definition of Tourism)/what is Tourism.* Retrieved from: www.tugberkugurlu.com/archive/definintion-of-tourism-unwto-definition-of-tourism-what-is-tourism (accessed: the 1st December 2019).

World Bank (2009). *Bangladesh transport policy note.* Retrieved from: https://bit.ly/202xOvp (accessed: the 30th July 2019).

World Tourism Organization (UNWTO) (2011). *Capacity building Asia workshop II.* Retrieved from: http://statistics.unwto.org/sites/all/files/pdf/unwto_tsa_1.pdf, p. 1 (accessed: the 1st December 2019).

World Travel and Tourism Council. (2019b). *Economic impact-2019.* Retrieved from: https://bit.ly/2m7L3k1 (accessed: the 5th September 2019).

Zulfiker, A. (2017). *Public transport in Bangladesh.* Retrieved from: www.researchgate.net/publication/318727300_Public_transport_in_Bangladesh (accessed: the 30th July 2019).

Part 3

The role of marketing in strategic delivery

6 Strategic analysis of competitiveness of travel and tourism in Bangladesh

Shah Alam Kabir Pramanik and
Md. Rakibul Hafiz Khan Rakib

Introduction

Even with distinctive tourist attractions like incredible scenic beauty, enthralling history, exciting cultural and archaeological heritages, the largest sandy sea beach, unique mangroves, wildlife, flora and fauna to provide better life and travel experiences to tourists (Majumder, 2015), Bangladesh could not create a distinctive image in the mind of tourists as a desired tourist destination until recently. While many countries have experienced a speedy growth in their inbound tourism (Dwyer and Forsyth, 1992), the Travel and Tourism Competitiveness Index of 2017 stated that international tourist arrivals in Bangladesh was only 125,000 and international tourism inbound receipts were US$148.4 million. Besides, Bangladesh has ranked 125th among 136 countries of the world and 5th among the five countries of South Asian region (Crotti and Misrahi, 2017). In the last two decades the growth rate of tourist arrival was 5.8% and 3.9%, respectively (Khondker and Ahsan, 2015). Falling tourist arrival is an ominous sign for the industry. Now tourism policy makers and destination marketers spotlight on improving competitiveness (Sotiriadis and Varvaressos, 2015) by creating a statutory framework to protect resources, and to monitor, control and enhance quality and efficiency in the industry (Soteriades, 2012; Goeldner and Ritchie, 2011).

To analyse and enhance the competitiveness of travel and tourism, plentiful models and frameworks (Travel and Tourism Competitiveness Index, Global Competitiveness Index, Porter's Five Forces Model, Porter's Diamond Model, DEA Travel Tourism Competitiveness Index etc.) have been undertaken (Martín et al., 2015; Dupeyras and MacCallum, 2013; Alonso, 2010; Crouch, 2010; Porter, 2008; Gooroochurn and Sugiyarto, 2005; Enright and Newton, 2004; Dwyer and Kim, 2003; Ritchie and Crouch, 2003). By taking two comprehensive models, namely the Travel and Tourism Competitiveness Index and Porter's Five Forces Model, this chapter aims at analysing the competitiveness of the travel and tourism industry of Bangladesh. Since a clear and precise tourism policy can contribute in strengthening the competitive position of the Bangladesh economy by solving tourism development issues, conserving manmade and natural heritage and achieving sustainable development (Hassan and

Kokkranikal, 2018; Edgell and Swanson, 2013) in the domestic as well as global tourism market, this work will help policy makers with information they need on improving competitiveness in tourism through formulating a better tourism policy.

Concept of competitiveness

When a firm has profitability over its rivals greater than the average profitability of all the firms in its industry then it is said that the firm has competitive advantage (Hill and Jones, 2013). To achieve competitive advantage in any industry, measuring the competitiveness of that industry acts as a strategic tool for future planning, management and development (Islam, 2014; Bhatia, 2006). Competitiveness is a relatively complex (Martín et al., 2015) and multidimensional (Spence and Hazard, 1988; Scott and Lodge, 1985) concept which encompasses all social, cultural and economic variables that affect a nation's performance in international markets (Dwyer and Kim, 2003). Despite the definitions of competitiveness offered from different (a firm, an industry or a nation) points of view, the concept of competitiveness is basically centred on human development, growth and increased standard of living or quality of life (Cho, 1998; Newall, 1992).

Destination competitiveness

Destination competitiveness is the ability of the destination to offer superior tourist experiences than the alternative destinations and create overall appeal so as to achieve competitive advantage. Goeldner and Ritchie (2011) defined destination competitiveness as the ability to compete efficiently and profitably in the tourism marketplace. Crouch and Ritchie (1999) defined competitiveness from a tourism perspective as a country's ability to boost its national wealth through creating additional value by managing its assets and processes, aggressiveness, attractiveness and closeness and by combining these relationships into a social and economic model. The World Economic Forum's (WEF) Global Competitiveness Report defines competitiveness as the set of factors, policies and institutions that determine a country's level of productivity (Blanke and Chiesa, 2011). Dwyer et al. (2000) affirmed that tourism competitiveness includes price differentials attached with the movement in exchange rate, productivity levels of the different components of the tourism industry and other qualitative factors that affect the attractiveness of a destination.

Key indicators of defining competitiveness in tourism industry

There is no single or unique set of competitiveness indicators that apply to all destinations. Under the guidance of OECD (Organization for Economic Co-operation and Development), Dupeyras and MacCallum (2013) explored some

Table 6.1 Key elements for defining the competitiveness in the tourism industry

Key Elements	Variables
Governance of tourism	Government support and tourism as a priority, regulations, a whole of government approach, a tourism strategy, safety and security, public/private partnerships, vertical cooperation, multilateral cooperation, institutions, budget allocated to tourism support
Product development	Product differentiation, innovation, investments, market share, provide unique experiences, increase the added value of tourism, develop high value segments, marketplace perspective including prospective travellers, tourism operators and small businesses
Quality of tourism services	Improve quality, welcome visitors, quality of life, social equity and cohesion, services to consumers
Price competitiveness	Prices, exchange rates, ratio price/quality, "value for money", taxation
Accessibility/connectivity	Infrastructure development, geostrategic position of the destination, proximity
Branding of the destination	Promotion and marketing, identity, image, awareness of the destination, breadth of appeal, market diversification
Natural and cultural resources	Sustainability, gastronomy, climate, biodiversity
Human resources development	Skills, education and training, labour productivity, tourism training centres

Source: Key elements for defining the competitiveness in tourism industry for OECD countries adapted from Dupeyras and MacCallum, 2013

key elements for defining the competitiveness in tourism industry. Those are shown in Table 6.1. This study also identified core indicators for measuring tourism competitiveness, namely tourism performance and impacts, ability of a destination to deliver quality and competitive tourism services, attractiveness, policy responses and economic opportunities.

Vengesayi (2003) measured competitiveness and attractiveness from two perspectives, namely destination perspective and tourist perspective, by providing a holistic approach called the Tourist Destination Competitiveness and Attractiveness (TDCA) dynamics (Figure 6.1).

The study measured the competitiveness on the basis of some pillars like intrinsic destination resources and mix of activities, experience environment – physical and social – and supporting services. The study also showed communication and promotion as moderating factors that influence tourism competitiveness. Moreover, Dwyer and Kim (2003) showed several indicators and sub indicators of destination competitiveness as shown in Table 6.2.

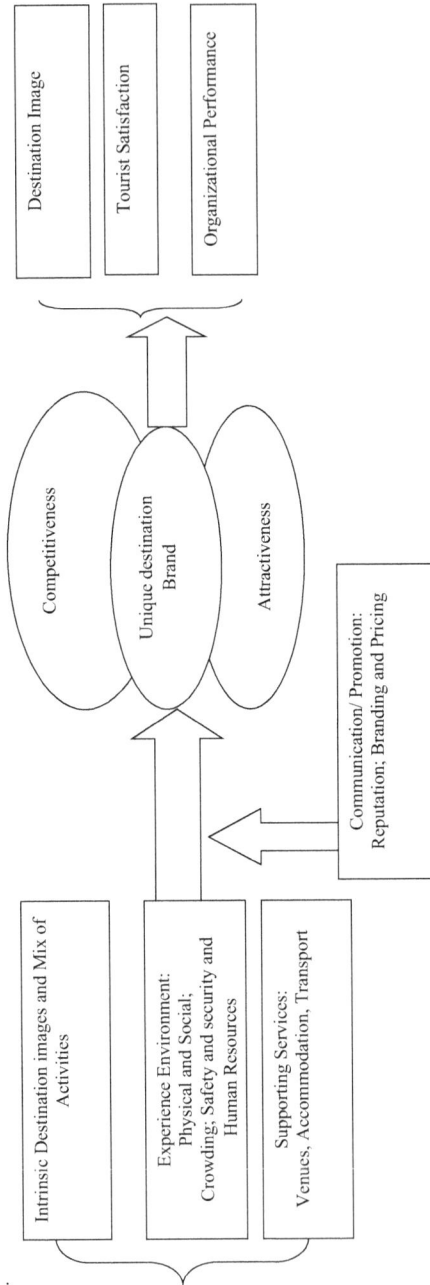

Figure 6.1 Conceptual model of Tourist Destination Competitiveness and Attractiveness (TDCA) dynamics (Vengesayi, 2003)

Table 6.2 Selected indicators of destination competitiveness

Endowed Resources	Created Resources	Supporting Factors	Destination Management	Situational Conditions	Demand Factors	Market Performance Indicators
a. Natural Resources: • Comfortable climate for tourism • Cleanliness/Sanitation • Natural wonders/scenery • Flora and fauna • Unspoiled nature • National parks/nature reserves b. Cultural/Heritage: • Historic/Heritage sites and museums • Artistic/architectural features • Traditional arts • Variety of cuisine • Cultural precincts and (folk) villages	a. Tourism Infrastructure: • Accommodation quality/variety • Airport efficiency/quality • Tourist guidance/information • Local transport efficiency/quality • Visitor accessibility to natural areas • Convention/exhibition facilities (capacity/quality) • Food services quality/variety b. Range of Activities: • Water based • Nature based • Adventure activities	a. General Infrastructure: • Adequacy of infrastructure to meet visitor needs • Health/medical facilities to serve tourists • Financial institution and currency exchange facilities • Telecommunication system for tourists • Security/safety for visitors • Local transport systems • Waste disposal • Electricity supply b. Quality of Services: • Tourism/hospitality firms which have well defined performance standards in service delivery • Firms have programmes to ensure/monitor visitor satisfaction	a. Destination management organization: • NTO (National Tourist Organization) acts as coordinating body for private and public sector tourism organizations • NTO effectively represents views of all tourism stakeholders in tourism development • NTO liaises effectively with private sector in tourism policy, planning and development • NTO provides statistical information as input to tourism policy, planning and development	a. Competitive (Micro) Environment: • Domestic business environment in destination • Management capabilities of tourism firms and organizations • Extent of competitive rivalry between firms in domestic tourism industry • Level of cooperation between firms in destination tourism industry • Links between tourism/hospitality firms and firms in other industrial sectors • Entrepreneurial qualities of local tourism stakeholders	a. Destination awareness b. Destination perception c. Destination preferences	a. Visitor statistics (numbers): • Number of foreign visitors • Growth rate of foreign visitors • Market share of destination – world, regional • Shifts in market share • Average length of stay • Rate of revisit b. Visitor statistics (expenditure): • Expenditure of foreign visitors (FX receipts) • Growth rate of expenditure of foreign visitors • Share of destination in total tourism expenditure – world, regional • Shifts in expenditure share

(Continued)

Table 6.2 (Continued)

Endowed Resources	Created Resources	Supporting Factors	Destination Management	Situational Conditions	Demand Factors	Market Performance Indicators
	• Recreation facilities • Sports facilities c. Shopping: • Variety of shopping items • Quality of shopping facilities • Quality of shopping items • Value for money of shopping items • Diversity of shopping experiences d. Entertainment: • Amusement/Theme parks • Entertainment quality/variety • Nightlife e. Special events/festivals	• Visitor satisfaction with quality of service • Industry appreciation of importance of service quality • Development of training programmes to enhance quality of service • Speed/delays through customs/immigration • Attitudes of customs/immigration officials c. Accessibility of Destination: • Distance/flying time to destination from key origins • Direct/Indirect flights to destination • Ease/cost of obtaining entry visa	• NTO strategically monitors and evaluates the nature and type of tourism development b. Destination marketing Management: • Reputation of NTO • Effectiveness of destination positioning • Strength/clarity of destination image • Efficient monitoring of destination marketing activities • Effective packaging of destination experiences • Links between destination tourism organizations and travel trade	• Access to venture capital • Tourism/hospitality firms operate in ethical manner • Firms use computer technology/commerce to achieve competitive advantage b. Destination Location: • Perceived 'exoticness' of location • Proximity to other destinations • Distance from major origin markets • Travel time from major origin markets c. Global (macro) environment: • The global business context • Political stability		• Foreign exchange earnings from tourism as percentage of total exports c. Contribution of tourism to economy: • Contribution of tourism to value added (absolute values and percentages, and rate of growth) • Domestic tourism • International tourism Contribution of tourism to employment (absolute numbers; percentage of total employment and rate of growth) • Domestic tourism • International tourism • Productivity of tourism industry sectors

- Ease of combining travel to destination with travel to other destinations
- Frequency/capacity of access transport to destination

d. Hospitality:
- Friendliness of residents towards tourists
- Existence of resident hospitality development programmes
- Resident support for tourism industry
- Ease of communication between tourists and residents

e. Market Ties:
- Business ties/trade links with major tourist origin markets
- Sporting links with major tourist origin markets
- Ethnic ties with major tourist origin markets

- NTO identification of target markets
- NTO strategic alliances with other NTO
- Destination marketing is based on knowledge of competitor products
- Present "fit" between destination products and visitor preferences.

c. Destination policy, planning, development:
- Existence of formal long-term "vision" for tourism industry development
- Destination "vision" reflects resident values

- Legal/regulatory environment
- Government policies for tourism development
- Economic conditions in origin markets
- Sociocultural environment
- Investment environment for tourism development
- Technology changes

d. Price competitiveness:
- Value for money in destination tourism
- Exchange rate
- Air ticket prices from major origin markets
- Accommodation prices
- Destination package tour prices

d. Indicators of Economic prosperity:
- Aggregate levels of employment
- Rate of economic growth
- Per capita income

e. Tourism investment:
- Investment in tourism industry from domestic sources
- Foreign direct investment in tourism industry
- Investment in tourism as percentage of total industry investment (and trend)

f. Price Competitiveness Indices:
- Aggregate price competitiveness indices
- By journey purpose
- By tourism sector

(*Continued*)

Table 6.2 (Continued)

Endowed Resources	Created Resources	Supporting Factors	Destination Management	Situational Conditions	Demand Factors	Market Performance Indicators
		• Religious ties with major tourist origin markets • Extent of foreign investment in local tourism industry	• Destination "vision" reflects tourism industry stakeholder values • Tourism policy conforms to a formal destination 'vision' • Tourism planning and development conforms to a formal destination 'vision' • Tourism development is integrated into overall industrial development • Ongoing tourism development is responsive to visitor needs • Extent to which research findings are integrated into tourism planning and development	• Price of destination visit relative to competitor destinations e. Safety/Security: • Level of visitor safety in destination • Incidence of crimes against tourists in destination		g. Government support for tourism: • Budget for tourism ministry • Budget for NTO • NTO expenditure on destination marketing (comparison with competitors) • Support for transport infrastructure • Industry programmes accessed by tourism industry • Tax concessions • Subsidies to industry' • Export marketing assistance • Vocational education skills/training for tourism industry

- Inventory of most significant attractors, facilities, services and experiences offered in destination
- Identification of major competitors and their product offerings
- Community support for special events

d. Human Resource Development:
- Public sector commitment to tourism/hospitality education and training
- Private sector commitment to tourism/hospitality education and training
- Training/education responsive to changing visitor needs

(*Continued*)

Table 6.2 (Continued)

Endowed Resources	Created Resources	Supporting Factors	Destination Management	Situational Conditions	Demand Factors	Market Performance Indicators
			• Range/quality of tourism/hospitality training programmes e. Environmental Management: • Public-sector recognition of importance of "sustainable" tourism development • Private sector recognition of importance of "sustainable" tourism development • Existence of laws and regulations protecting the environment and heritage • Research and monitoring of environmental impacts of tourism			

Source: Dwyer and Kim, 2003, pp. 400–405

Measuring destination competitiveness

The authors use the Travel and Tourism Competitiveness Index and Porter's Five Forces Model to measure and enhance the competitiveness of the travel and tourism industry of Bangladesh.

Measuring competitiveness of the tourism industry of Bangladesh through the Travel and Tourism Competitiveness Index

The Travel and Tourism Competitiveness Index (TTCI) is a comprehensive and recognized model for analysing competitiveness in the tourism industry. The TTCI was developed by WTTC in 2005 to formalize a model that could be used to analyse the competitiveness in any destination worldwide as well as that could be a useful tool in policy analysis (Martín et al., 2015; Bandura, 2008). The TTCI has been used extensively by different scholars (Sotiriadis and Varvaressos, 2015; Blanke et al., 2013) to measure the competitiveness and to formulate the policy for tourism development in different countries. The TTCI 2017 framework is shown in Figure 6.2.

The TTCI 2017 is composed of four broad categories of variables that facilitate or drive travel and tourism (T&T) competitiveness. The four broad categories are

i   Enabling environment
ii   T&T policy and enabling conditions
iii Infrastructure
iv   Natural and cultural resources

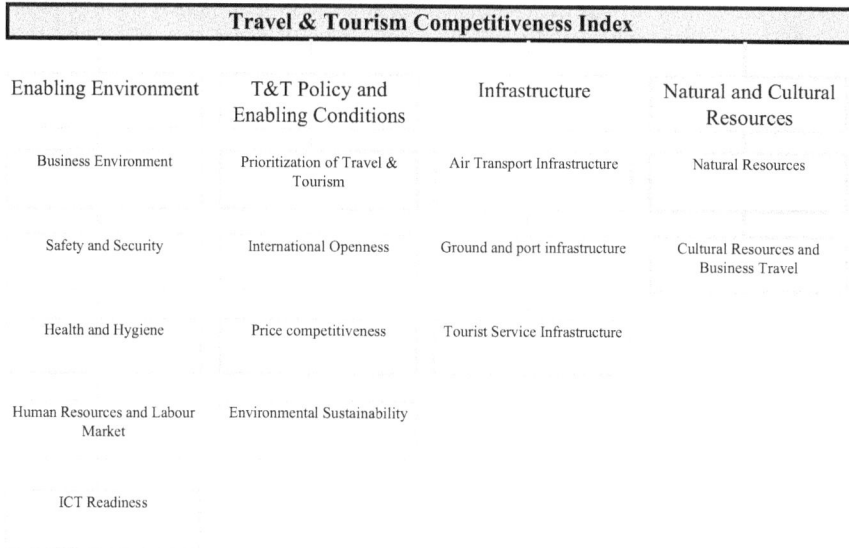

Travel & Tourism Competitiveness Index			
Enabling Environment	T&T Policy and Enabling Conditions	Infrastructure	Natural and Cultural Resources
Business Environment	Prioritization of Travel & Tourism	Air Transport Infrastructure	Natural Resources
Safety and Security	International Openness	Ground and port infrastructure	Cultural Resources and Business Travel
Health and Hygiene	Price competitiveness	Tourist Service Infrastructure	
Human Resources and Labour Market	Environmental Sustainability		
ICT Readiness			

Figure 6.2 The Travel and Tourism Competitiveness Index 2017 framework

Source: Adapted from Travel and Tourism Competitiveness Index, World Economic Forum (2017)

Among the four broad categories there are 14 pillars under which the tourism competitiveness is measured among 136 countries. Among the 136 countries Bangladesh ranked 125th position with average score 2.89 only. The detailed of competitiveness index among the South Asian countries are shown below.

In Table 6.3 we have found that among the five South Asian countries Bangladesh has achieved the lowest score, 2.89, and the 5th position. In the enabling environment category there are five pillars namely business environment, safety and security, health and hygiene, human resource and labour market and information and communication technology (ICT) readiness. In all these pillars except ICT readiness, Bangladesh is underperforming and the degree is severe for safety and security measures. This indicates that Bangladesh has to improve safety and security measures to enhance its competitiveness in the global market.

In the category of T&T policy and enabling conditions there are four pillars: prioritization of T&T, international openness, price competitiveness and environmental sustainability. Bangladesh is currently in below par position in all these pillars. The Bangladesh government has to take some measures to attract both domestic and foreign tourists. Besides, the government and other private destination marketers should rationalize the price of our tourism products and services in order to sustain the industry in the long run. International openness strategies of Bangladesh like visa requirements and openness of bilateral air service agreements should be smooth to gain a competitive position in global market.

The infrastructure category consists of three pillars: air transport infrastructure, ground and port infrastructure and tourist service infrastructure. As air transport infrastructure and tourist services infrastructure in Bangladesh are very poor, more international and domestic flight, flexible flight schedules, hassle-free arrival and departure facilities and preventing tourists' harassment in our airports should be ensured. Bangladesh is doing well in ground and port infrastructure.

In the natural and cultural resources index, the position of Bangladesh is quite dissatisfactory. Bangladesh government should emphasize the protection of its existing natural and cultural resources and heritages. Formalized attention should be given to the unexplored natural attractions. Every large corporate and international business meeting should be complimented with pleasure trips to our different attractive destinations.

Measuring competitiveness of the tourism industry of Bangladesh through Michael Porter's Five Forces Model

To analyse the competitiveness of tourism industry many researchers have used Porter's Five Forces Model (Dobrivojevic, 2013; Ali and Parvin, 2010). Porter (1980) outlined five forces which affect the competitiveness and industry profitability, namely threats of new entrants, bargaining power of buyers, bargaining power of suppliers, rivalry among the existing firms and impact of substitutes.

Table 6.3 The Travel and Tourism Competitiveness Index 2017: South Asian Region

South Asian Countries	Global Position	Average Score	Enabling Environment					T&T Policy and Enabling Conditions				Infrastructure			Natural and Cultural Resources	
			Business environment	Safety and security	Health and hygiene	Human resource and labour market	ICT readiness	Prioritization of T&T	International Openness	Price Competitiveness	Environmental Sustainability	Air Transport Infrastructure	Ground and port infrastructure	Tourist Service Infrastructure	Natural Resources	Cultural Resources & Business Travel
Bangladesh	125th	2.89	4.1	3.7	4.3	3.8	3.1	3.2	2.5	4.7	3.4	1.9	3.1	1.9	2.4	1.6
India	40th	4.18	4.3	4.1	4.4	4.4	3.2	3.9	3.7	5.8	3.1	3.9	4.5	2.7	4.4	5.3
Bhutan	78th	3.61	4.7	6.1	4.6	4.3	3.9	5.0	2.9	6.0	4.6	2.7	2.5	2.7	3.5	1.3
Nepal	103rd	3.28	4.1	4.8	5.0	4.2	2.6	4.8	2.8	5.6	3.4	2.0	1.9	2.3	4.2	1.3
Pakistan	124th	2.89	3.9	3.1	4.5	3.1	2.5	3.4	2.2	5.4	3.1	2.1	3.0	2.3	2.2	1.9
South Asian Average Score			4.2	4.4	4.6	4.0	3.1	4.1	2.8	5.5	3.5	2.5	3.0	2.4	3.3	2.3

Source: The Travel and Tourism Competitiveness Index 2017, adapted from World Economic Forum (2017)

Threats of new entrants

New entrants or potential competitors are those companies that are not currently operating their business in an industry but if they want, they have the capability to do so. Whether a new firm enter into an industry or not it depends on entry barriers. The tourism industry of Bangladesh is neither fully formalized nor strictly regulated by the government and tourism authority, so anyone can enter into this industry at any time. This industry is described by emergence of new entrepreneurs. In Bangladesh, tourists' destinations like Cox's Bazar, Saint Martin, Bandarban, Rangamati Hill Tracts and different natural and heritages sites enjoy strong brand loyalty. Hence, newcomers will have to face strong barriers to enter into the tourism industry of Bangladesh.

Bargaining power of buyers

Bargaining power of buyers is their ability to bargain down the prices that companies charge in the industry or to lift up costs of the companies by claiming better quality products and services (Hill and Jones, 2013). As there are many travel agents, tour operators and corporate travel consultants operating in Bangladesh, tourists' bargaining power is high here. There are some travel and tour operators who have online facilities and websites where tourists can collect information and which enable the tourists to wrangle down the prices. Young tourists are technology-driven and very savvy (Hence, 2018) and can easily express their feelings and observations in different digital platforms, thus making themselves stronger in taking destination selection and visit decisions than ever before.

Bargaining power of suppliers

Bargaining power of suppliers is the ability of suppliers to increase the prices of the input they provide, or to lift up the costs of the industry by providing low-quality inputs or poor services. In the tourism industry suppliers include transportation, airlines, hotels, travel agents, tour operators, ancillary services providers and so on. In Bangladesh, a strong syndicate of transportation service providers sometimes raises the cost of tourism and destination marketers through charging unreasonable ticket prices. Suppliers also squeeze profits out of the tourism industry by increasing the cost of destination marketers as these suppliers have profound impact on both the backward and forward linkages of the destination marketers' value chain.

Rivalry among existing firms

Rivalry among existing firms refers to the intensity of competition among the existing firms who are currently doing business within the industry. The intensity of competition among the existing firms depends on the industry's competitive structure. The competitive structure of the tourism industry of Bangladesh is fragmented because there are many small destinations available all over the country that hold small shares. As the industry's competitive structure is not consolidated, here competition is not fierce. The demand of tourism is increasing in Bangladesh; hence, the authors believe competition among the existing firms will be low because they will not engage in a cutthroat approach.

Impacts of substitutes

Substitute products refer to those products offered by different businesses or industries that have the ability to meet the same customer needs. In the tourism and hospitality industry people visit different destinations for refreshment, enjoyment, learning and so on. There are ample opportunities and probabilities to bring new tourist spots, destinations or products which will provide refreshment, enjoyment, learning and entertainment to tourists and visitors. Anyone can bring close substitutes with little variations in the levels of services and amenities. This industry has a constant challenge of substitute products.

   The competitiveness of the tourism industry of Bangladesh through Porter's five forces model is summarized in Figure 6.3.

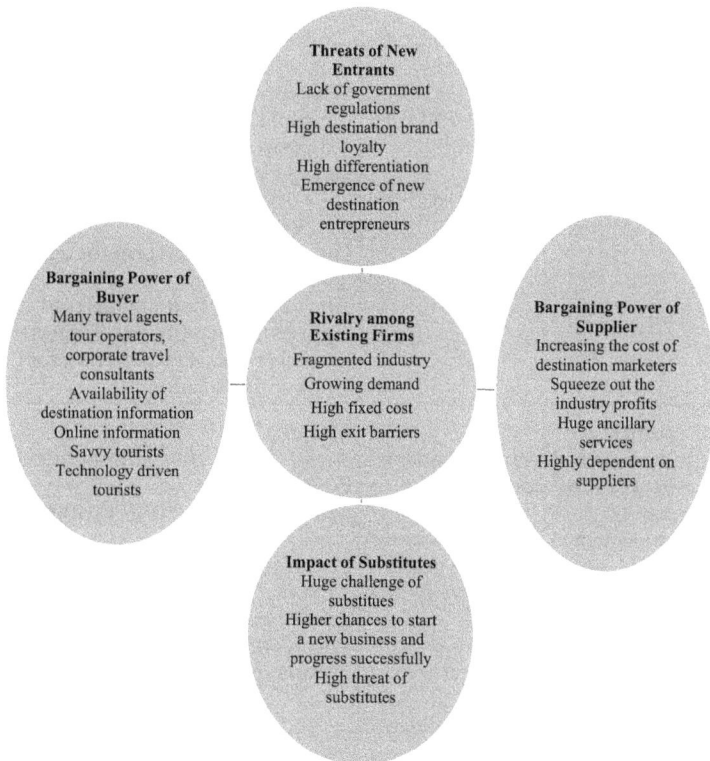

Figure 6.3 Porter's Five Forces Model applied to the tourism industry of Bangladesh
Source: Authors' own construction with the help of Porter's Model proposed in 1980

SWOT analysis of the tourism industry of Bangladesh

SWOT analysis is a strategic planning technique of overall evaluation of strength, weakness, opportunities and threats of a company (Kotler and Armstrong, 2012). Before determining the objectives and goals of a company, SWOT analysis is must. Strengths are capabilities, resources and other positive factors that are internal to the destination which may assist the destination serve its customers and attain its objectives. Weaknesses are limitations and negative factors that are internal to the destination which may hamper the destinations' expected performance. Opportunities are positive external environmental factors or trends that the destination can be able to develop to its advantage. Threats are negative external factors or trends the destination faces which may present obstacles to performance. SWOT analysis of tourism industry of Bangladesh is expressed in Figure 6.4.

	Internal		
Positive	**Strengths:** Huge natural, archaeological and historical sites Age-old cultural and religious heritages and events Huge unexplored destinations Number of World Heritage Sites Availability of human resources Hospitality of the residents of BD Largest sea beach in the world Availability of gastronomy Unique mangrove forest like Sundarbans and swamp forest like Ratargul	**Weaknesses:** No formal authoritative rules, regulations and associations Inefficiency in managing the destinations Lack of appropriate coordination among the destination' stakeholders Lack of private and government investment Inappropriate marketing strategies Improper management in supplies and support services Lack of formal safety and security measures Lack of government prioritization of tourism sector Limited budget for promoting tourism overseas Poor digitalization initiatives from the government and private sectors Lack of strategic information for policy formulation	Negative
	Opportunities: Relatively comfortable weather conditions Young generations' zest for new destinations More engagement in international events e.g. World Cup cricket, SAAF games Huge scope of business tourism Mega-project-based tourism New information technology Privatization of the public sector in tourism Increased popularity of short holidays People's tendency to escape from urban and stressful life	**Threats:** Strong competition within the region Poor political image Unplanned extraction of natural resources Language barrier of the Bangladeshi people Lack of public awareness Threats of security Natural disasters	
	External		

Figure 6.4 SWOT analysis of the tourism industry of Bangladesh

Source: Authors' own constructions

Strategic situation analysis for the Bangladesh tourism industry

In strategic situation analysis, we will focus on parameters like growth of tourism industry, international tourist arrivals, international tourism receipts and contribution to GDP and employment.

Tourism is seen as a growing overall economic activity and an emerging industry in developed, developing and underdeveloped countries (Tasci and Knutson, 2004), and this increase in such activity is greatly desirable (Dwyer et al., 2004). But the performance of our country's tourism industry is not satisfactory until now. Table 6.4 shows international tourist arrivals in our country was only 125,000, the lowest among neighbouring and competing countries. Simultaneously, a large number of tourists from Bangladesh are going to India, Nepal or Bhutan for enjoying their vacation during Eid, Puja or some special events (Hamid, 2019). Although the average receipts per arrival of tourists is acceptable, due to low tourist arrival, total international tourism inbound receipts are not satisfactory. Moreover, the T&T industry contribution to country's GDP and employment are also poor in comparison to the other neighbouring countries.

Strategic situation analysis of our tourism industry based on the data provided by BPC and WEF is quite difficult, as these data differ significantly and are very often found incomplete. For example, tourist arrivals in 2013 as per BPC for the first six months is 278,780 whereas the total tourist arrivals in 2013 as per WEF (2015) is 148,000! Thus, making a conclusion based on this type of data on same indicator from different sources is tough. The number of tourist arrivals in our country is decreasing as reflected by the reduced growth rate (–3.59%) in 2016 compared to 2015. Besides, earnings from tourism is also showing negative growth rate (–28.98%) in 2016 as compared to 2015. All these discussions demand a prioritized comprehensive national tourism policy for our country.

Structural problems and issues of the tourism industry of Bangladesh

The tourism industry of Bangladesh has faced challenges for decades. There is a lack of appropriate and updated tourism policy which makes our tourism industry underperforming. Non-existence of formal authoritative rules, regulations and associations is the major problem of the industry. The industry is supplier driven rather than market driven (Ishtiaque, 2012). The suppliers squeeze out the industry profits. Besides, the competitive market structure of this industry itself creates problem as the market structure is fragmented. Further, to attract tourists from home and abroad integrated marketing communication programmes are absent which leads to poor growth of the tourism industry.

Table 6.4 Key performance of the tourism industry of Bangladesh in 2017 in comparison with neighbouring countries

Key Indicators	Bangladesh	Bhutan	India	Nepal	Pakistan
International tourist arrivals	125,000	155,121	8,027,133	538,970	965,498
International tourism inbound receipts	US $148.4 million	US $71.2 million	US $21,012.7 million	US $481.3 million	US $317.0 million
Average receipts per arrival	US $1,187.2	US $458.7	US $2,617.7	US $892.9	US $328.3
T&T industry GDP	US $5,193.0 million (2.4% of total GDP)	US $0.0 million (0.0% of total GDP)	US $41,582.4 million (2.0% of total GDP)	US $804.9 million (4.0% of total GDP)	US $7,362.0 million (2.8% of total GDP)
T&T industry employment	1,138,690 jobs (2.0% of total employment)	0 jobs (0.0% of total employment)	23,454,400 jobs (5.5% of total employment)	426,395 jobs (3.2% of total employment)	1,429,580 jobs (2.4% of total employment)

Source: Travel and Tourism Competitiveness Index 2017, adapted from World Economic Forum (2017)

Table 6.5 Year-wise total tourist arrivals and foreign earnings in Bangladesh

Year	Tourists Arrivals		Earnings	
	Number	*Growth Rate (%)*	*Amount (in Crore TK)*	*Growth Rate (%)*
2012	588,193	−0.92	825.40	–
2013	278,780 (first six months)	–	949.56	15.04
2014	–	–	1227.30	29.24
2015	643,094	–	1136.91	−7.36
2016	620,000	−3.59	807.32	−28.98

Source: Bangladesh Parjatan Corporation (BPC), n.d.

Lessons from other countries

India's incredible journey – Bangladesh can follow

India is an ideal example of affluent natural attractions, rich history and cultural heritage which has been successfully preserving, developing and marketing its different tourism products and services to the world. India has undertaken its tourism branding strategy under the slogan "Incredible India" in order to develop and promote its tourist attractions to the tourists of western countries and became successful in drawing their attention. In addition to natural, heritage, cultural and archaeological tourism, the government is also trying to develop rural tourism, golf tourism, polo tourism, cruise tourism and medical tourism under the "Incredible India" campaign through a very strategic approach. Bangladesh, having many natural, geographical, cultural and historical attractions like India, can follow India's strategy and develop alternative tourism products such as rural tourism, religious tourism, eco-tourism, tribal tourism and so on.

Malaysian tourism – a case for Bangladesh to imitate

Malaysia entered into the tourism arena in the early 1990s and adopted their first National Tourism Policy in 1992. They also set up a separate tourism ministry in 2004. Throughout this time, they have tried several branding strategies from "Beautiful Malaysia" to "Only Malaysia" followed by "Fascinating Malaysia" and lastly settled on the current campaign of "Malaysia, Truly Asia". Now they are hoping to grow their tourism industry three times by 2020. Bangladesh can also imitate this success sequence of Malaysia. It has tried only one branding campaign namely "Beautiful Bangladesh (School of Life)". But, to be competitive in the global tourism market, additional campaigns should be launched by adopting a structured national tourism policy.

Conclusion and recommendations

Tourism is increasingly considered as a promising business sector in Bangladesh not only for its capability to earn foreign exchange but also for its ability to act as a medium for representing our country's rich natural, cultural and age-old unique heritages to the world. Government first should formulate a specific tourism development policy on a priority basis and make short-term and long-term master plans for developing the country's tourism industry. Moreover, standard transportation, accommodation, eating and shopping facilities and ancillary services need to be developed in the tourist spots. Digital marketing and online technology such as websites, social media, community blogs and so on should be developed by the destination marketers so that prospective tourists can easily get necessary information about destination and assistance like booking a ticket and hotel reservation online. Besides, the formalities for foreigners from abroad to come to Bangladesh for pleasure trip should be made easy. More police stations and boxes need to set up close to the tourist spots and the coverage of tourist police should also be spread in order to ensure sufficient safety and security to the tourists and visitors. Special attention should be given to preserve rivers, sea beaches, water resources, forests and wildlife resources and related flora and fauna in order to uphold ecological balance besides maintenance of historical and heritage sites. Repetitive promotional activities should also be taken to continuously and persuasively inform domestic as well as foreign tourists to make Bangladesh competitive and an attractive tourist destination.

References

Ali, M. M. and Parvin, R. (2010). *Strategic management of tourism sector in Bangladesh to raise gross domestic product: An analysis*. Retrieved from: https://bit.ly/2nd0qsm (accessed: the 25th May 2019).

Alonso, V. (2010). Factores críticos de éxito y evaluación de la competitividad de los destinos turísticos. *Estudios y perspectivas en turismo*, 19(2), pp. 201–220.

Bandura, R. (2008). *A survey of composite indices measuring country performance: 2008 update*. Retrieved from: https://bit.ly/2oCKQ9S (accessed: the 22nd May 2019).

Bangladesh Parjatan Corporation (n.d.). *Annual report 2016–2017*. Retrieved from: https://bit.ly/2msaSfu (accessed: the 30th May 2019).

Bhatia, A. K. (2006). *The business of tourism: Concepts and strategies*. Auckland: Sterling Publishers Private Limited.

Blanke, J. and Chiesa, T. (2011). *The travel & tourism competitiveness index 2011: Assessing industry drivers in the wake of the crisis*. Geneva: World Economic Forum, pp. 3–33.

Blanke, J., Chiesa, T. and Crotti, R. (2013). *The travel & tourism competitiveness index 2013: Contributing to national growth and employment*. Geneva: World Economic Forum, pp. 3–41.

Cho, D. S. (1998). From national competitiveness to bloc and global competitiveness. *Competitiveness Review*, 8(1), pp. 11–23.

Crotti, R. and Misrahi, T. (2017). *The travel & tourism competitiveness index: Travel & tourism as an enabler of inclusive and sustainable growth*. Retrieved from: https://bit.ly/2o5O4zU (accessed: the 10th May 2019).

Crouch, G. I. (2010). Destination competitiveness: An analysis of determinant attributes. *Journal of Travel Research*, 50(1), pp. 27–45.

Crouch, G. I. and Ritchie, J. R. (1999). Tourism, competitiveness, and societal prosperity. *Journal of Business Research*, 44(3), pp. 137–152.

Dobrivojevic, G. (2013). Analysis of the competitive environment of tourist destinations aiming at attracting FDI by applying Porter's Five Forces Model. *British Journal of Economics, Management & Trade*, 3(4), pp. 359–371.

Dupeyras, A. and MacCallum, N. (2013). Indicators for measuring competitiveness in tourism: A guidance document. *OECD Tourism Papers*, 2013/2, OECD Publishing, pp. 1–62.

Dwyer, L. and Forsyth, P. (1992). The case for tourism promotion: An economic analysis. *The Tourist Review*, 47(3), pp. 16–26.

Dwyer, L., Forsyth, P. and Rao, P. (2000). The price competitiveness of travel and tourism: A comparison of 19 destinations. *Tourism Management*, 21(1), pp. 9–22.

Dwyer, L. and Kim, C. (2003). Destination competitiveness: Determinants and indicators. *Current Issues in Tourism*, 6(5), pp. 369–414.

Dwyer, L., Mellor, R., Livaic, Z., Edwards, D. and Kim, C. (2004). Attributes of destination competitiveness: A factor analysis. Tourism Analysis, 9(1), pp. 91–101.

Edgell, D. L. and Swanson, J. (2013). *Tourism policy and planning: Yesterday, today, and tomorrow*. Oxon: Routledge.

Enright, M. and Newton, J. (2004). Tourism destination competitiveness: A quantitative approach. *Tourism Management*, 25(6), pp. 777–788.

Goeldner, C. R. and Ritchie, J. R. B. (2011). *Tourism: Principles, practices, philosophies* (12th ed.). Hoboken, NJ: John Wiley & Sons, Inc.

Gooroochurn, N. and Sugiyarto, G. (2005). Competitiveness indicators in the travel and tourism industry. *Tourism Economics*, 11(1), pp. 25–43.

Hamid, M. A. (2019). *Biswo-Prekkhapote Bangladesher Porjoton* (in Bengali). Dhaka: Dibyaprakash.

Hassan, A. and Kokkranikal, J. (2018). Tourism policy planning in Bangladesh: Background and some steps forward. *e-Review of Tourism Research (eRTR)*, 15(1), pp. 79–87.

Hence, B. G. (2018). Urban experiential tourism marketing: Use of social media as communication tools by the food markets of Madrid. *Journal of Tourism Analysis: Revista de Análisis Turístico*, 25(1), pp. 2–22.

Hill, C. W. L. and Jones, G. R. (2013). *Strategic management: An integrated approach* (9th ed.). New Delhi: Cengage Learning India Private Limited.

Ishtiaque, A. N. A. (2012). A schema of tourism development model: A case of Bangladesh. *Dhaka University Journal of Marketing*, 15(June), pp. 175–186.

Islam, M. K. (2014). A study on development strategies of tourism in Bangladesh. *Unpublished PhD Thesis*. Dhaka: University of Dhaka.

Khondker, B. H. and Ahsan, T. (2015). *Background paper on tourism sector*. Retrieved from: https://bit.ly/2XpAbza (accessed: the 20th May 2019).

Kotler, P. and Armstrong, G. (2012). *Principles of marketing* (14th ed.). Upper Saddle River, NJ: Pearson Education, Inc.

Majumder, D. (2015). Contributions and loopholes of tourism sector in Bangladesh. *Jagannath University Journal of Social Sciences*, 3(1–2), pp. 1–19.

Martín, J. C., Mendoza, C. and Román, C. (2015). A DEA travel – tourism competitiveness index. *Social Indicators Research*, 130(3), pp. 937–957.

Newall, J. E. (1992). The challenge of competitiveness. *Business Quarterly*, 56(4), pp. 94–100.

Porter, M. E. (1980). *Competitive strategy: Techniques for analyzing industries and competitors*. New York, NY: The Free Press.

Porter, M. E. (2008). The five competitive forces that shape strategy. *Harvard Business Review*, 86(1), pp. 57–71.

Ritchie, B. and Crouch, G. (2003). *The competitive destination: A sustainable tourism perspective*. Oxfordshire: CABI Publishing.

Scott, B. R. and Lodge, G. C. (1985). *U.S. competitiveness in the world economy*. Boston, MA: Harvard Business School Press.

Soteriades, M. (2012). Tourism destination marketing: Approaches improving effectiveness and efficiency. *Journal of Hospitality and Tourism Technology*, 3(2), pp. 107–120.

Sotiriadis, M. and Varvaressos, S. (2015). A strategic analysis of Greek tourism: Competitive position, issues and lessons. *African Journal of Hospitality, Tourism and Leisure*, 4(2), pp. 1–14.

Spence, A. M. and Hazard, H. A. (1988). *International competitiveness*. Cambridge: Ballinger Publishing Company.

Tasci, A. D. A. and Knutson, B. J. (2004). An argument for providing authenticity and familiarity in tourism destinations. *Journal of Hospitality and Leisure Marketing*, 17(1), pp. 73–82.

Vengesayi, S. (2003). A conceptual model of tourism destination competitiveness and attractiveness. In *ANZMAC conference proceedings*. Adelaide, University of South Australia, 1–3 December, pp. 637–647.

World Economic Forum (2015). *The travel & tourism competitiveness report 2015: Growth through Shocks*. Geneva: World Economic Forum.

World Economic Forum (2017). *The travel & tourism competitiveness report 2017: Paving the way for a more sustainable and inclusive future*. Geneva: World Economic Forum, p. xiv.

7 Conceptual analysis on tourism product and service promotion with special reference to Bangladesh

Shah Alam Kabir Pramanik and
Md. Rakibul Hafiz Khan Rakib

Introduction

Bangladesh is a South Asian country, endowed with rich history, archaeological sites, religious places, traditional events and natural beauty (Majumder, 2015; Ahmad, 2013). All these attractions have made Bangladesh a suitable destination for tourists (Hassan and Kokkranikal, 2018). But surprisingly, the country fails to attract a sufficient number of tourists. One of the prime reasons for this failure is insufficient and ineffective promotional activities performed by our destination marketers (Hossain, 2015).

The tourism promotion mix consists of the activities such as advertising, personal selling, sales promotion, direct marketing, events and experience marketing, public relations, digital marketing, sponsorships, social media, online marketing, e-word of mouth communication and so on that a destination marketer uses to communicate its information, features and benefits to its present and potential tourists (Bao, 2018) with a view to informing, reminding and persuading them to visit or revisit the destination (Goeldner and Ritchie, 2011; Kotler et al., 2010). But the destination marketers of Bangladesh have failed to recognize the critical role of promotion in developing the tourism industry. This necessitates the concerned tourism authority and destination marketer to initiate some measures for improving the country's destination image to the potential tourists for drawing their attention and growing their interest (Eccles, 1995; Dwyer and Forsyth, 1992) in selecting Bangladesh as a tourist destination and visit or revisit the same (Hossain, 2015). To attract tourists from home and abroad, integrated marketing promotion, popularly known as integrated marketing communication (IMC), is required (Aronsson and Tengling, 1995). Although many tools of marketing promotion are available to destination marketers, the question is which tools are effective for the promotion of tourism products and services? Is it possible to attract both domestic and foreign tourists with the same promotional tools? This chapter will try to answer those questions.

The concept of promotion

Promotion is one of the most important tools of the marketing mix that helps to create awareness, improve image and stimulate sales of any product or services. A company's promotional tools consist of the specific blend of advertising, personal selling, public relations, sales promotion and direct marketing in order to provide clear and persuasive message to audience (Kotler et al., 2010). Several prior studies on marketing recognize the importance of finding out the consumers' needs, wants and demands through marketing communication to provide them with their desired goods and services (Al-Dmour et al., 2016), helped in suggesting distinguished and varied products (Jeong and Lee, 2017; Bahri-Ammari and Nusair, 2015) and assisted to identify the most effective methods of marketing communication with a view to influence the behaviours and purchase decisions of consumers (Erdem and Jiang, 2016; Chen et al., 2011). Thus, promotion is considered as one of the most important tools in communicating and connecting with consumers (Van Waterschoot and Van den Bulte, 1992).

Tourism promotion

In the tourism industry, promotions can play a pivotal role for creating destination awareness, image, satisfaction and loyalty. The tourism promotion mix, also known as tourism communication mix, consists of a range of media, tools and techniques that the destination service providers use to convey the message and attract tourists to purchase the offering and visit the destination including websites, information kits, advertising, personal selling, travel shows, sales promotion and public relations (Bao, 2018). Tourism marketing promotions, by carefully mixing and matching different tools, can contribute to build destination brand equity by establishing the destination in tourists' memory with a positive destination image.

Hospitality organization utilizes different promotional tools and elements in formulating promotional strategies (Al-Debi and Mustafa, 2014). Through providing information to the visitors and tourists about the new products and services, it arouses desirability, boosts the demand and affects the final visit decision, promoting establishment of its inevitability (Chiu and Ananzeh, 2012). For creating greater tourism awareness to both the domestic and international tourists, large-scale advertising campaigns should be undertaken to improve the image of the destination as well as its country for the development of the tourism industry (Rao et al., 1990). Advertisements, websites and direct marketing are very important promotional tools for the tourist product whose demand is declining (Ali, 2016; Al-Azzam, 2016). Besides advertising, more public relations efforts and publicity programmes should be devoted to ensure more media coverage from a tourism perspective. Vogt and Stewart (1998) confirmed that travellers need precise information in taking the decision to travel to a particular destination and then, for all the decisions they made while actually travelling. Murphy et al. (2007) validated this argument and emphasized the importance of word of mouth (WOM) information sources like friends, family members, relatives and so on in the travel decision-making process. With the dynamic change in the online world, new IT-related communication formats like online promotion or social media are gaining more effectiveness over the traditional communication channels (Nail, 2005; cited in Trusov et al., 2009), which leads to a reinterpretation of some traditional promotional tools like word-of-mouth communication as electronic word-of-mouth (eWOM). Information and communication technology (ICT) and internet-based platforms lessen consumer's information search effort, reduce uncertainty and related perceived risks (Mackay and Vogt, 2012; Gretzel and Yoo, 2008), as what is shared among the users creates greater confidence and credibility (Litvin et al., 2008).

Objectives of tourism promotion

The core objective of any promotion activity is to generate demand for a particular product or service (Goeldner and Ritchie, 2011). Usually, the objective of tourism promotional campaign is to raise visitation in a destination for facilitating growth and prosperity of the country's tourism industry. In general, the objectives of tourism promotion include those discussed next.

Providing information about the destination

Through tourism promotion, destination marketers and promoters can provide necessary information about the destination like what the destination actually is, its key attractions, availability of related services associated with the destination, perfect time for a visit and so on.

Creating brand/destination awareness

Brand awareness is reflected by consumers' brand recognition or recall activity (Kotler and Keller, 2006). Destination marketers can keep the destination's name in front of the tourists and positively reinforce its image by providing a consistent promotional message.

Increase demand

Tourism marketing promotions are used primarily to increase demand for tourism products or services through attracting new visitors to a specific destination.

Increase sales

Once tourists become customers of a tourism attractions or site, destination marketer should encourage them to increase their spending. Hotels, for example, can use sales promotion tools like a "customer rewards card" for their long stay with reduced occupancy charge.

Encourage re-visit behaviour

Destination marketers or service providers can employ retention marketing promotion to turn first-time customers into repeat customers through spreading positive word of mouth information about the tourism products and services;

collecting customer information and regularly contacting them with updated information, special offers, advance notice of discounts and sales; and encouraging them coming back again and again.

Effective tools for tourism products and services promotion

The tourism promotion mix consists of the modes of communication discussed next.

Advertising

Advertising has been defined as any paid form of non-personal promotion and presentation of goods, services or ideas by an identified sponsor (Goeldner and Ritchie, 2011; Kotler and Keller, 2006). Advertising is one of the widely used promotional tools for creating destination awareness and identity (Hossain, 2015) through the use of numerous advertising media such as print and broadcast ads, the internet, newspapers, brochures and booklets, magazines, directories, television, outdoor, billboards, motion pictures or radio. Among the different advertising media, print ads has become a primary vehicle in promoting state tourism and "selling" a wide range of nation's destinations and modes of travel through creating positive destination imagery (Motes and Hilton, 2002).

Personal selling

Personal selling involves face-to-face interaction with prospective buyers for the purpose of making presentations, responding to their questions and securing orders (Kotler and Keller, 2006). From a tourism perspective, personal selling entails direct contact between destination marketers and tourists (Goeldner and Ritchie, 2011), either face-to-face or through some sort of telecommunications such as telephone contact for securing immediate feedback. It plays an important role to make the potential tourists aware of the tourism products and services through personal presentation and promotion and persuade them to visit the destination (Hossain, 2015). Personal selling includes sales presentations, sales meetings, trade shows, incentive programmes and samples.

Sales promotion

Sales promotion is the use of a wide range of short-term incentives to encourage experimental use or purchase of a product or service (Kotler and Keller, 2006) or to stimulate a desired result from prospective customers, trade intermediaries and the salesforce (Cooper et al., 2005). Through boosting sales, sales promotion can play critical role in travel and tourism marketing. Tourism sales promotions are the short-term incentives offered to tourists for inducement to purchase

(Mill and Morrison, 2002), which also covers the salesforce and channel of distribution as well. Sales promotion includes discounts, coupons, contests, games, sweepstakes, tourism fairs, trade shows, displays, exhibits, premiums and gifts and demonstrations.

Events and experiences

Kotler and Keller (2006) define events and experiences as company-sponsored actions and programmes intended to create day-by-day or special brand-related interactions. Events can serve as one of the key marketing propositions in place promotion by functioning as animators of destination attractiveness (Getz and Page, 2016). Events like the 2011 ICC Cricket World Cup held in Bangladesh bring lots of excitement and memorable experiences to both domestic and international tourists. In tourism, the concept of experience focuses on emotions and feelings consumers experience during their consumption activities by contributing value addition for tourists and improved satisfaction (Tahar et al., 2018). Events and experiences include common promotional platforms such as sports, entertainment, arts, festivals, street activities and so on.

Public relations and publicity

Public relations can be defined as the social conscience of the company that treated public interest as first priority while making any decisions through building good relations with its publics including employees, suppliers, visitors and the community (Goeldner and Ritchie, 2011). Kotler and Armstrong (2012) define public relations as the activities planned to build good relations with the company's different publics by attaining positive publicity, building up, promoting and protecting a good company image and handling or heading off adverse events, rumours and stories. Publicity refers to non-personal and nonpaid communications by an identified sponsor about an organization, product, service or idea (Belch and Belch, 2005). Public relations include press releases, news conference, speeches, sponsorships, company-sponsored events, special events, press relations, corporate communications, donations, lobbying, counselling and so on.

Direct marketing

With the rapidly changing customer needs and wants, the mass marketing–oriented promotion models are being widely replaced by individual-oriented marketing promotion models. Direct marketing as a component of the IMC mix is getting popular as the fastest growing of all other tools (Solomon, 2010), and enjoys remarkable growth in recent years (Katzenstein et al., 1994). It involves connecting with the targeted customers directly with a view to securing immediate response and building long-term relationships (Kotler et al., 2010). Direct marketing uses a number of tools such as telephone marketing, catalogue marketing, e-mail marketing, mobile marketing, electronic shopping, face to face marketing, fax mail, voice mail and so on to connect with carefully targeted visitors.

Word of mouth communication

Word of mouth (WOM) communication can be termed as interpersonal communication among consumers relating to their personal experiences with a product or firm (Richins, 1983). It has long been acknowledged as one of the credible and important external sources of information for travel planning (Hwang et al., 2006; Kotler et al., 2006). WOM information from friends and relatives is considered as one of the most relied upon sources of information for selecting a particular destination (Murphy et al., 2007; Beiger and Laesser, 2004). As the use of the internet for travel and tourism planning becomes ever more established, travel decision-making processes are increasingly influenced by eWOM (Gretzel and Yoo, 2008), thus traditional WOM is being widely replaced by eWOM. Besides relatives, friends or colleagues, consumer opinion platforms such as blogs and social media pages have established themselves as important spaces for eWOM (Hennig-Thurau et al., 2004).

Internet marketing/online promotion

Internet marketing, also known as i-marketing, online marketing, web-based marketing or e-marketing, has changed the travel and tourism industry forever (Bao, 2018; Goeldner and Ritchie, 2011). Tourism service providers like airlines, hotels, tour operators and travel agencies develop digital solutions to improve user experience and boost online sales (Papetti et al., 2018). Tourists get instant and updated information from the internet on tourism products or services to ease their travel decision making (Middleton et al., 2009; Morrison, 2002). Continuous growth of the Internet and World Wide Web has created a new marketing potential for tourist destinations (Flynn, 1995), including for Bangladeshi destination marketers. Online promotion includes activities like websites, e-mail marketing messages, search engine marketing, newsletters, banner advertising, mobile applications, e-partnerships and Web 2.0 strategies.

Social media marketing

Social media is an online network of relationship among the people, sometimes termed as "consumer-generated media" (Goeldner and Ritchie, 2011). Recently, a segment of young travellers emerged who prefer exploring new zones and destinations like Bichanakandi or Sada Pathor of Sylhet, which are apparently less known but attractive, to mass tourism. These young travellers share almost similar profiles and are frequent users of social media with the capacity of influencing other users by initiating a new trend (Hence, 2018). So, for Bangladeshi destination marketers, ensuring their destination's presence in social media like Facebook or YouTube becomes mandatory to reach these large segments of tourists worldwide. Social media includes various types of social networks like Facebook, YouTube, Myspace, blogs, microblogging sites like Twitter, Qaiku, Google Buzz, wikis, podcasting, forums like Trinet and content communities.

IMC for domestic tourists and foreign tourists

IMC is the specific mix of advertising, personal promotion, public relation, sales promotion, direct marketing, events and experiences marketing and so on that are combined to construct and maintain long-term profitable customer relationships by delivering a clear, consistent and inducing message of market offerings (Kotler et al., 2010). Destination marketers should spend more time on marketing promotion so that potential tourists can recognize and accept their tourism products and services (Eccles, 1995). For creating a clear and long-lasting impact, most of the tourism organizations use a blending of different promotional activities. Although the IMC tools used to attract domestic tourists considerably differ from the tools used to attract foreign tourists, our destination marketers failed to recognize the difference. For this reason, the authors suggest separate IMC mixes for domestic and foreign tourists with some modification in the existing tools by including a special international communication focus.

IMC for domestic tourists

For domestic tourists, all the promotional tools that we have described in the previous section can be used. However, the authors suggest using a combination of tools rather using a single one, because the combined impact of all or some of the IMC tools is always greater than the impact generated by a single promotional tool in attracting and retaining tourists.

IMC for foreign tourists

Almost all the major IMC tools that are used for domestic tourists can also be used for foreign tourists, but the authors dropped personal selling as an IMC tool for foreign tourists because of its poor reach in the international arena from a Bangladeshi tourism standpoint. Moreover, all the tools must be used with special attention to foreign perspectives. Special attention should also be given to broadcast tourism promotion programmes on television and the Internet. Also, billboards in airports, features in tourism magazines and advertisements on both public and private vehicles can attract foreign tourists. International tourism fairs and travel shows, airport demonstrations and special discounts should be focused on while designing sales promotion programme. International mega sports events like the Cricket World Cup, carnivals, international tourists' activation programme like "Beautiful Bangladesh" or "Visit Bangladesh" should get careful consideration for designing events and experience programmes. For designing public relations and publicity programmes, international conferences,

Advertising	Personal Selling	Sales Promotion	Events and Experiences	Public Relations and Publicity	Direct Marketing	Word of Mouth (WOM)	Online Promotion	Social Media Marketing
Print and broadcast ads, The internet, Newspapers, Brochures and booklets, Magazines, Directories, Television, Outdoor, Billboards, Motion pictures, Radio.	Sales presentations, Sales meetings, Trade shows, Incentive programs, Samples like free visit.	Discounts, Coupons, Contests, Games, Sweepstakes, Tourism fairs, Trade shows, Displays, Exhibits, Premiums and gifts, Demonstration.	Sports, Entertainment, Arts, Festivals, Street activities.	Press releases, News, Conference, Speeches, Sponsorships, Company-sponsored events, Special events, Press relations, Corporate communication, Donations.	Telephone marketing, Catalog marketing, E-mail marketing, Mobile marketing, Face to face marketing, Voice mail.	Friends, Family members, Colleagues, Relatives, Consumer opinion platforms such as blogs.	Websites, E-mail marketing messages, Search engine marketing, Newsletters, Banner advertising, Mobile applications, E-partnerships, Web 2.0 strategies.	Social networks, Blogs, Microblogging site, Wikis, Podcasting, Forums, Content communities.

(IMC for Domestic Tourists)

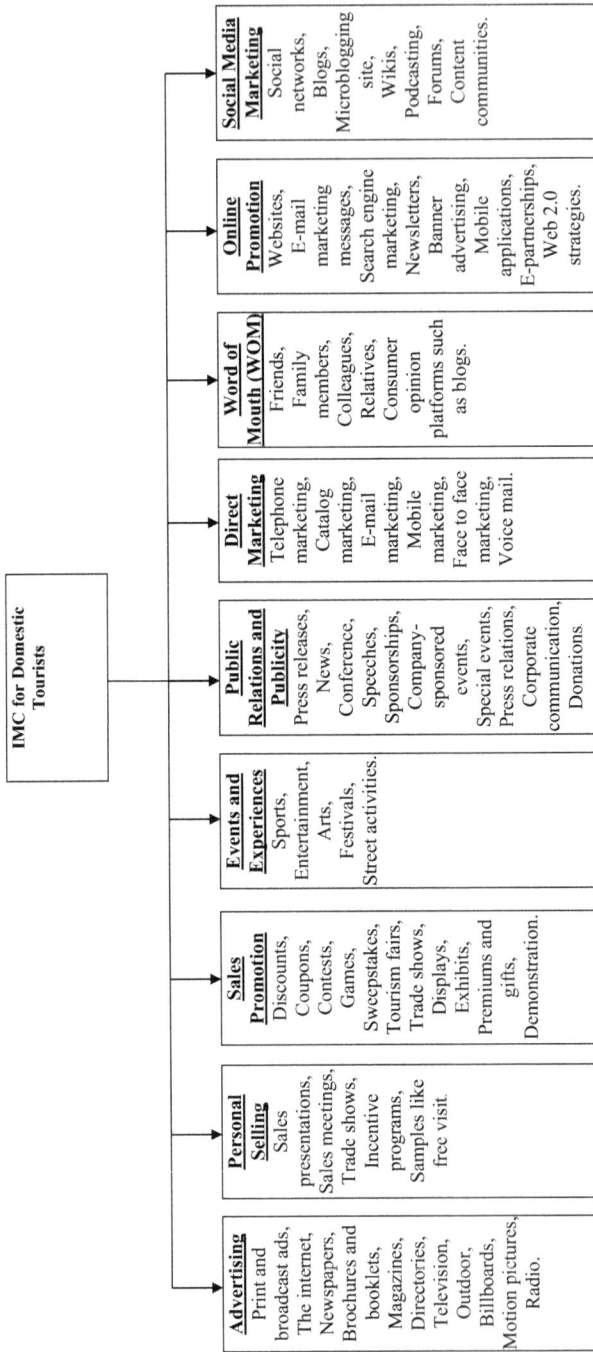

Figure 7.1 IMC tools for domestic tourists

Source: Authors' own compilation

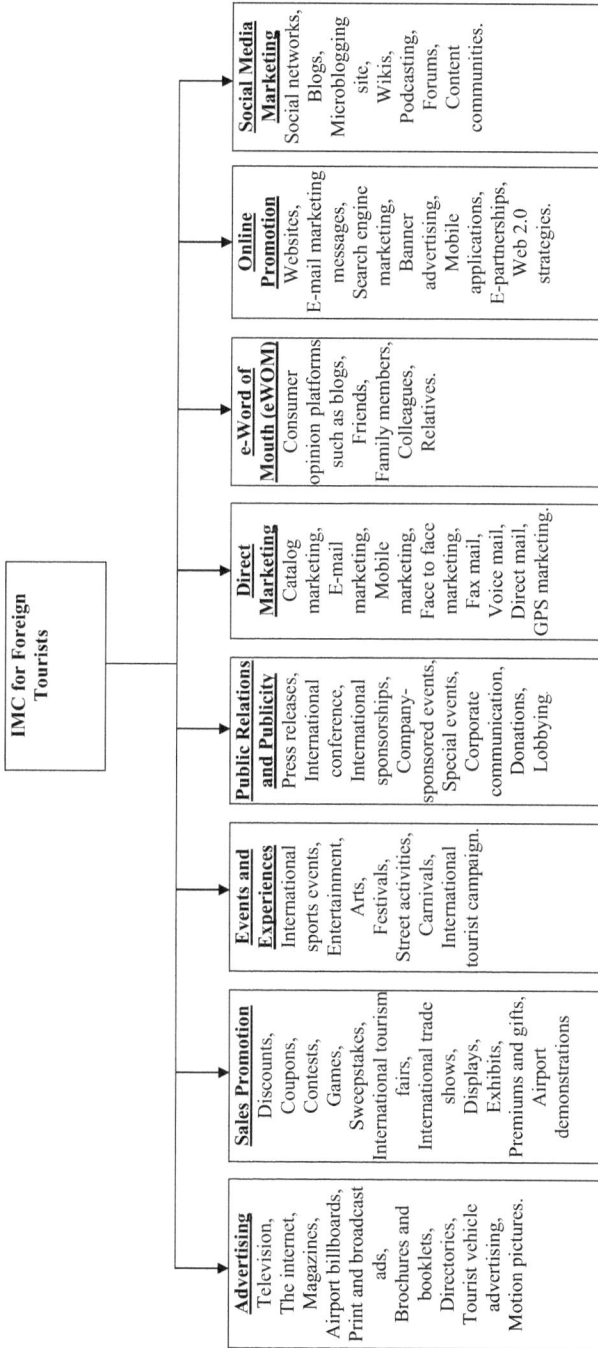

IMC for Foreign Tourists

Advertising	Sales Promotion	Events and Experiences	Public Relations and Publicity	Direct Marketing	e-Word of Mouth (eWOM)	Online Promotion	Social Media Marketing
Television, The internet, Magazines, Airport billboards, Print and broadcast ads, Brochures and booklets, Directories, Tourist vehicle advertising, Motion pictures.	Discounts, Coupons, Contests, Games, Sweepstakes, International tourism fairs, International trade shows, Displays, Exhibits, Premiums and gifts, Airport demonstrations	International sports events, Entertainment, Arts, Festivals, Street activities, Carnivals, International tourist campaign.	Press releases, International conference, International sponsorships, Company-sponsored events, Special events, Corporate communication, Donations, Lobbying	Catalog marketing, E-mail marketing, Mobile marketing, Face to face marketing, Fax mail, Voice mail, Direct mail, GPS marketing.	Consumer opinion platforms such as blogs, Friends, Family members, Colleagues, Relatives.	Websites, E-mail marketing messages, Search engine marketing, Banner advertising, Mobile applications, E-partnerships, Web 2.0 strategies.	Social networks, Blogs, Microblogging site, Wikis, Podcasting, Forums, Content communities.

Figure 7.2 IMC tools for foreign tourists

Source: Authors' own compilation

international sponsorships, donations and lobbying should be given special attention. All the activities involved in direct marketing programmes for domestic tourists can also be used for attracting foreign tourists. For foreign tourists, eWOM communication is of vital importance. Hence, consumer opinion platforms such as blogs should get special dealing. Online promotion is one of the most impactful tools for attracting foreign tourists and all the activities involved in internet marketing programmes for domestic tourists equally apply here. Finally, in line with today's trends, social media promotion should be given highest priority while designing IMC programmes. Social networking sites like Facebook, YouTube and so on, microblogging sites like Twitter and other consumer opinion blogs should be managed carefully for informing, attracting and retaining tourists.

Setting the promotional budget for tourism product and services

Marketers usually follow sophisticated approaches for setting the promotional budget. Commonly used methods for setting the promotional budget for tourism product and services are described next.

Affordable or available fund method

Under the affordable or available fund method, destination marketers set the promotional budget at the level the company has the capacity to spend or afford. This method is the simplest of all, but as this method leads to an uncertain annual budget, it makes the long-range planning of marketers very difficult (Kotler and Armstrong, 2012).

Percentage of sales method

Destination marketers can also set their promotional budget at a specific percentage of the destination's past, present or anticipated sales revenue (Kotler and Armstrong, 2012). This method is simple to use as it assumes a positive correlation between sales and promotion expenditure, but it is not a scientific method as there is no specific basis for determining the exact percentage, meaning the long-term planning is still very difficult.

Competitive parity method

Under the competitive parity method, destination marketers set their promotion budget by considering the competitors' promotional activities and costs to match competitors' budget spending (Kotler and Armstrong, 2012). Although this method may help to avoid promotion war, the budget suitable to one business may not be right for the other.

Objective and task method

Destination marketers set promotional budget under the objectives and task method by defining the specific objectives to be attained through promotion, deciding on the tasks to be performed to realize these objectives, measuring the cost of completing these tasks and making the sum of the costs of all these tasks (Kotler and Armstrong, 2012). This is believed to be the most scientific method of promotional budget-setting, but at the same time, it is the most complex method to use.

Expert opinion method

Under this method, destination marketers ask experts (like marketing managers, academicians and tourism marketing consultants) from both internal and external sources to estimate the amount to be spent for promotion of the destination for a specific period of time. Although this method usually provides bias-free estimates, it is not scientific as it depends on the expert's experience, judgment, intuition, intelligence and attitude.

Tools for measuring tourism promotion effectiveness

Developing and implementing a promotional programme is always costly. There-fore, destination marketers must examine how much effect the promotion has created on the target audience. This involves *inquiry tests*, measuring the number of inquiries generated from the promotional message through customer queries or phone calls; *recall tests*, asking the tourists whether they remember the promo-tional message and the points they can recall; *recognition tests*, testing the extent to which a tourist can rightly identify or recognize a particular destination just by viewing the destination's attributes, logo or promotional campaign; *attitude tests*, measuring the past and present attitudes of the tourists to find out any change in their thoughts and feelings toward the destination; and *behaviour tests*, measuring tourists' behaviour resulting from the promotional campaign such as number of people who visited or talked about the destination to others (Kotler and Arm-strong, 2012; Belch and Belch, 2005).

Destination marketers would also like to measure overall sales and revenue; sales per month, quarter or year; sales during promotional and non-promotional periods; return on investment (ROI); tracking the visitors to the destination web-site; and measuring the commenting, shares and likes on social media to evaluate the effectiveness of their promotional programmes.

Framework for developing tourism campaigns

This is a simple framework based on the basic system approach consisting of three components: input, process and output. While designing a tourism pro-motion framework, as input, destination marketers must consider the objectives

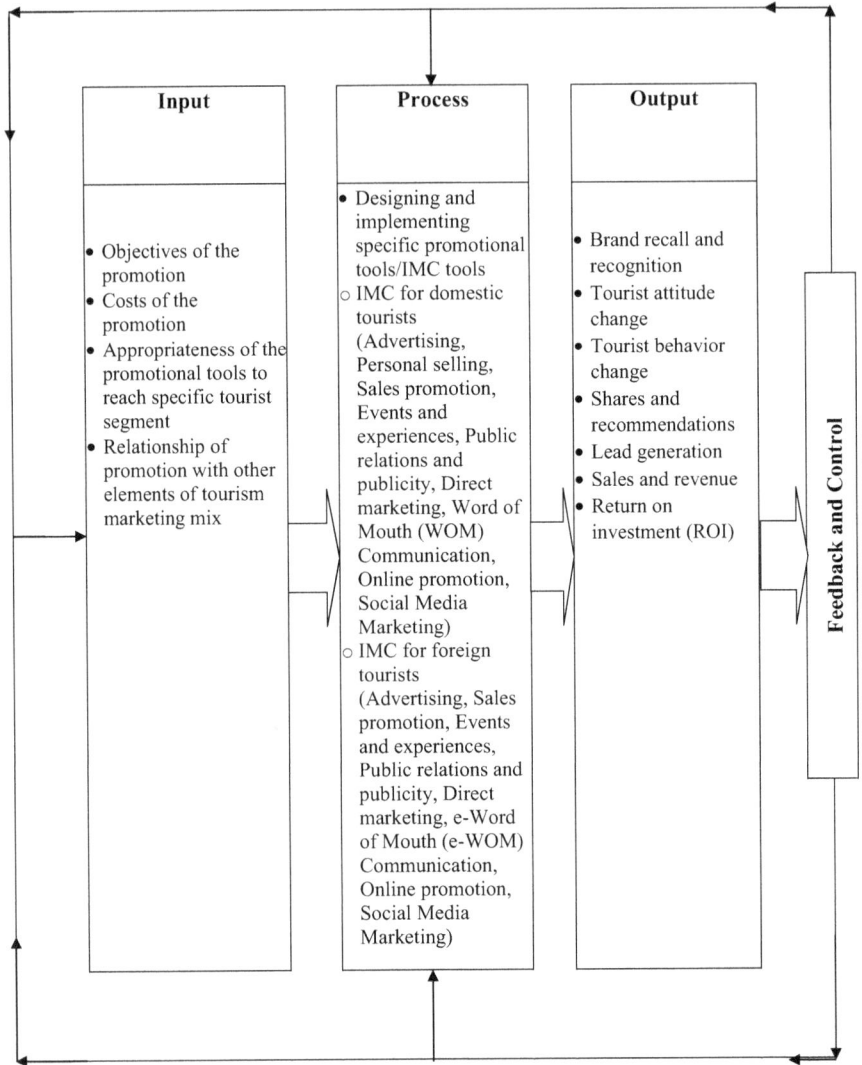

Figure 7.3 Conceptual framework for developing tourism promotion campaigns

Source: Authors' own construction with the help of basic tourism promotion framework proposed by Kreck (1972)

of the promotion, costs, appropriateness of the promotional tools and relationship of the promotion with other elements of the tourism marketing mix. The processing component of our proposed model is concerned with designing and implementing the IMC programme by considering what to say, how to say it, when to say it, through which channel the message should be carried and so on. The output component involves the outcome of a destination's promotional effort consisting of such results as increased destination brand recall and recognition, increased social shares and WOM recommendations, improved sales and revenue position, increased return on investment (ROI) and so on. If the promotional campaign fails to bring desired results or any distortion from the predetermined objectives are found, destination marketers must quickly read that feedback, implement necessary controlling mechanisms and take corrective measures.

Present status of marketing promotion in the Bangladesh tourism industry

Bangladesh Parjatan Corporation (BPC) acts as the national tourism organization (NTO) of Bangladesh (Hossain, 2015). As NTO, BPC tries to promote Bangladesh as a dreamy tourist destination among the prospective tourism markets. BPC and two other state-owned tourism-related authorities, namely Bangladesh Tourism Board (BTB) and Department of Archaeology, along with other private destination marketers and tour operators, use almost all the traditional promotional tools including distribution brochures, booklets, leaflets, souvenirs, posters, maps, guides and so on for spreading information and stimulating demand among the prospective tourists. Moreover, BPC, BTB, the Department of Archaeology and other private tour operators publish newspaper, magazine and television advertisements in local media. They also participate at tourism fairs, shows, festivals or exhibitions arranged locally and internationally. But the use of international media and digital marketing tools are very limited. Recently, numerous campaigns have been taken by the BPC and BTB to attract both domestic and foreign tourists which include TVC on "Beautiful Bangladesh (School of Life)"; billboards and pictorial ads during ICC Cricket World Cup 2011; "Visit Bangladesh 2016 – Life Happens Here" campaign on digital and social platforms like state-owned websites, Facebook and YouTube; the Bangladesh Folk Festival 2011; co-sponsoring Bangladesh Premier League 2011; arranging the Asian Tourism Fair 2011 and more (Bangladesh Tourism Board, n.d.).

To capture the present state of tourism promotion, the authors have studied 25 destination authorities, marketers and tour operators. The findings from this study are summarized in Table 7.1.

The study found that all the destination marketers, authorities and tour operators use almost all types of promotional tools for promoting their tourism products and services, and none of them rely upon any single tool. But most of them (72%) have no separate IMC programmes for domestic tourists and international tourists. The majority (52%) of them set their promotional budget by following affordable or available fund methods. Besides, 80% of them do not measure the effectiveness of their promotional tools or campaigns.

Table 7.1 Present scenario of tourism promotion in Bangladesh on some selected dimensions

Statement	Options	Frequency	Percent	Valid Percent
Types of promotional tools used	Use of more than one promotional tools	25	100.0	100.0
	Total	25	100.0	100.0
Separate tools to attract the domestic and foreign tourists	Yes	7	28.0	28.0
	No	18	72.0	72.0
	Total	25	100.0	100.0
Methods of setting the promotional budget for tourism product and services	Affordable or available fund method	13	52.0	52.0
	Percentage of sales method	5	20.0	20.0
	No formal method	7	28.0	28.0
	Total	25	100.0	100.0
Tools for measuring the tourism promotion effectiveness	Behaviour test	1	4.0	4.0
	Other measures such as sales, revenue, ROI etc.	4	16.0	16.0
	Effectiveness of promotional tools are not measured	20	80.0	80.0
	Total	25	100.0	100.0

Source: Field Survey, 2019

The challenges of implementing promotion tools in tourism industry of Bangladesh

Although the success of any destination largely depends on successful implementation of integrated marketing promotion tools, there are also some challenges that the destination marketers in Bangladesh have been facing in implementing the programmes. Determination of promotion objective is always a difficult task for destination marketers. For different types of tourism and destinations, different types of promotional strategies are also required. By going with digital trends, addressing consumers' digitalization acceptance in selecting destinations, marketing has become a tough job. Hence, dealing with the impact of the ever-changing social media landscape and getting access to a number of content marketing tactics is always challenging. Moreover, filling seasonality gaps and managing fluctuating demand in visitation through promotion is challenging (Destination Think, 2016). Besides, in a Bangladesh context, managing and measuring word-of-mouth communication about a particular destination, collecting and implementing tourist feedback and deciding on what to measure across all promotional tools is not easy.

Conclusion and recommendations

Tourism is now considered one of the fastest growing industries in the world. Having numerous natural, archaeological, historical and religious attractions, Bangladesh is an emerging tourism market. But the growth rate of this country's tourism industry is not satisfactory. To reverse its declining fortune, Bangladesh should formulate aggressive promotional strategies so that they can communicate and promote the attractions of its different destinations and tourism facilities to potential tourists. Besides, tourism marketers in Bangladesh should devote more money to tourism promotion so that they can run their promotional activities in international media such as TV channels like Discovery, National Geographic, BBC Earth and so on, plus world-renowned newspapers, magazines and other media. Moreover, tourism marketers should pay special attention to digital marketing and social media marketing to provide the tourists with user friendly, updated and eye-catching information and offers. Besides, cooperation among the government authorities and private tour operators should be increased so that tourism promotion can be placed as a national priority in the country's overall development goals.

References

Ahmad, S. (2013). *Tourism industry in Bangladesh*. Retrieved from: www.thedailystar.net/news/tourism-industry-in-bangladesh (accessed: the 25th May 2019).

Al-Azzam, A. F. M. (2016). A study of the impact of marketing mix for attracting medical tourism in Jordan. *International Journal of Marketing Studies*, 8(1), pp. 139–149.

Al-Debi, H. A. and Mustafa, A. (2014). The impact of services marketing mix 7P's in competitive advantage to five stars hotel – case study Amman, Jordan. In *The Clute institute international academic conference*. Orlando, FL, pp. 39–48.

Al-Dmour, H., Al-Madani, S., Alansari, I., Tarhini, A. and Al-Dmour, R. (2016). Factors affecting the effectiveness of cause-Related marketing campaign: Moderating effect of sponsor-cause congruence. *International Journal of Marketing Studies*, 8(5), pp. 114–127.

Ali, F. (2016). Hotel website quality, perceived flow, customer satisfaction and purchase intention. *Journal of Hospitality and Tourism Technology*, 7(2), pp. 213–228.

Aronsson, L. and Tengling, M. (1995). *Turism-Världens Största Näring: Turism Och Reseservice*, Malmö: Liber-Hermod.

Bahri-Ammari, N. and Nusair, K. (2015). Key factors for a successful implementation of a customer relationship management technology in the Tunisian hotel sector. *Journal of Hospitality and Tourism Technology*, 6(3), pp. 271–287.

Bangladesh Tourism Board (n.d.). *The role of tourism board in promotion and marketing* (in Bengali). Retrieved from: https://bit.ly/2N3ome2 (accessed: the 15th May 2019).

Bao, H. (2018). Marketing of Tourism Services/Experiences. In M. Sotiriadis (ed.), *The emerald handbook of entrepreneurship in tourism, travel and hospitality*. Bingley: Emerald Publishing Limited, pp. 261–275.

Beiger, T. and Laesser, C. (2004). Information source for travel decision: Towards a source process model. *Journal of Travel Research*, 42(4), pp. 357–371.

Belch, G. E. and Belch, M. A. (2005). *Advertising and promotion: An integrated marketing communications perspective* (6th ed.). New Delhi: TATA McGraw-Hill Publishing Company Ltd.

Chen, P-J., Okumus, F., Hua, N. and Nusair, K. (Khal). (2011). Developing effective communication strategies for the Spanish and Haitian-Creole-speaking workforce in hotel companies. *Worldwide Hospitality and Tourism Themes*, 3(4), pp. 335–353.

Chiu, L. K. and Ananzeh, O. A. (2012). Evaluating the relationship between the role of promotion tools in MICE tourism and the formation of the touristic image of Jordan. *Academica Turistica*, 5(1), pp. 59–73.

Cooper, C., Fletcher, J., Fyall, A., Gilbert, D. and Wanhill, S. (2005). *Tourism: Principles and practice* (3rd ed.). London: Pearson Education Ltd.

Destination Think (2016). *50 challenges that 50 destination marketers told us they face.* Retrieved from: https://destinationthink.com/50-challenges-destination-marketers-2016/ (accessed: the 27th May 2019).

Dwyer, L. and Forsyth, P. (1992). The case for tourism promotion: An economic analysis. *The Tourist Review*, 47(3), pp. 16–26.

Eccles, G. (1995). Marketing, sustainable development and international tourism. *International Journal of Contemporary Hospitality Management*, 7(7), pp. 20–26.

Erdem, M. and Jiang, L. (2016). An overview of hotel revenue management research and emerging key patterns in the third millennium. *Journal of Hospitality and Tourism Technology*, 7(3), pp. 300–312.

Flynn, P. (1995). *The worldwide web handbook: A guide for users, authors and publishers.* London: International Thomson Computer Press.

Getz, D. and Page, S. J. (2016). Progress and prospects for event tourism research. *Tourism Management*, 52, pp. 593–631,

Goeldner, C. R. and Ritchie, J. R. B. (2011). *Tourism: Principles, practices, philosophies* (12th ed.). Hoboken, NJ: John Wiley & Sons, Inc.

Gretzel, U. and Yoo, K. H. (2008). Use and impact of online travel reviews. In P. O'Connors, W. Hopken and U. Gretzel (eds.), *Information and communication technologies in tourism 2008.* Vienna: Springer, pp. 35–46.

Hassan, A. and Kokkranikal, J. (2018). Tourism policy planning in Bangladesh: Background and some steps forward. *e-Review of Tourism Research (eRTR)*, 15(1), pp. 79–87.

Hence, B. G. (2018). Urban experiential tourism marketing: Use of social media as communication tools by the food markets of Madrid. *Journal of Tourism Analysis: Revista de Análisis Turístico*, 25(1), pp. 2–22.

Hennig-Thurau, T., Gwinner, K. P., Walsh, G. and Gremler, D. D. (2004). Electronic word-of- mouth via consumer-opinion platforms: What motivates consumers to articulate themselves on the internet? *Journal of Interactive Marketing*, 18(1), pp. 38–52.

Hossain, M. J. (2015). Promotional measures in developing tourism industry in Bangladesh: An empirical study. *Unpublished PhD Thesis.* Dhaka: University of Dhaka.

Hwang, Y., Gretzel, U., Xiang, Z. and Fesenmaier, D (2006). Information search for travel decisions. In D. R. Fesenmaier, H. Werthner and K. W. Wober (eds.), *Destination recommendation systems: Behavioural foundations and applications.* Cambridge, MA: CAB International, pp. 3–16.

Jeong, M. and Lee, S. A. (2017). Do customers care about types of hotel service recovery efforts? An example of consumer-generated review sites. *Journal of Hospitality and Tourism Technology*, 8(1), pp. 5–18.

Katzenstein, H., Kavil, S., Mummalaneni, V. and Dubas, K. (1994). Design of an ideal direct marketing course from the students' perspective. *Journal of Interactive Marketing*, 8(2), pp. 62–72.

Kotler, P. and Armstrong, G. (2012). *Principles of marketing* (14th ed.). Upper Saddle River, NJ: Pearson Education, Inc.

Kotler, P., Armstrong, G., Agnihotri, P. Y. and Haque, E. U. (2010). *Principles of marketing: A South Asian perspective* (13th ed.). New Delhi: Dorling Kindersley (India) Pvt. Ltd.

Kotler, P., Bowen, J. and Maken, J. (2006). *Marketing for hospitality and tourism* (4th ed.). Englewood Cliffs, NJ: Prentice Hall.

Kotler, P. and Keller, K. L. (2006). *Marketing management* (12th ed.). Upper Saddle River, NJ: Pearson Education, Inc.

Kreck, L. A. (1972). A Dimension of tourism promotion. *The Tourist Review*, 27(3), pp. 86–92.

Litvin, S. W., Goldsmith, R. E. and Pan, B. (2008). Electronic word of mouth in hospitality and tourism management. *Tourism Management*, 29(3), pp. 458–468.

Mackay, K. and Vogt, C. (2012). Information technology in everyday and vacation contexts. *Annals of Tourism Research*, 39(3), pp. 1380–1401.

Majumder, D. (2015). Contributions and loopholes of tourism sector in Bangladesh. *Jagannath University Journal of Social Sciences*, 3(1–2), pp. 1–19.

Middleton, V. T. C., Fyall, A., Morgan, M. and Ranchhod, A. (2009). *Marketing in travel and tourism* (4th ed.). Oxford: Elsevier.

Mill, R. C. and Morrison, A. M. (2002). *The tourism system* (4th ed.). USA: Kendall/Hunt Publishing Company.

Morrison, A. (2002). *Hospitality and travel marketing* (3rd ed.). Boston, MA: Cengage Learning.

Motes, W. H. and Hilton, C. B. (2002). Promoting state tourism: Exploring perpetual and behavioural effects of syntactical construction in print advertisements. *Journal of Travel and Tourism Marketing*, 13(3), pp. 1–18.

Murphy, L., Mascardo, G. and Benckendorff, P. (2007). Exploring word-of-mouth influences on travel decisions: Friends and relatives vs. other travellers. *International Journal of Consumer Studies*, 31(5), pp. 517–527.

Nail, J. (2005). *What's the buzz on worth of mouth marketing? Social computer and consumer control put momentum into viral marketing*. Retrieved from: www.forrester.com/Research/Document/Excerpt/0.7211.36916.00.html (accessed: the 19th April 2019).

Papetti, C., Christofle, S. and Guerrier-Buisine, V. (2018). Digital tools: Their value and use for marketing purposes. In M. Sotiriadis (ed.), *The emerald handbook of entrepreneurship in tourism, travel and hospitality*. Bingley: Emerald Publishing Limited, pp. 277–295.

Rao, S. R., Thomas, E. G. and Javalgi, R. G. (1990). *Project findings and recommendations for strategic master plan for Bangladesh tourism*. WTO and UNDP: Report prepared for the Government of the People's Republic of Bangladesh.

Richins, M. L. (1983). Negative word-of-mouth by dissatisfied consumers: A pilot study. *Journal of Marketing*, 47(1), pp. 68–78.

Solomon, M. R. (2010). *Consumer behaviour: Buying, having, and being* (9th ed.). Upper Saddle River, NJ: Prentice Hall.

Tahar, Y. B., Haller, C., Massa, C. and Bédé, S. (2018). Designing and creating tourism experiences: Adding value for tourists. In M. Sotiriadis (ed.), *The emerald handbook of entrepreneurship in tourism, travel and hospitality*. Bingley: Emerald Publishing Limited, pp. 313–328.

Trusov, M., Bucklin, R. E. and Pauwels, K. (2009). Effects of word-of-mouth versus traditional marketing: Findings from an internet social networking site. *Journal of Marketing*, 73(5), pp. 90–102.

Van Waterschoot, W. and Van den Bulte, C (1992). The 4P classification of the marketing mix revisited. *Journal of Marketing*, 56(4), pp. 83–93.

Vogt, C. and Stewart, S. (1998). Affective and cognitive effects of information use over the course of a vacation. *Journal of Leisure Research*, 30(4), pp. 498–520.

8 Promoting tourism in Bangladesh

Policies, issues and challenges

Md Aslam Mia

Introduction

Bangladesh is one of the most densely populated countries located in South Asia and has a border with India, Myanmar and Bay of Bengal. Being a riverine, low-lying and located in a strategic geographical area, Bangladesh has been bestowed with treasures of natural beauty, hundreds of significant relics from old and middle age civilizations (Rahman, 2017) and thousands of man-made establishments (e.g. the tea garden). Despite having plenty of tourist spots, the tourism industry is still in its nascent stage even after the elapse of 45 years of independence.

Undoubtedly, tourism is one of the important components of the service industry for most countries nowadays, and many East Asian countries (e.g. Thailand, Malaysia, Indonesia etc.) have used this platform to develop and diversify their economy. Ample studies have investigated the link between tourism and various aspects of an economy and found a positive effect, such as growth, employment, foreign exchange earnings, productivity, community development, welfare of the residents and so on (Arslanturk et al., 2011; Chou, 2013; Fang et al., 2016; Gokovali, 2010; Jin, 2011; Kim et al., 2013; Nissan et al., 2011; Schubert et al., 2011).

Despite having tremendous potential, the tourism industry in Bangladesh did not perform well when compared with some of its South Asian counterparts. For example, Figure 8.1 shows the tourist arrivals in Bangladesh and some of the selected South Asian countries from 2000 to 2014. While the arrivals of foreign tourists have substantially increased for the countries like Sri Lanka, Bhutan and Nepal from 2009, Bangladesh has observed rather a slight decline since then. This has not only raised the question of tourism policies undertaken by the government of Bangladesh, but also raised the concern of future development of the tourism industry.

On top of that, the tourism receipts as a percentage of total exports also showed a declining trend from 2000 to 2015, which also indicates that Bangladesh has failed to enjoy the maximum economic benefits of tourism. The potential outcome of this declining trend also revealed that the economic activities of the tourism industry also suffered an economic slump and poses a credible threat to the employment of thousands of people involved in the industry.

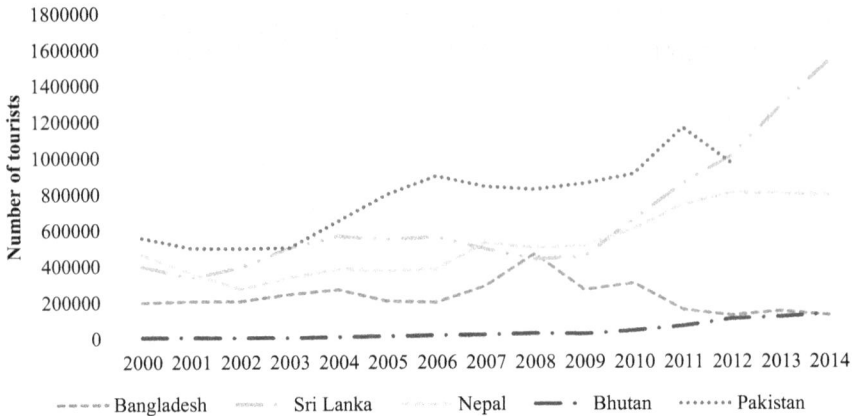

Figure 8.1 Number of international tourist arrivals in selected South Asian countries (2000–2014)

Source: Author's estimate based on the World Bank data

While many countries are now transforming and diversifying their economy and economic activities, Bangladesh should also reduce its dependency on apparel export as a core contributor to her export earnings to cope with various external shocks. Recently, oil-dependent Saudi Arabia has taken major economic reforms to reduce its dependency on oil exports and to engage in other economic activities and focus on tourism. As a result, the Saudi government has taken several initiatives that include visa-free travel for some nationalities and exemptions of various rules for foreign tourists, among others.

In a report published by World Economic Forum (WEF) (2019), Bangladesh has improved in travel and tourism competitiveness index (TTCI) from 125th position to 120th among 140 countries in 2019. While the recent report may bring some hope to many, however, it leads to a conclusion that the tourism industry in Bangladesh has certainly disappointed many stakeholders due to its slow pace. It also raises concern on why Bangladesh has done so poorly in tourism growth and what would be the solution to overcome those challenges.

While there is no shortcut and one-size-fits-all strategy to revive the tourism industry overnight, however, there are certainly ways that the government can undertake to promote tourism in the country. Thus, the study aims to analyse various factors that promote tourism growth and development as well as tourism policies undertaken in the last few decades in Bangladesh. The findings of the study would shed lights to relevant stakeholders to make the tourism sector an important contributor to the economy of Bangladesh.

The rest of the chapter is as follows. The next section briefly discusses the factors affecting tourism growth and development. Then there is a thorough review of tourism policies undertaken by the government of Bangladesh and

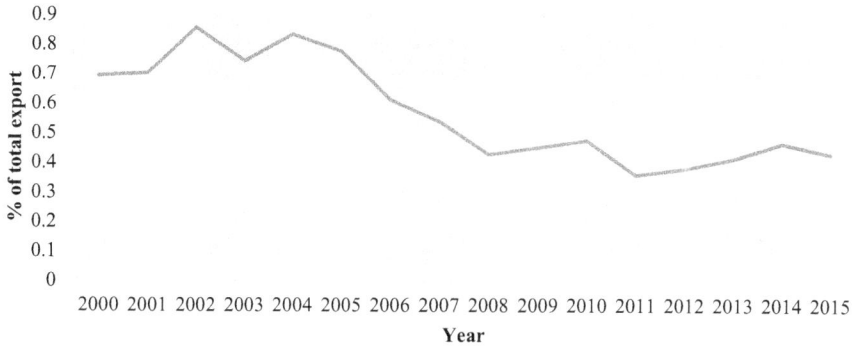

Figure 8.2 International tourism receipts (% of total exports) in Bangladesh (2000–2015)

Source: Author's estimate based on the World Bank data

various actors in implementing those policies. Finally, the chapter concludes with some highlights of the policy implications to revive the tourism industry in Bangladesh.

Factors affecting tourism growth and development

Generally, not all countries will succeed in promoting tourism. Some countries are naturally blessed while others are not. Despite natural blessings, there are some other factors that also lead to tourism growth and development. Patil (2017) has succinctly discussed some of the major factors that affect the tourism of a country, namely, environmental factors, socio-economic factors, historical and cultural factors, religious factors, institutional factors (this is proposed by the authors) and others. Each of the factors are depicted in Figure 8.3.

First, environmental factors include both the weather and scenic beauty of a country, which could stimulate tourist arrivals. For example, good weather, such as pleasant temperature and ample sunshine, could attract tourists from colder regions. On the other hand, people from relatively hot weather will likely to travel to countries with colder weather. For example, during winter season, people in the colder regions will tend to travel to a country with warmth weather and sunshine. In support of this statement, Malaysia receives relatively large number of tourists from other countries in the months of December, January and February (the winter period in most of the colder region) due to its warm weather (Malaysia Tourism, 2019). The highest number of tourists during these months is believed to be pushed by the weather factor.

Additionally, scenic beauty also stimulates tourist arrivals, such as long and clean beaches, sunset and sunrise points, hilly regions, lakes and waterfalls (Mendoza-González et al., 2018; Picken, 2017). Countries with the best sea beaches, such

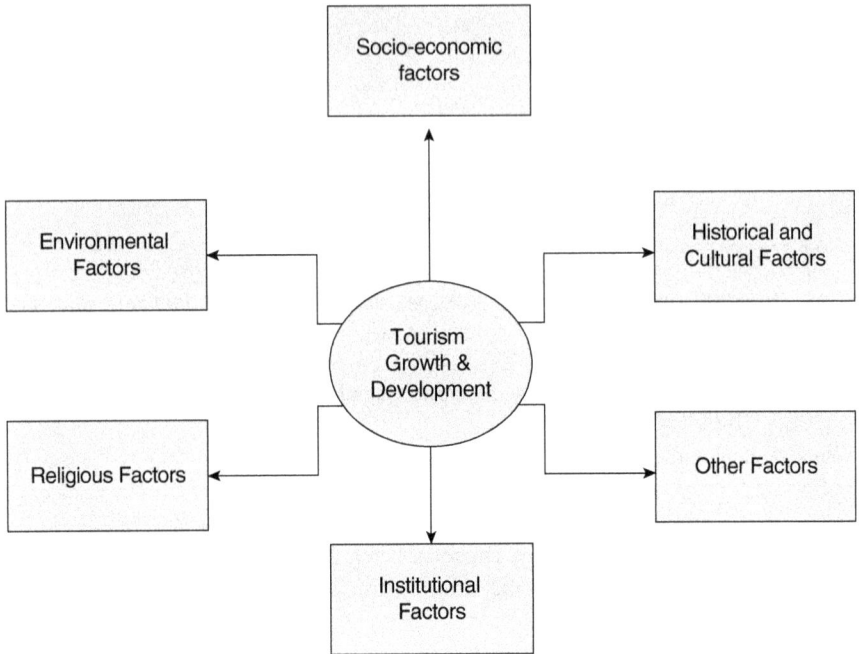

Figure 8.3 Factors affecting tourism development and growth

Source: Patil, 2017 and the author*

*redrawn by the author and added institutional factor in the original model.

as the Philippines accommodate many tourist arrivals to these destinations (Ong et al., 2011; Porter et al., 2015). Fortunately, the weather in Bangladesh is a sub-tropical monsoon climate and one can expect it not to be so cold during winter (no snowfall) and not so hot in summer. The average temperature in winter is around 10°C while it is on average around 25–30°C during summer.[1] In terms of scenic beauty, Bangladesh is also blessed to have one of the world's largest unbroken sea beaches (approximately 120 km long) and a mangrove forest in the southern part of the country (Hoque et al., 2018).

Patil (2017) has highlighted four different sub-factors under the category of socio-economic factors, namely, accessibility, accommodation, amenities and ancillary service. Accessibility means how well the tourist sites are linked with rest of the country. To be more specific, accessibility may be coined with the term "tourism transport system". Prideaux (2018) defined tourism transport system as the "organization and operation of passengers transport services by rail, road air and sea transport mood to facilitate travel by tourists from their home to a destination, between destination and within destination".

Most countries promote their domestic transportation to tourist spots with well-connected railways and roadways. For example, China, South Korea, Spain and Japan promote high-speed train service to most of their tourist spots with reasonable prices. These tickets are also available to buy online, which makes it even more convenient for the tourists.

The existing literature also support that easy connectivity/transportation not only promote tourism, but also boost the domestic transportation industry (Albalate and Fageda, 2016; Dinu, 2018; Rehman Khan et al., 2017; Spasojevic et al., 2018; Sharpley, 2006). Moreover, some of the rural parts or disintegrated islands are also connected by air or water transport systems so that the tourists could have a hassle-free journey to their intended destination.

While good accommodation facilities remain one of the prerequisites for tourists to visit any tourist spots (Sharpley, 2000; Taylan Dortyol, 2014), friendly and inexpensive accommodation options could trigger tourist arrivals. Nonetheless, in the era of sustainable development goals, some tourists are also interested to see how well tourist spots are maintained and what are the facilities available at the spot. Tourists are very sensitive to these services before they make an intended journey. Some of the ancillary services are also important to promote tourism in an area. For example, easily available and accessible banking facilities, mobile and internet connections, hospitals and insurance could motivate tourists to travel (Rajan et al., 2016; Ranjan Debata et al., 2013; Reihanian et al., 2012; Turner, 2010).

Third, historical and culturally significant tourist spots always receive special attention from potential tourists (both domestic and foreign) (Greenwood, 2011; Hospers, 2011; Yang et al., 2010). Tourists have a preference to visit old heritage sites, ancient monuments and palaces to observe and understand the history. This has been supported by the theory of "the tourist gaze" developed by Urry (1990, 1992) that says people visit certain places because they are not normally seen in their workplace or home. Although the main activity of tourists is "gazing at signs", places gazed at by tourists are not random in most cases (Hospers, 2011).

In the context of India, visiting historic sites remains third among other types of tourism activities in the country (Shankar, 2015). On top of that, Penang (one of the largest island cities in Malaysia) was named the second best tourist spot in 2017 by CNN due to its centuries-old historical sites, favourite cuisines and inexpensive accommodation options (Dermawan, 2017). Moreover, historical spots like Taj Mahal (in India), the pyramids (in Egypt), Machu Picchu (in Peru) and so on are considered as the top most historical attractions worldwide.

Fourth, religious factors also contribute to tourism growth (Bond et al., 2015; Raj and Griffin, 2015; Shepherd, 2016). There are some spots that are significant from the various religious viewpoints and visiting those places are a priority or obligation for different faiths. Mecca and Medina are the two famous religious spots for Muslims, Varanasi and Amritsar are famous for Hindus and Sikhs, respectively, and Putuoshan in China and Borobudur temples in Indonesia for Buddhists (Eid, 2012; Fatimah, 2015; Henderson, 2011; Wong et al., 2013). Every year, thousands of people make a way to visit those places due to their religious rituals. Apart from that, the overall religion of a country may also influence the tourist arrivals, particularly from a similar faith.

Fifth, institutional factors, which may include the overall economic situation of the host country, absence of violence, corruption, security and so on substantially contribute to a country's overall tourist arrivals (Das and Dirienzo, 2010; Erkuş-Öztürk, 2011; Hadwen et al., 2011; Snieška et al., 2014; Wang and Ap, 2013). While other factors in tourism may not be easy to amend or change, the institutional factors are often the result of government tourism policies.

Evaluation of tourism organizations and policies in Bangladesh

Despite gaining independence in 1971, Bangladesh had no comprehensive tourism policies until the early 1990s. However, Bangladesh Parjatan Corporation (BPC), also known as Bangladesh Tourism Organization (BTO), was established with the purview of promoting tourism in the country in 1972.[2] While the governance mechanism of BPC is autonomous, however, it is under the jurisdiction of Ministry of Civil Aviation and Tourism now. BPC plays dual role to promote tourism in the country. First, they oversee infrastructure and product development, and second, undertake promotional activities about tourism in Bangladesh.

Since tourism is a service industry and it requires skilful human resources to support the growth, the National Hotel and Tourism Training Institute (NHTTI) was established in 1974 to provide training to the relevant manpower. Since then, there was not much development in the tourism industry in Bangladesh as there was no comprehensive policy framework, guidelines or rules. The only visible changes seen during this time period was the shift of civil aviation and tourism to different ministries until 1986. This poor development in tourism in Bangladesh might be due to the political upheaval and lack of financial resources that did not allow this industry to grow at its full swing.

The first master plan about tourism in Bangladesh was enacted in 1992 with the help of United Nations Development Programme (UNDP) and World Trade Organization (WTO). Since the sector was paralyzed by decades of ignorance from various governments, it needed huge investment. Thus, one of the core objectives of this policy was to attract both domestic and foreign investment particularly in the infrastructure-related areas.

Several other incentives were also announced including tax exemptions, land allotment and concessionary funds/loans so as to encourage private sectors to come forward and contribute to the development of the industry. Foreign investment rules were also relaxed, such as new establishments could be fully owned by foreigners or jointly with domestic partners. Implementations of these policies were coordinated through various departments and ministries, overseas missions/embassies, cabinets and prime minister's office.

While the sector started to gain its momentum after the enactment of the first tourism policy in 1992, a formal diploma course in hotel management was offered by the NHTTI to supply the required human resources. Although there is no exact number of graduates available from the respective authority, they are now offering three diploma courses, one professional chef course and several

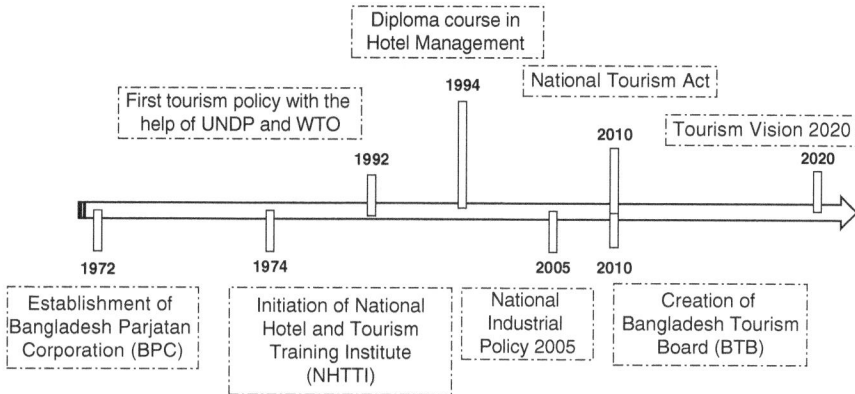

Figure 8.4 Timeline of tourism development in Bangladesh
Source: The author

certificate courses.[3] Certainly, the courses offered by the NHTTI are not only meeting the domestic demand and enhancing tourism quality, but also exporting human resource to other countries that increase export earnings.

Another significant policy initiative taken by the Bangladesh government was to award tourism as an industry status in 2005 to push the sector to grow at a faster pace (Tasnim, 2019). That was much needed to retain the continuous flow of domestic and foreign investment towards the tourism industry. Moreover, the recognition of tourism as an industry provided many benefits to foreign investors such as tax-exempted salaries for a number of years with technical expertise relevant to the industry, flexible ownership transfers and so on.

So far, one of the most comprehensive policies in the history of tourism in Bangladesh was the National Tourism Act 2010 (referred to as the Act hereafter). The Act includes total 31 objectives and goals targeted at various segments of the tourism industry in Bangladesh (Hassan and Burns, 2014). To implement the Act effectively and efficiently, a new organization named Bangladesh Tourism Board (BTB) was founded in 2010 with a vision to make Bangladesh one of the top tourist destinations in South Asia. While the new Act was welcomed by all the stakeholders, there are issues that remain unclear and ambiguous. For example, the two main actors of tourism organization in Bangladesh, BPC and BTB, were not even sure about the two different tourism acts (1992 and 2010) (Hassan and Burns, 2014; Ishtiaque, 2013).

The "Tourism Vision 2020" campaign in Bangladesh aims to attract a greater number of tourists, create a positive tourism image, ensure political stability, increase tourism contribution to GDP by 4–5% and liberate inbound and outbound tourism (Ishtiaque, 2013). Despite the fact that there are many opportunities to achieve these objectives by tapping the vast historical heritage, there remain many weaknesses. For example, Ishtiaque (2013) highlighted that lack of coordination among various tourism organizations, tendency of overcharging

```
              ┌─────────────────────────┐
              │  Tourism Administration   │
              │  Authority in Bangladesh  │
              └─────────────────────────┘
                            │
                            ▼
                  ┌───────────────────┐
                  │  Ministry of Civil │
      ┌───────────│   Aviation and     │───────────┐
      │           │      Tourism       │           │
      │           └───────────────────┘           │
      ▼                                            ▼
┌─────────────────────┐              ┌─────────────────────┐
│ Bangladesh Parjantan │              │  Bangladesh Tourism  │
│     Corporation      │              │        Board         │
└─────────────────────┘              └─────────────────────┘
      │
      ▼
┌─────────────────────┐
│  National Hotel and  │
│Tourism Training Institute│
└─────────────────────┘
```

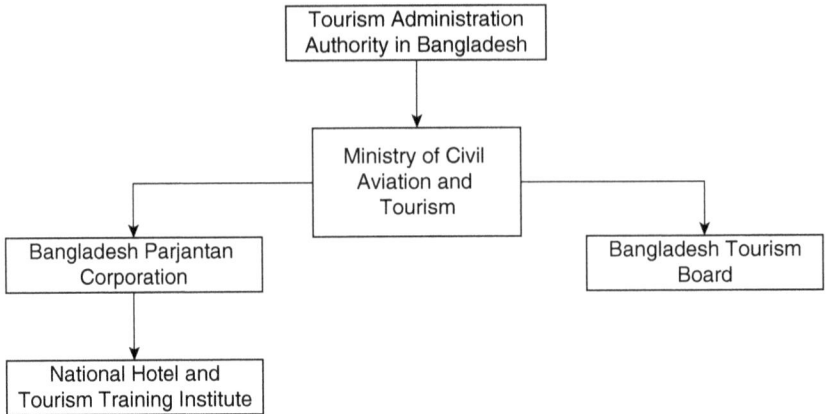

Figure 8.5 Direct stakeholders of tourism industry/policy in Bangladesh
Source: The author

foreign tourist, restrictive airline policies by the Bangladesh government, lack of quality transportation systems and low-yield foreign tourists.

Additionally, the Bangladesh government also imposes relatively a high departure tax that ranges from US$5.9 (around BDT 500) to US$30 (around BDT 2500) (Australian High Commission, 2019), which is considered to be relatively high compared to other countries in the region. The high airport tax not only increases the overall cost to travel, but also demotivates the potential tourists from entering the country.

For implementing tourism policies, the Ministry of Civil Aviation and Tourism coordinates with the two main stakeholders, BPC and BTB, along with NHTTI to ensure that the policies are taken seriously and implemented efficiently (Figure 8.5). Despite the comprehensive tourism policies that are in place, there are many deficiencies observed in terms of manpower, tourism branding, lack of marketing strategy, visionary mindset and insufficient budgetary allocations (Howladar, 2015). Since the tourism industry is not standalone but rather linked with various other ministries/departments, efficient coordination among those different sectors remains a crucial factor to promote tourism.

Other factors that lead to the declining tourist arrivals in the case of Bangladesh are the overall political stability and rule of law. It is evident from Figure 8.6 that the condition of those two important factors are relatively poor and that reflects a risky choice for the foreign tourist. Bangladesh stands at the lower percentile of those two indicators, which can be used to gauge the overall safety condition.

Ample studies have investigated how political instability (Kebede, 2018) and terrorism (Lutz and Lutz, 2018) affect the tourism industry and image of a country (Sönmez, 1998), and the majority of the studies have observed an inverse relationship between them (Ingram et al., 2013; Issa and Altinay, 2006; Seddighi

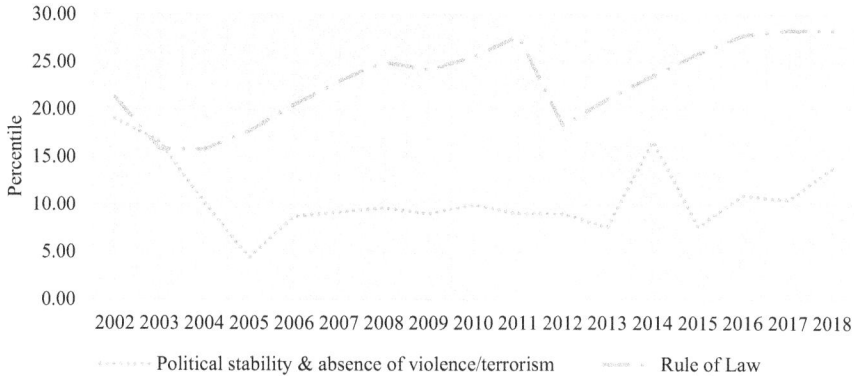

Figure 8.6 Political stability and rule of law percentile rank of Bangladesh among all countries (2002–2018)

Source: The author's estimate based on the World Bank data*

*percentile ranges from 0 (lowest) to 100 (highest) rank. Higher percentile ranks represent better condition of the country.

et al., 2002; Yaya, 2009). Meaning to say, political instability will reduce inbound tourists, thus a negative consequence in the tourism-related activities. Bangladesh has seen frequent political unrest, several terrorist attacks and potential threats from different radical organizations since early 2007 until late 2015 that might partly explain the poor tourist arrivals during this time.

Another institutional deterrence behind poor tourist arrivals is the visa policy. Visas are one of the important tools to control the movement of people across borders. Countries with friendly visa policies generally receive a large number of tourists compared to their counterparts (Balli et al., 2013; Cheng, 2012). In recent years, many countries are trying to minimize the hassle of visa application, minimum requirements and processing time to ensure that a visa does not become a burden to the potential tourists.

Some of the countries became successful in attracting tourists by eliminating visa requirements, such as Indonesia. Indonesia observed a sharp rise in tourist arrivals after it started issuing free visas (literally no visa is required for hundreds of countries) and aggressive promotion (The Jakarta Post, 2016). Countries like Cambodia, Malaysia, Sri Lanka and Myanmar started issuing e-visas to many countries to ensure hassle-free travel. Issuing e-visas has a couple of implications. For example, it can save resources for the issuing country and earn US$ by minimizing the issuing cost.

While Bangladesh has recently taken several initiatives to issue visas to foreign tourists, the majority of the ordinary people still have to apply for a visa manually through the embassy/consulate in different countries. While most officials and diplomats can travel to Bangladesh without a visa, there are also countries that are entitled to get on-arrival visas or e-visas. However, if Bangladesh really wishes to attract more foreign tourists, we have to liberalize the visa policy by allowing more countries to get on-arrival or e-visas.

E-visas can be a suitable option rather than issuing on-arrival visas to ordinary people. First, issuing e-visas can identify potential tourists that are blacklisted in other countries or have restricted travel permits. Second, the government can reject any visa application before the tourist arrives in the country to ensure that the tourist does not have to go through the hassle if on-arrival visa is denied at the port of entry. Third, the government can also earn a sizable amount of visa fees from the tourist. Thus, the study strongly suggested that increasing the coverage of e-visas to many countries and ordinary people, particularly the neighbouring countries and charging a modest visa fee in line with other countries could stimulate tourist arrivals. In this case, successful examples of Sri Lanka, Cambodia and Malaysia could be taken into account for an effective policy implementation. Moreover, bilateral government policies towards visas can also ensure hassle-free travel both for inbound and outbound tourists.

Having said that, the successful example of Bhutan, which has made tourism visas free for some countries but maintains a visa quota for western nations to ensure low-volume but high-yield tourism return, can also be explored as an option (Nyaupane and Timothy, 2010). By doing this, Bhutan managed to reduce the number of low-yield tourists and backpackers that do not contribute much to the economy. This is to ensure that the tourism benefits can be maximized without potential negative consequences on natural and environmental resources.

Conclusion

After evaluating the factors of tourism growth and development as well as the tourism policies of Bangladesh, it is evident that Bangladesh has many tourist spots and capacity to host thousands of tourists. Unfortunately, the tourist arrivals and export earnings from the tourism industry is rather in a declining trend, which exposes the inefficiency of marketing and promoting tourism to the outside world. Moreover, the infrastructure development in the country may not be at the standard level yet, hence, transportation facilities within the country should receive special attention from the respective authorities. Moreover, coordination among various ministries/departments should also be enhanced to ensure that there is no ambiguity in implementing the tourism policies effectively.

It is understood that for any sector or any country, a well-planned execution of policies through strong institutions are prerequisite for modest growth and development. In support of this, Hassan and Burns (2014) also reiterated the importance of proper and effective tourism policies to cope with many global challenges. To formulate effective tourism policies and enhance institutional strength, a visionary mindset is significantly important due to rising geo-political issues. Well-equipped and resourceful tourism organizations in Bangladesh could be the stepping stone to support future exponential growth in the industry.

Notes

1 The specific temperature varies based on different locations. For an overview of the temperature countrywide, kindly see www.climatestotravel.com/climate/bangladesh
2 BPC started its activity from the year 1973.
3 For details of the diploma course, chef course and professional certificate, kindly see www.nhtti.org/index-2.html

References

Albalate, D. and Fageda, X. (2016). High speed rail and tourism: Empirical evidence from Spain. *Transportation Research Part A: Policy and Practice*, 85, pp. 174–185.
Arslanturk, Y., Balcilar, M. and Ozdemir, Z. A. (2011). Time-varying linkages between tourism receipts and economic growth in a small open economy. *Economic Modelling*, 28(1), pp. 664–671.
Australian High Commission (2019). *Departure travel tax for foreign nationals.* Retrieved from: https://bangladesh.embassy.gov.au/daca/Departure_Travel_Tax.html (accessed: the 16th November 2019).
Balli, F., Balli, H. O. and Cebeci, K. (2013). Impacts of exported Turkish soap operas and visa-free entry on inbound tourism to Turkey. *Tourism Management*, 37, pp. 186–192.
Bond, N., Packer, J. and Ballantyne, R. (2015). Exploring visitor experiences, activities and benefits at three religious tourism sites. *International Journal of Tourism Research*, 17(5), pp. 471–481.
Cheng, K. M. (2012). Tourism demand in Hong Kong: Income, prices, and visa restrictions. *Current Issues in Tourism*, 15(3), pp. 167–181.
Chou, M. C. (2013). Does tourism development promote economic growth in transition countries? A panel data analysis. *Economic Modelling*, 33, pp. 226–232.
Das, J. and Dirienzo, C. (2010). Tourism competitiveness and corruption: A cross-country analysis. *Tourism Economics*, 16(3), pp. 477–492.
Dermawan. (2017). *Penang named 2nd top tourist destination for 2017 by CNN.* Retrieved from: www.nst.com.my/news/2017/01/202954/penang-named-2nd-top-tourist-destination-2017-cnn (accessed: the 16th November 2019).
Dinu, A.-M. (2018). The importance of transportation to tourism development. *Academic Journal of Economic Studies*, 4(4), pp. 183–187.
Eid, R. (2012). Towards a high-quality religious tourism marketing: The case of Hajj service in Saudi Arabia. *Tourism Analysis*, 17(4), pp. 509–522.
Erkuş-Oztürk, H. (2011). Emerging importance of institutional capacity for the growth of tourism clusters: The case of Antalya. *European Planning Studies*, 19(10), pp. 1735–1753.
Fang, B., Ye, Q. and Law, R. (2016). Effect of sharing economy on tourism industry employment. *Annals of Tourism Research*, 57, pp. 264–267.
Fatimah, T. (2015). The impacts of rural tourism initiatives on cultural landscape sustainability in Borobudur area. *Procedia Environmental Sciences*, 28, pp. 567–577.
Gokovali, U. (2010). Contribution of tourism to economic growth in Turkey. *Anatolia*, 21(1), pp. 139–153.
Greenwood, J. (2011). Driving through history: The car, the open road, and the making of history tourism in Australia, 1920–1940. *Journal of Tourism History*, 3(1), pp. 21–37.

Hadwen, W. L., Arthington, A. H., Boon, P. I., Taylor, B. and Fellows, C. S. (2011). Do climatic or institutional factors drive seasonal patterns of tourism visitation to protected areas across diverse climate zones in Eastern Australia? *Tourism Geographies*, 13(2), pp. 187–208.

Hassan, A. and Burns, P. (2014). Tourism policies of Bangladesh – a contextual analysis. *Tourism Planning and Development*, 11(4), pp. 463–466.

Henderson, J. C. (2011). Religious tourism and its management: The hajj in Saudi Arabia. *International Journal of Tourism Research*, 13(6), pp. 541–552.

Hoque, M. A., Ara, E. and Shoeb-Ur-Rahman, M. (2018). Forest-based Tourism in Bangladesh: Challenges Unveiled for the Sundarbans. *Forest*, 10(36), pp. 97–107.

Hospers, G.-J. (2011). City branding and the tourist gaze. In K. Dinnie (ed.), *City branding: Theory and cases*. London: Palgrave Macmillan, pp. 27–35.

Howladar, Z. H. (2015). *Tourism in Bangladesh: Problems and prospects*. Retrieved from: www.theindependentbd.com/arcprint/details/21969/2015-11-06 (accessed: the 16th November 2019).

Ingram, H., Grieve, D., Ingram, H., Tabari, S. and Watthanakhomprathip, W. (2013). The impact of political instability on tourism: Case of Thailand. *Worldwide Hospitality and Tourism Themes*, 5(1), pp. 92–103.

Ishtiaque, A. N. A. (2013). Tourism vision 2020: A case of Bangladesh tourism with special emphasis on international tourist arrivals and tourism receipts. *Journal of Business*, 34(2), pp. 13–36.

Issa, I. A. and Altinay, L. (2006). Impacts of political instability on tourism planning and development: The case of Lebanon. *Tourism Economics*, 12(3), pp. 361–381.

Jin, J. C. (2011). The effects of tourism on economic growth in Hong Kong. *Cornell Hospitality Quarterly*, 52(3), pp. 333–340.

Kebede, N. (2018). The fate of tourism during and in the aftermath of political instability: Ethiopia tourism in focus. *Journal of Tourism Hospitality*, 7, p. 337.

Kim, K., Uysal, M. and Sirgy, M. J. (2013). How does tourism in a community impact the quality of life of community residents? *Tourism Management*, 36, pp. 527–540.

Lutz, B. J. and Lutz, J. M. (2018). Terrorism and tourism in the Caribbean: A regional analysis. *Behavioural Sciences of Terrorism and Political Aggression*, pp. 1–17.

Malaysia Tourism (2019). *Malaysia tourism statistics in brief*. Retrieved from: www.tourism.gov.my/statistics (accessed: the 16th November 2019).

Mendoza-González, G., Martínez, M. L., Guevara, R., Pérez-Maqueo, O., Garza-Lagler, M. C. and Howard, A. (2018). Towards a sustainable sun, sea, and sand tourism: The value of ocean view and proximity to the coast. *Sustainability*, 10(4), p. 1012.

Nissan, E., Galindo, M.-A. and Méndez, M. T. (2011). Relationship between tourism and economic growth. *The Service Industries Journal*, 31(10), pp. 1567–1572.

Nyaupane, G. P. and Timothy, D. J. (2010). Power, regionalism and tourism policy in Bhutan. *Annals of Tourism Research*, 37(4), pp. 969–988.

Ong, L. T. J., Storey, D. and Minnery, J. (2011). Beyond the beach: Balancing environmental and socio-cultural sustainability in Boracay, the Philippines. *Tourism Geographies*, 13(4), PP. 549–569.

Patil, S. (2017). *Five main factors influencing the growth of tourism with diagram*. Retrieved from: http://articles-junction.blogspot.com/2013/07/five-main-factors-influencing-growth-of.html (accessed: the 16th November 2019).

Picken, F. (2017). Beach tourism. In L. Lowry (ed.), *The sage international encyclopedia of travel and tourism*. Thousand Oaks, CA: SAGE, pp. 135–136.

Porter, B. A., Orams, M. B. and Lück, M. (2015). Surf-riding tourism in coastal fishing communities: A comparative case study of two projects from the Philippines. *Ocean and Coastal Management*, 116, pp. 169–176.

Prideaux, B. (2018). Tourism and surface transport. In C. Cooper, S. Volo and W. C. Gartner (eds.), *The Sage handbook of tourism management*. London: SAGE, pp. 297–313.

Rahman, A. (2017). *Tourism in Bangladesh: Problems and prospects*. Retrieved from: www.daily-sun.com/post/234275/2017/06/16/Tourism-in-Bangladesh:-Problems-and-Prospects (accessed: the 16th November 2019).

Raj, R. and Griffin, K. A. (2015). *Religious tourism and pilgrimage management: An international perspective*. Wallingford: CABI.

Rajan, A., Manuel, B., Sankar, K. H. and Gunasekar, S. (2016). Technological advancement and tourism: A panel data analysis of ASEAN countries. Paper presented at *the 2016 IEEE International Conference on Computational Intelligence and Computing Research (ICCIC)*. Madras: IEEE. The 15th – 17th December 2016.

Ranjan Debata, B., Sree, K., Patnaik, B. and Sankar Mahapatra, S. (2013). Evaluating medical tourism enablers with interpretive structural modeling. *Benchmarking: An International Journal*, 20(6), pp. 716–743.

Rehman Khan, S. A., Qianli, D., SongBo, W., Zaman, K. and Zhang, Y. (2017). Travel and tourism competitiveness index: The impact of air transportation, railways transportation, travel and transport services on international inbound and outbound tourism. *Journal of Air Transport Management*, 58, pp. 125–134.

Reihanian, A., Mahmood, N. Z. B., Kahrom, E. and Hin, T. W. (2012). Sustainable tourism development strategy by SWOT analysis: Boujagh National Park, Iran. *Tourism Management Perspectives*, 4, pp. 223–228.

Schubert, S. F., Brida, J. G. and Risso, W. A. (2011). The impacts of international tourism demand on economic growth of small economies dependent on tourism. *Tourism Management*, 32(2), pp. 377–385.

Seddighi, H. R., Theocharous, A. L. and Nuttall, M. W. (2002). Political instability and tourism. *International Journal of Hospitality and Tourism Administration*, 3(1), pp. 61–84.

Shankar, S. (2015). Impact of heritage tourism in India: A case study. *International Journal of Innovative Research in Information Security (IJIRIS) India*, 2(6), pp. 59–61.

Sharpley, R. (2000). The influence of the accommodation sector on tourism development: Lessons from Cyprus. *International Journal of Hospitality Management*, 19(3), pp. 275–293.

Sharpley, R. (2006). Transport for travel and tourism. In R. Sharpley (ed.), *Travel and tourism*. London: SAGE, pp. 44–50.

Shepherd, R. J. (2016). *Faith in heritage: Displacement, development, and religious tourism in contemporary China*. Oxon: Routledge.

Snieška, V., Barkauskienė, K. and Barkauskas, V. (2014). The impact of economic factors on the development of rural tourism: Lithuanian case. *Procedia – Social and Behavioural Sciences*, 156, pp. 280–285.

Sönmez, S. F. (1998). Tourism, terrorism, and political instability. *Annals of Tourism Research*, 25(2), pp. 416–456.

Spasojevic, B., Lohmann, G. and Scott, N. (2018). Air transport and tourism – a systematic literature review (2000–2014). *Current Issues in Tourism*, 21(9), pp. 975–997.

Tasnim, M. N. (2019). A review of tourism policy and performance: Bangladesh perspective. Project Paper, United International University, Dhaka, Bangladesh. pp. 1–43.

Taylan Dortyol, I. (2014). How do international tourists perceive hotel quality? *International Journal of Contemporary Hospitality Management*, 26(3), pp. 470–495.

The Jakarta Post (2016). Free visas, promotion lead to sharp rise in tourist numbers. Retrieved from: www.thejakartapost.com/news/2016/07/02/free-visas-promotion-lead-to-sharp-rise-in-tourist-numbers.html (accessed: the 16th November 2019).

Turner, L. (2010). "Medical tourism" and the global marketplace in health services: US patients, international hospitals, and the search for affordable health care. *International Journal of Health Services*, 40(3), pp. 443–467.

Urry, J. (1990). *The tourist gaze*. London: Sage.

Urry, J. (1992). The tourist gaze "revisited". *American Behavioural Scientist*, 36(2), pp. 172–186.

Wang, D. and Ap, J. (2013). Factors affecting tourism policy implementation: A conceptual framework and a case study in China. *Tourism Management*, 36, pp. 221–233.

Wong, C. U. I., Ryan, C. and McIntosh, A. (2013). The Monasteries of Putuoshan, China: Sites of secular or religious tourism? *Journal of Travel and Tourism Marketing*, 30(6), pp. 577–594.

World Economic Forum (2019). *The travel and tourism competitiveness report 2019: Travel and tourism at a tipping point*. Retrieved from: www.weforum.org/reports/the-travel-tourism-competitiveness-report-2019 (accessed: the 16th November 2019).

Yang, C.-H., Lin, H.-L. and Han, C.-C. (2010). Analysis of international tourist arrivals in China: The role of World Heritage Sites. *Tourism Management*, 31(6), pp. 827–837.

Yaya, M. E. (2009). Terrorism and tourism: The case of turkey. *Defence and Peace Economics*, 20(6), pp. 477–497.

Part 4
Technology in tourism marketing

9 Innovative technology application in tourism marketing

Azizul Hassan and Célia M. Q. Ramos

Introduction

Technological gadgets are transformed to have capacities in influencing the interaction that consumers make with digital information, content, media, feedback, intelligence and product or service offers to experience tourism, to contribute to involvement and create memories associated with the trip (Bilgihan et al., 2016). Tourism is established as one of the world's capable economic industries where the application of innovative technologies is playing crucial roles, which triggered the redesign of the tourist distribution chain across the globe. In tourism marketing, consumers are prioritized in terms of their requirements and the capacity to meet those requirements, to achieve the aims of increased sales and loyalty. Innovative technologies become powerful and indispensable elements to manage, distribute and market tourism (Benckendorff et al., 2019). The application of augmented reality (AR) can make the tourists able to add value travel experiences and engage in local culture and values more deeply. The service generated from AR application can also be extended to online visa processing, flight booking, online check in, issuing boarding passes on their smartphones, passing through the automated clearance gates and even electronically validating their boarding passes for boarding the plane. All of these jointly can improve the facilitation of travel, ensuring maximum security, quality products and services and accessibility as well as the elimination of mobility issues, while increasing the flow of how business transaction data is performed (Kazmer et al., 2018). An innovative technology like AR can possibly persuade consumers of tourism products and services. This research analyses the attachment of an innovative technology as AR with tourism from the postmodernism context emphasizing how digital marketing can be applied in this context.

Innovation technology example: augmented reality

AR application in tourism as an element of postmodern and digital marketing has been described as follows:

> Augmented reality "augments" the viewer's surroundings with new digital imagery and information. . . . To that extent, AR may or may not mean much to the average consumer – a cool image is great, but it doesn't create a

connection. Thus, augmented reality in and of itself usually isn't the goal of any marketing campaign; the goal is to use AR in such a way that it creates an interactive experience, engaging the customer through a rich and rewarding experience.

(Vong, 2015, p. 1)

Buchholz (2014, p. 2) explains AR in a more expressive way:

By Augmented Reality, abbreviated to AR, we understand the computer assisted augmenting of perception by means of additional interactive information levels in real time. The distinction between AR and Virtual Reality: in the case of Virtual Reality, the user is totally immersed in a virtual world that has no connection with reality.

One of the latest digital technology trends is AR, which is incorporated in tourism marketing and advertising strategies. AR superimposes information of a real-world image and offers added content when viewing an object (Nayyar et al., 2018). Thus, AR is able to offer its users with unique and complete experiences that combine "real" and virtual elements. In AR, users can interact in real time using their smartphone, laptop or tablet computer. The advantages generated from the incorporation of AR into digital marketing strategy include first, tourism product or service brand awareness enhancement (Yung and Khoo-Lattimore, 2019). AR campaigns support increased awareness of a specific tourism brand or product through the cultivation of a "wow" factor encouraging both the media and the tourists talking about the brand. AR is specifically effective for awareness raising as the technology is rather new and the interaction experience with a brand through AR still stays as a unique experience for most tourists. The use of AR differentiates a particular tourism brand from the competitors by applying innovative techniques and generating expectation for promoting a tourism product or service. Second, AR helps increase user engagement, because when tourists receive a new experience, they tend to get attracted to the tourism brand (Rauschnabel et al., 2019). A useful link between that specific brand and tourists is established and the brand memory gets reinforced. This is important mainly for the saturation of advertising to which a tourist becomes exposed. The integration of AR with specific tourism product or service marketing campaigns can possibly lead to improved experience: when tourists have more information at their disposal, it deepens their knowledge of what they are experiencing. Tourists with the support of AR can make relatively quicker choices and complete transactions at a greater speed, once they are more knowledgeable about the product they can get. Third, AR supports bringing digital technology to the real world. One of the key advantages that AR can provide is the capacity to create a link between the digital marketing campaign to the tourists' physical experience in more innovative, new and inventive ways (Chaffey and Ellis-Chadwick, 2019). AR links the conventional marketing channels with the digital that allows tourism product and service marketers to turn a brochure or static physical advert into a digital experience that can have easier linkage with a digital marketing campaign. Fourth, AR transforms the tourism B2B vendor/customer experience in many different ways. Sales processes in such a business type have been typically challenging on the context of tourist expectations and the restrictions of tourist product or service providers in terms of the actual capacity. AR is believed to have the potential for creating a crucial improvement in the complete sales process chain. AR significantly improves the creation of lively materials for sales promotion, which when complemented with technological devices that awaken the senses of the human being, contribute to the increase of personal satisfaction and well-being.

Digital marketing

In general, digital marketing refers to the type of marketing that relies on technology use as the most basic way to reach and interact with target consumer bases (Ryan, 2014). Digital marketing relies on technology application and deals with marketing in technology-backed market settings. This marketing demands technology use as a valid requirement for its application in a business system. Perhaps digital marketing is feasible in changed market situations where the dynamism and efficiency of marketers requires enhanced capacities to interact with target consumer bases. Such assumptions are particularly more generalisable in market settings with advanced technology use. Digital marketing in certain cases can appear as a common solution to overcome identified issues related to restricted geographical market.

The use of the Internet is seen as more effective to reach diverse consumer segments and thus to persuade them to get access or purchase available offers, services or products (Chaffey and Ellis-Chadwick, 2019). Also, the Internet is the generic platform for digital marketing while television, radio, smartphones and computing devices are tools or means to serve customized and personal purposes. Tools used in digital marketing can be diverse. Also, these tools are used in certain perspectives followed by careful understanding of market demand. Different consumers can use different marketing tools aiming to communicate with certain interests and persuade people to purchase. Basically, digital marketing is divided into two types: pull digital marketing and push digital marketing.

Pull digital marketing

Pull digital marketing is the type of marketing where consumers actively search for marketing contents. Examples include e-mail marketing as the most common and conventional pull digital marketing tool, where target consumers are sent emails with details of the offered products or services. Website marketing is one of the most recent tools of digital marketing, which helps to deepen the relationship between the customer and the brand or product and consequently increase their loyalty (Buss and Foley, 2019).

Blog marketing is a widely used pull digital marketing tool and typically comes with descriptive features of a certain product or service. These descriptions are posted on different blogs to attract the attention of general consumers. With the help of these blogs, consumers then search for more in-depth information to reach a purchase decision for a particular product or service, which results in increased sales and brand awareness.

Push digital marketing

On the other side, push digital marketing is the form of marketing where marketers send messages to consumers with detailed contents of a particular product or service, without considering what they are searching for (Chaffey and Ellis-Chadwick, 2019). This type of marketing tool most prominently includes social media marketing.

Social media marketing is a form of marketing using social networking websites like Facebook, Twitter, Google Plus, Pinterest, LinkedIn and so on. The unprecedented popularity and growth of social networks have helped to widen the scope of such marketing and to transform communication between brands and consumers. Augmented reality is the other tool of such marketing that is viewed as having promise in coming decades (Bjork and Guss, 1999). This type of marketing blends technology with reality to create interest among consumers, while increasing brand awareness and presentation of associated products.

Keyword marketing is sometimes called search engine marketing and focuses on advertisement on search engines relating to a specific product or service promotion (Yamamoto, 2018). The very basic and popular concept of this kind of marketing is pay-per-click marketing that attaches clicks by customer.

Neuromarketing is a relatively complex marketing tool that necessitates high expertise and skills in analysing neurology-associated marketing behaviour of a consumer. This is involved with studying affective and cognitive responses and sensorimotor of a potential consumer that stimulates marketing behaviour.

Other common push digital marketing tools include the following (Chaffey and Ellis-Chadwick, 2019): Inbound marketing is a sort of marketing that depends on making and sharing important information to attract a large number of consumer groups; content marketing includes infographics, articles, videos or podcasts to attract potential consumer bases; article marketing is the posting of certain products or services on different websites that is quite similar to blog marketing; affiliate marketing is marketing associated with a third party "affiliate". This is a form of referral marketing. Video marketing is a form of video streaming on the Internet where a consumer can get a clear view of a product or service. This does not necessarily involve a full-length video of an exact product or service. Rather, videos are mostly posted on media like YouTube.

Five distinct roles of interactive technology are identified by Deighton and Kornfeld (2009): thought tracing, activity tracing, property exchanges, social exchanges and cultural exchanges. However, a changing shift is largely influenced by technology-powered capacity enhancements of consumers leading to improved communications and information search. The availability of applicable technologies is one of the reasons for such a shift. In addition, technology supports consumers to get more information on selected products and services that can influence their purchase decisions. It is obvious that consumers behold stronger capacities and amenities for digital marketing. Their abilities to adapt with updated technologies are also developed by the acquisition of more operational skills as an outcome of innovative technologies. Thus, the shift outlines both similarities and dissimilarities between traditional and updated concepts and frameworks in digital marketing. Such a shift allows newer theories or models in relation to technology adoption and purchase decision making by consumers.

The application of traditional theories in digital marketing is interesting when marketing dynamics are seen as inconsistent. Generational theory is relevant to digital marketing where each generation communicates on online platforms to allow marketers to reach them. This theory accepts that consumer behaviours and attitudes are general focusing mainly on shared experiences. These consumers are born in a same generation. They manipulate their early days followed by their worldviews. Millennials (24–39 age group) are the wisest users of the Internet, Generation X (40–55 age group) is more active than the earlier group in areas of financial research while pre-Baby Boomers are becoming the fastest user of social networks as supported by smartphone gadgets (Pew Internet Generation, 2015). For all age groups, general activities include shopping, travel arrangements, health information searches or podcast downloading.

Game theory is a mathematical concept. This analyses strategic relations between individuals and agents to be generated on the basis of agent's preference. Supposedly, the agents can have contradictory precedence. The realistic application of this theory is outlined by gamification. The use of gamification in market research is evidenced by Pepsi-Cola particularly, for studying vending sales. With a specially designed game, Pepsi-Cola managed to receive consumer reactions on vending machine placement as well as the required number of machines in a building (Marketing Society, 2019).

Collective intelligence theory argues that age groups are relatively more creative than simply collecting their requirements. Crowd sourcing is an advanced phenomenon of the Internet that performs on the basis of collective intelligence. In crowd sourcing, projects are broken down into smaller assignments and distributed among individuals to reach completion. A popular mid-Atlantic resort town, Ocean City in Maryland, employs crowd sourcing for promoting special events on Facebook (Small Business, 2015).

Network theory is concerned with all types of relationships including both subjective and objective knowledge. Social network analysis as a learning tool is involved in outlining inter-societal networks as well as behaviour influences. Digital marketing channels as Groupon, Foursquare, Facebook or Twitter are changing as networking platforms between consumers and marketers listen to consumer opinions and allow marketers to spread messages within their networks. Networking becomes influential; recent research reveals that the success of entertainment products is not only the outcome of acting power or story but the influence of consumer decisions (Harvard Business Review, 2012). In digital marketing, theorists apply diverse approaches of both a scientific and non-scientific nature to clarify consumer purchase decision making (Karimi, 2013). Conventional understanding about consumers gets changed in contexts where more functions are added with the support of technology use than simply product or service purchase.

On the basis of postmodernism and digital marketing, an innovative technology application like AR in tourism can be valid, once the leveraging of business and experience stirs a new disruption in tourism. The interaction between technology and product or service consumers becomes functional, and such interactivity is a required element of digital marketing. Consumers in tourism use technology to get informed and communicate about a target product or service. For tourism, digital marketing offers numerous usable tools to meet consumer demands (Weber and Roehl, 1999) and contribute to reach their preferences. For tourism, digital marketing appears in forms like brand marketing, B2B marketing, referral marketing, mobile marketing, digital marketing and many more (Bennett and Radburn, 1991). This classification signifies that any other marketing tool like AR can be experimental and subject of intensive research, justifying validity of this study and that will be essential for the activity.

AR and digital marketing

The expectations of tourists and global trends are continually pushing the tourism industry to adopt digitalization. This adoption can offer better performances that are both economically and socially sustainable, plus environment-friendly marketing strategies followed by the satisfaction of tourists. This is one of the reasons the tourism industry needs to innovate and adopt a workable and better digital transformation process. All of the beneficiaries and stakeholders need to collaborate to enhance the capacity of the tourism industry for facing global market competition and challenges (Kasliwal and Agarwal, 2019).

As a transversal economic sector, tourism can have relevancy in technology applications. The impact generated from digitalization comes with innovations in tourism product or service manufacturing and resource management followed by effective marketing on digital platforms. Technologies have become a powerful and indispensable tool for tourism resource management, distribution and marketing. The harnessing of digital and innovation advances offers the tourism industry opportunities for improving community empowerment, inclusiveness and efficient resource management and a sustainable development agenda. As a result, tourism needs to adopt technological innovation for realizing its potential contributors. Technological innovations have made changes when AR can ensure smart travel facilitation through offering experience sharing that is rather unique and exclusive (Yung and Khoo-Lattimore, 2019).

The common efforts of digital platforms, feedback, information content, intelligence, media and tourism services can possibly make the tourists plan their travel and experience tourism in more meaningful ways. The use of personal computing devices, smartphones, the Internet and wireless technologies have put tourism in a position to adopt any innovative technologies. In general, tourists seek to have experiences than merely visiting an attraction. Tourist destinations have always been a subject of enjoyment for the tourists. With the application of AR, these tourists will be able to enjoy their desired destination digitally. The digitalization of tourism marketing has theoretically transformed the decision-making process to purchase tourism products or services as well as the planning of travel, booking the hotel, confirming tickets or even managing tourism activities (Yung and Khoo-Lattimore, 2019).

The benefits generated from the application of AR for tourism marketing need to spread over the relevant investors, stakeholders and the governmental agencies. Thus, digital transformation can provide benefits to the tourists also. The application of AR has connected the tourists on an international level helping the most vulnerable tourist segments. A tourism destination that is smart with the support of AR can transform the tourism industry as a whole (Hsu et al., 2018). An AR-supported smart destination adopts innovative technologies, accepts innovation, creates accessibility and ensures sustainability and inclusivity throughout the complete travel cycle including the pre-, during and post-travel phases. AR technology–applied smart destinations can turn conventional tourism marketing more inclusive with the inclusion of entities for tourist communication or tourist retention.

AR marketing for the Bangladesh tourism industry

For many countries in the world including Bangladesh, tourism is the key industry in their economy. In a competitive tourism business industry, innovative technologies application remains important for meeting the demand of time and consumers. Thus, the tourism industry in Bangladesh has to be able to adopt new and innovative technologies as its foremost concern. In the present as well as in the upcoming years, millennials largely dominate the tourism industry and AR has huge potential for enhancing tourists' experiences. The newer types of AR mobile apps offer beneficial data and information, guides, navigation, translation and many other features. In Bangladesh, in particular, the tourism industry stays on four specific service offers: transportation, accommodation, catering and tourist attractions. AR has the capacity to assist and improve each of these service offers (Dinçsoy, 2020; Gottlieb, 2018).

Accommodations

In terms of accommodation in Bangladesh, the hospitality sector was the pioneer in implementing AR. The ways through which AR can support a hotel is amazing that offers a range of offers in different areas of operation. For accommodation, AR can support with information and advertising. The application of AR for creating all-around room tours with details of accommodation, prices and relevant information are useful. This is an effective way for advertising the hotel and for engaging tourists for trying out a wide range of services of the hotel. The application of AR can possibly turn one-time guests into loyal and regular guests.

More convenient and reliable data and information about the accommodation service offers can be possible from using AR. Tourists require solid information about the housekeeping, services, sightseeing, food and so on. Getting reliable and good data and information has never been easy. However, times have changed and tourists now with the use of smartphones or tablets can open the specific hotel app, point to a particular marker and find all of the relevant information and data. AR has turned this simple and easy. Hotels in Bangladesh can have interactive wall maps in all rooms, the tourists can point their phone at and are thus able to check local tourist attractions nearby.

Tourists in these days are technology savvy. They do not tend to and are unwilling to observe things that they used to do even a few years early. Tourists using AR have different ways for observing issues. Tourists in these days do not necessarily pay attention to pamphlets that are simple. AR makes it possible to discover information on a relatively new level by getting 3D animations from printed flyers. Hotels in Bangladesh can collaborate with AR service providers for producing such interactive advertisements in their magazines. Also, hotel app users can scan the advertisements for unlocking a presentation video.

AR can turn a simple hotel into an interactive one. The success of AR has motivated and inspired many businesses for utilizing AR that includes the hospitality industry also. Young millennials are keener to stay in particular hotels having AR access. Bangladesh can have hotels that display each site of their hotels with added AR objects and offer rewards to the competition winners.

Transport

Moving to a city and transport from one place to another by car, train or bus can be seriously difficult for tourists in most cases because the place is unfamiliar to them. In such situations, tourists also wonder and become confused how to travel and where to travel. However, travelling with an AR app can help and guide tourists to get route, direction, next stop and even show the places of interest. Tourists from other countries can enjoy the AR app as this can meet their demands. Bangladesh can have AR apps for tourists that turn a bus or metro map or other similar items into an attractive guide with multiple languages. Cities in Bangladesh can have the example of New York with Tunnel Vision. The application of AR can certainly help change a city's look by engaging them to AR apps.

Restaurants/catering

Restaurants, diners, bars and relevant other catering facilities can apply AR to benefit their business. This can be performed in many ways, for example a brand new restaurant menu can have an interactive 360-degree view of each dish, ingredients and accurate portion size. These are currently performed by AR application Kabaq (Kabaq, 2019). Tourists can enjoy AR games in restaurants while their order is being served. These AR games can attract tourists in an easy way. Navigation help can be offered with the special mobile apps that use AR for providing data and information about the cafes, restaurants and bars nearby.

Attraction

Tourists mostly travel to explore tourist attractions, which the use of AR can help. This can transform a good city into an extraordinary one. With the support of AR, tourists can view the evolution of landmark buildings in the time context. Tourists can also enjoy 3D place models, get fun tour guides and so on. Cities in Bangladesh can surely adopt AR for this purpose. AR can be applied in tourism for serving many other purposes (Hassan et al., 2018; Jomsri, 2019), as discussed next.

Marketing

In Bangladesh, almost all of the agencies engaged in tourism business use paper catalogues and brochures weighing to tons. The potential of AR with print media is abundant in the virtual world. The access of animated visualization can attract tourists in an increased number where AR marketing in Bangladesh can of course offer more and better data and information and thus persuade tourists to take quicker decisions.

Navigation

The foremost purpose of tourists while travelling is to discover destinations and sights that are new. This can be highly challenging and almost impossible in unknown places. In such situations, AR apps or maps with AR access to navigate turn this job easier and accessible.

Language barriers

AR can eliminate the trouble and barrier of language in an unfamiliar place. A tourist would definitely be keen to learn a language or signs that are local. AR apps with innovative features like Google Translate can help avoid miscommunication or misunderstanding by availing prompt translation service to the tourists' native language.

In almost all cases, tourism brings new experiences. AR can obviously add experiences to tourists in far more useful way than ever imagined. AR is a useful technology for tourists in terms of hotel booking and tours, accessible travel information, eliminating language barriers and advanced navigation.

In Bangladesh, the application of AR in the tourism industry is a possibility. This application can be critical and challenging at the same time due to the lack of knowledge and institutional capacities. For the application of AR for marketing to become possible, both the public and private institutions need to be involved, including the Ministry of Civil Aviation and Tourism, Bangladesh Tourism Board, Bangladesh Parjatan Corporation, experts from the relevant academic institutions and other private and public organizations. Their effective and efficient participation can make the application of AR for tourism marketing in Bangladesh possible. This is one of the reasons the World Tourism Day theme was chosen as "Tourism and the Digital Transformation". This is one of the most important aspects of tourism that innovative technologies are adopted for: marketing aimed at making changes and supporting digital development trends. Observing this day with the theme would only be able to bring changes in the Bangladesh tourism industry when the relevant authorities, beneficiaries and stakeholders can offer new and innovative ideas, projecting them within the observance of this day, and adopt the digital transformation that is better and workable. Furthermore, that digital transformation will help to support and provide the means to ensure the conditions for tourist destinations to meet the challenges associated with the sustainable development goals proposed by the United Nations.

Conclusion

A better way to interact with consumers can be digital platforms, which are relatively easier for communication, information exchange, transaction execution and presenting messages fast and at low cost, and that can be accessed anywhere and anytime. An innovative technology like AR can help promote tourism activities in a destination to both local and foreign tourists. A tourism industry that is technologically advanced can also benefit the development of tourism entrepreneurs, the enhancement of community interests and the management of tourism resources efficiently. In line with the sustainable development goals, innovative technologies can ensue the rapid, continued and inclusive growth of tourism, while contributing to reducing poverty, improving quality of life and health services, reducing inequality and improving access to education, among other things, whose main beneficiaries are the local population.

This innovative technology has connected tourists to global digital platforms and some important allies to ensure sustainable tourism development. This is the reason for which a technology as AR will be able to bring a momentum for the economy and the society of Bangladesh. Thus, the potential of an innovative technology like AR can advance the demand of soft skills in effective planning and managing the resources. The massive and enormous effects of AR on tourism marketing will have positive effects on the tourism workforce.

References

Benckendorff, P. J., Xiang, Z. and Sheldon, P. J. (2019). *Tourism information technology*. Wallingford: CABI.
Bennett, M. and Radburn, M. (1991). Information technology in tourism: The impact on the industry and supply of holidays. In M. T. Sinclair and M. J. Stabler (eds.), *The tourism industry: An international analysis*. Wallingford: CAB International, pp. 45–65.

Bilgihan, A., Smith, S., Ricci, P. and Bujisic, M. (2016). Hotel guest preferences of in-room technology amenities. *Journal of Hospitality and Tourism Technology*, 7(2), pp. 118–134.

Bjork, P. and Guss, T. (1999). The internet as a marketspace – the perception of the consumers. In D. Buhalis and W. Schertler (eds.), *Information and communication technologies in tourism 1999*. Vienna: Springer, pp. 54–65.

Buchholz, R. (2014). *Augmented reality: New opportunities for marketing and sales*. Retrieved from: http://bit.ly/1nMCLYO, p. 2 (accessed: the 1st October 2019).

Buss, D. M. and Foley, P. (2019). Mating and marketing. *Journal of Business Research*. Retrieved from: https://doi.org/10.1016/j.jbusres.2019.01.034.

Chaffey, D. and Ellis-Chadwick, F. (2019). *Digital marketing*. London: Pearson.

Deighton, J. and Kornfeld, L. (2009). Interactivity's unanticipated consequences for markets and marketing. *Journal of Interactive Marketing*, 23(1), pp. 2–12.

Dinçsoy, M. O. (2020). Sustainable development and industry revolutions. In K. T. Çalıyurt (ed.), *New approaches to CSR, sustainability and accountability*. Singapore: Springer, pp. 61–79.

Gottlieb, O. (2018). Time travel, labour history, and the null curriculum: New design knowledge for mobile augmented reality history games. *International Journal of Heritage Studies*, 24(3), pp. 287–299.

Harvard Business Review (2012). *Before they were stars*. Retrieved from: https://hbr.org/2012/01/before-they-were-stars (accessed: the 1st August 2019).

Hassan, A., Ekiz, A., Dadwal, S. and Lancaster, G. (2018). Augmented reality adoption by tourism product and service consumers: Some empirical findings. In T. Jung and tom Dieck, M. C. (eds.), *Augmented reality and virtual reality: Empowering human, place and business*. New York, NY: Springer, pp. 47–64.

Hsu, C. C., Tsaih, R. H. and Yen, D. (2018). The evolving role of IT departments in digital transformation. *Sustainability*, 10(10), p. 3706.

Jomsri, P. (2019). Creative innovation of augmented reality for promote sustainable tourism of Chiang Mai Moat. *Journal of Physics: Conference Series*, 1335(012010), pp. 1–6.

Kabaq (2019). *Augmented reality food*. Retrieved from: www.kabaq.io/ (accessed: the 1st August 2019).

Karimi, S. (2013). A purchase decision-making process model of online consumers and its influential factor across sector analysis. *PhD Thesis*. Manchester: Manchester Business School.

Kasliwal, N. and Agarwal, S. (2019). Green marketing initiatives and sustainable issues in hotel industry. In U. S. Panwar, R. Kumar, N. Ray (eds.), *Handbook of research on promotional strategies and consumer influence in the service sector*. Hershey, PA: IGI Global, pp. 197–214.

Kazmer, M. M., Glueckauf, R. L., Schettini, G., Ma, J. and Silva, M. (2018). Qualitative analysis of faith community nurse–led cognitive-behavioral and spiritual counseling for dementia caregivers. *Qualitative Health Research*, 28(4), pp. 633–647.

Marketing Society (2019). *Pepsi pushes augmented reality to the MAX*. Retrieved from: www.marketingsociety.com/the-library/pepsi-pushes-augmented-reality-max (accessed: the 1st August 2019).

Nayyar, A., Mahapatra, B., Le, D. and Suseendran, G. (2018). Virtual reality (VR) and augmented reality (AR) technologies for tourism and hospitality industry. *International Journal of Engineering and Technology*, 7(2), pp. 156–160.

Pew Internet Generation (2015). *Millennials*. Retrieved from: www.pewInternet.org/topics/millennials/ (accessed: the 1st August 2019).

Rauschnabel, P. A., Felix, R. and Hinsch, C. (2019). Augmented reality marketing: How mobile AR-apps can improve brands through inspiration. *Journal of Retailing and Consumer Services*, 49, pp. 43–53.

Ryan, D. (2014). *Understanding digital marketing: Marketing strategies for engaging the digital generation*. London: Kogan Page Limited.

Small Business (2015). *Theories of digital marketing*. Retrieved from: http://small business.chron.com/theories-digital-marketing-36397.html (accessed: the 1st August 2019).

Vong, K. (2015). *How brands are using augmented reality in marketing to engage customers*. Retrieved from: http://bit.ly/SrBNTq, p. 1 (accessed: the 1st September 2019).

Weber, K. and Roehl, W. S. (1999). Profiling people searching for and purchasing travel products on the World Wide Web. *Journal of Travel Research*, 3, pp. 291–298.

Yamamoto, M. (2018). Comparative study on the effectiveness of keyword search advertising to provide tourists information. *Journal of Tourism and Hospitality Management*, 5(4), pp. 135–143.

Yung, R. and Khoo-Lattimore, C. (2019). New realities: A systematic literature review on virtual reality and augmented reality in tourism research. *Current Issues in Tourism*, 22(17), pp. 2056–2081.

Part 5

Globalization and tourism marketing

10 Globalization effects on tourism marketing in Bangladesh

Md. Sohel Rana, Muhammad Khalilur Rahman, Mohammad Fakhrul Islam and Azizul Hassan

Introduction

Globalization is a web that connects people across geographical limitations through the means of economics, technology, politics, tourism, trade, transactions and so on. Over the years, the buzzword of globalization has become a controversial subject because of its nature of bringing more openness in many diversified areas. However, globalization has created numerous opportunities for many nations to grow faster, taking a competitive advantage. Every country is rigorously putting its best effort to internationalize its domestic products and services according to the demand from the global market. The globalization process has also facilitated the movement of people from one country to another for multiple reasons. Recently, tourism is one of the dynamic service sectors that has been contributing largely to the national economies of many countries. Globalization and tourism have a deep connection with each other. Globalization has made tourism spread beyond the national borders. The tourism sector is not just involved in the movement of people across the borders but it facilitates capital flows from one country to another, cultural flows, political collaboration, free flow of ideas and education, healthcare facilities and many implications on our social and human lives including environmental pollutions, criminal behaviour, diseases and terrorist activities (Hjalager, 2007). The contribution of tourism in any country's economic development is huge and tourism is considered and recognized as a source of economic growth and development (Brida and Risso, 2009; Seetanah, 2011).

Ivanov and Webster (2013) have identified that globalization has a stimulus impact on inbound tourism, which significantly contributes to the country's gross domestic product (GDP). However, a previous study conducted by Ivanov (2005) has identified that tourism may also create leakages in GDP since many organizations will tend to import goods and services from foreign countries for the quality consumption of the tourists during their stay in the host countries. Hence, many countries are focusing on globalizing their tourism sector through a number of contracts and agreements bilaterally and multilaterally among the different countries around the world. The globalization of the tourism industry of a country may require undergoing an irrevocable globalizing process and a set of activities that depends on manifold communication among the countries. For example, there remain many issues that require countries to sort out how the limitations of outsourcing, transnational ownership structures, cross-border marketing collaboration and investment and so on will be resolved and implemented. Many scholars and academics have studied in the field of tourism and discussed the impacts of globalization upon tourism industries (Ivanov and Webster, 2013; Scheyvens, 2011). Bangladesh has been emerging as a developing nation which has attracted many people around the world as a surprising land in the South Asian region. Consequently, Bangladesh should take this as an opportunity to promote its historical, traditional and cultural heritage throughout the world to make a long-lasting impression about beautiful Bangladesh. However, before drawing any policy framework for Bangladesh it is important to understand how the global tourism sector has been attracting people in different parts of the world. The understanding of a global overview of tourism will help set an effective tourism framework for Bangladesh on the global stage.

Overview of global tourism

Tourism has been developed into a global phenomenon. Global tourism has contributed to the economic sectors widely and social activities in the current era. It contributes around 5% of the world's GDP. Global tourism is a key export sector for many developing and developed counties. According to World Tourism Organization (UNWTO) (2019), international tourist arrivals worldwide are estimated to increase an average of 3.9% a year during the period between 1995 and 2010. However, international tourism arrivals are slackening down gradually from 3.8% in 2011 and it is predicted that it will be 2.5% in 2030 (see Figure 10.1) due to the higher base volume, lower GDP growth, low elasticity of travel to GDP and shift from falling transport costs to increased ones.

Based on Figure 10.1, international tourist arrivals in the world will increase by 43 million a year on average between 2010 and 2030. It is crucial to consider all estimates presented here in complete terms, rather than just in terms of an average annual growth rate (UNWTO, 2018). As the base volume is increasing, a lower step of increase still indicates a greater growth in total members. The increase rate between 2010 and 2030 signifies an increase of 43 million global tourist arrivals a year compared to an increase of 28 million in 1995–2010 (UNWTO, 2019). This growth of 43 million is equivalent to the absolute international tourist arrivals in a major destination.

According to the report of UNWTO (2018), international tourist arrivals increased by 6% to 1.4 billion in 2018. International tourist arrivals in Europe were 672 million in 2018, whereas Asia and the Pacific recorded 323 million, America 211 million, Africa estimated 63 million arrivals and the Middle East showed 58 million international tourist arrivals (UNWTO, 2018).

International Tourist Arrivals (mil)

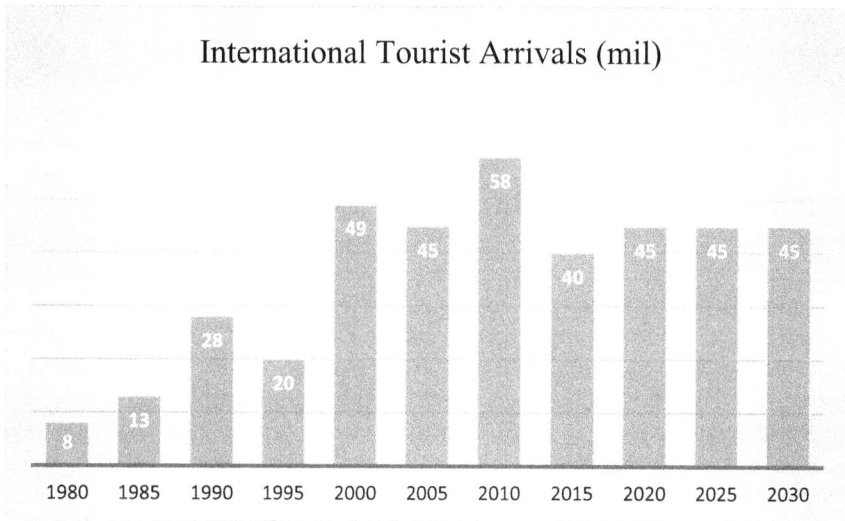

Figure 10.1 International tourist arrivals change over the previous year (millions)
Source: UNWTO, 2019

A brief historical background of tourism in Bangladesh

Bangladesh came into being as an independent country in the global map in 1971 through a bloody war that lingered for nine months against what was then West Pakistan. The pride of the nation glorified several times when the great leader of the country, Bangabandhu Sheikh Mujibur Rahman, took the responsibility to reconstruct the war-ravaged country and started to take developing initiatives. Geographically Bangladesh is situated at the northeastern part of South Asia and lies between 20°34′ and 26°36′ north latitude and 88°01′ and 92°41′ east longitude (Bangladesh Tourism, 2009). Meanwhile, the major parts of Himalayas lie at the northern belt and the southern part of Bangladesh is widely surrounded by the Bay of Bengal. Bangladesh entitles to about 147,570 square kilometres of low-lying alluvial and plain land which is blessed by numerous rivers criss-crossed from the north to the south and east to the west. Bangladesh is a blessed country where people of diversified races, languages, cultures and religions find themselves peaceful and happy living with each other. The majority are Muslims, but they are harmonious and sympathetic to the people of other religions like Hindus, Christians and Buddhists that gives a unique cultural integration in the peoples' livelihoods which is vividly reflected in their behavioural traits.

  Bengal once was regarded as one of the affluent and enriched regions until the 16th century in the subcontinent. Due to its wealth and alluvial land, Bengal was a cherished land to occupy for many empires. For example, the great warrior Mohmmad Bakhtiar Khalzhi from Turkistan captured Bengal with only 20 men

(Bangladesh Tourism, 2009). Subsequently, many traders came to this land and started reigning different parts of Bengal. In the early 16th century, the Europeans started to come through the Bay of Bengal and set up their trading posts. However, eventually the British started to rule the country as a part of British India and ruled over 200 years until 1947 (Bangladesh Tourism, 2009). During the Pakistan regime Bangladesh remained underdeveloped and continuously overlooked from many developing activities.

After the independence of Bangladesh, Bangabandhu Sheikh Mujibur Rahman initiated to establish Bangladesh Parjatan Corporation (BPC) by Presidential Order 143 in 1972. The purpose of the ordinance was to develop tourism and hospitality facilities across the country, create a positive image about the country, generate employment opportunities, establish tourism institutions and tourism parks and to establish foreign liaison (Daily Sun, 2017). However, the government of Bangladesh took initiatives to reform the national tourism policy in 2010 aiming to create employment opportunities, contribute to the national economic development and ensure environmental sustainability (Alam et al., 2009). Since Bangladesh contains a rich historical background and diversified cultural harmony, the country attracts many tourists to visit the archaeological sites, historical mosques, monument, resorts, beaches, rivers, dams, forests, tribal peoples' livelihoods and diverse wildlife. Bangladesh has immense potential in the field of tourism development due to its geographical location, cultural integration and natural resources. This particular sector has been showing possible contribution to economic growth, employment generation and marginal and infrastructural development. Therefore, effective policy planning is badly needed to make the tourism sector viable economically in Bangladesh.

This chapter focuses on exploring globalization factors that may affect tourism marketing of Bangladesh. Subsequently, the second section of the chapter discusses globalization factors, tourism marketing plans, cross-border marketing collaboration, outsourcing transnational ownership, foreign representation, digital connectivity, data security, transport networks and challenges of global tourism marketing. The third section shows methodological approaches including data collection and data analysis. The fourth section of the chapter discusses the results of the study and finally the chapter concludes with the policy implications for global tourism marketing and conclusion remarks.

Literature review

Globalization factors affecting tourism in Bangladesh

The natural beauty of Bangladesh consists of beaches, rivers, coasts, religious places, mosques, pagodas, waterfalls, hills, forests and tea gardens. The historical Mosque City of Bagerhat, the Sunderbans and the Buddhist Vihara at Paharpur are the major heritage sites in the county. Many international and local tourists visit the country's tourist attraction sites to look at the beauty of nature of this country. Bangladesh can be a prime tourism destination in the world if the government concentrates on what is required for tourist attractions. The World Travel and Tourism Council (WTTC) reported that the travel and tourism industry has contributed $5.3 billion, which is approximately (2.2%) of the total GDP in the country in 2017 and it is expected to rise 6.8% per year to $5.7 billion in 2028 (Dhaka Tribune, 2018; The Financial Express, 2019). Every year a greater number of tourists come to Bangladesh to sightsee and enjoy its natural beauty. However, due to some limitations, the country has failed to introduce itself as a tourism destination country in the world. But it is assumed that the country can be one of the top tourist attraction destinations in the world soon (Roy and Roy, 2015). *The Financial Express* (2019) reported that the country's growth in the tourism industry has a significant impact on the development of rural communities. The tourism industry in Bangladesh is increasingly driven by micro, small and medium enterprises (MSMEs) particularly in the region-based hotels, restaurants, recreational activities and tour operators.

There are many types of tourism in the country that attract people to travel to the country's destinations. Domestic tourism involves residents travelling within the country. International tourism involves inbound and outbound tourism. However, inbound tourism involves non-residents visiting within the country whereas outbound tourism comprises residents travelling within another country. Based on these purposes of travel, tourism in Bangladesh can be classified into mainly leisure tourism, religious tourism, family tourism, health tourism, sports tourism, educational tourism and business tourism, which may attract both domestic and international tourists.

Tourism marketing plans at global stages

The global tourism marketing poses a huge potential. It generates more than $1.3 trillion in revenue per year. In 2019, digital travel sales are set to reach $755 billion (Digital Marketing Institute, 2019). The tourism and travel sector are competitive. Thus, brands must use new technological marketing strategies to stand out. There are five major key strategies for making travel and tourism marketing successful at the global stages: put the target audience front and centre, utilize all aspects of social media, identify key moments on the booking journey, know which devices they use and when and welcome social proof (tourists are routinely using review sites such as Google, TripAdvisor, and Yelp to guide their buying decisions). Hjalager's (2007) study explains the various indicators of globalization of the tourism marketing industry and indicates that it has huge business potential in the globalization process, for instance, investments, outsourcing, marketing collaborations, multinational ownership structure, selling and purchasing knowledge. The free movement of labour developments is not only confined to manufacturing sectors but also very much relevant for the modernization of tourism.

The ASEAN region has been successful in attracting new tourists. The success is well illustrated in Table 10.1, which reveals that some countries experienced significant growth between the years 2009 and 2010.

Table 10.1 Growth of international visitor arrival 2009–2010

No.	Member Country	2009	2010	Growth (%)
1	Brunei Darussalam	157,464	214,290	36.09
2	Cambodia	2,161,577	2,508,289	16.04
3	Indonesia	6,323,730	7,002,944	10.74
4	Lao PDR	2,008,363	2,513,944	25.13
5	Malaysia	23,646,191	24,577,196	3.94
6	Myanmar	762,547	791,505	3.80
7	Philippines	3,017,099	3,520,471	16.68
8	Singapore	9,681,259	11,638,663	20.22
9	Thailand	14,149,841	15,936,400	12.63
10	Viet Nam	3,772,259	5,049,855	33.87
	Total	65,680,330	73,752,641	12.29

Source: ASEAN, 2019

Cross-border marketing collaboration

The globalization process not only facilitates economic growth, expansion of trades and businesses, development of a global communication system and rapid technological advancement but also accelerates cross-border collaboration in different fields. Tourism is such a sector which is known as a hyper-globalizer (Held et al., 1999), and that demands global communication and collaboration to attract tourists around the world. Hence, tourism enterprises in different countries must seek international marketing collaboration to sell their tourism market positions and brand expansions by import and export of tourism business concepts through franchising and licensing which require flexible human resourcing and enhancing of the international labour market (Hjalager, 2007). The tourism market consists of a large number of small firms which are characterized and influenced by a large number of individuals' purchasing decisions. In such circumstances, joint marketing can provide the best advantages for both the small tourism firms and tourists, since it has been proved that joint marketing campaigns have been an effective part of promotion over the years (Middleton, 1988). Tourism marketing campaigns are mostly destination-based and often seek collaboration from regional associations, whose marketing activities are sometimes partly publicly financed. However, there remains a number of privately financed marketing collaborations (Hjalager, 2007). Nevertheless, joint venture marketing collaborations are relatively cost-effective initiatives. Cross-border marketing collaboration not only conducts joint venture marketing but it also shares knowledge, culture, know-how and technologies to attract potential tourists of the world at low cost. Sometimes meetings, incentives, conferences and events like tourism fairs are organized in a professional and specialized manner which create new categories of agents that offer promotional services to the enterprises (Swarbrooke and Horner, 2001). Hence, cross-border marketing collaboration is a potential promotional initiative that equips a country with required competencies, capacities and interlinkages to become a successful tourism campaigner at the global stage.

Outsourcing transnational ownership

For a long time, tourism business owners have invested in foreign countries (Johanson, and Vahlne, 1977). Globalization not only connects one nation to another, but it also transfers ideas, resources, capital, manpower and many productive elements across the geographical boundaries. As mentioned previously, the tourism sector is greatly influenced by the globalization process. The resourceful countries get the opportunities to invest in some of the countries where tourism is attractive to the people around the world and has potential. In such circumstances, the host country's individual firms or tourism destinations may embrace globalization in different stages through controlling costs, gaining market shares and competencies and gaining access to the important resources, international tourism networks and so on (Hjalager, 2007). In the meantime, global tourism companies willing to invest in the host countries go through local rules regulations, government policies in utilizing local resources, taxes, share of ownership, infrastructures, communication technologies, transportation opportunities and so on scrupulously (Hjalager, 2007). Lately, Bangladesh has been much opened for foreign investors. The government has taken a long-term plan to brand the tourism sector in Bangladesh. Hence, the development of infrastructure and institutional restructure should be put in place to grab the global opportunities to make the tourism sector thrive seamlessly. Foreign investment through transnational ownership opens up new markets for the products of the host country and the raw materials and physical assets are manoeuvred in a controlled manner (Hjalager, 2007). Investment in the tourism business may come in diversified forms. For example, in the early 1960s and '70s, hotel capacity was given much importance and the tour operators invested heavily in this segment and thus they made them resourceful, not just dependent on local players. Investment from northern Europe particularly concentrated on hotels in metropolitan areas which incurred high expenses and generated low earnings. On the other hand, when ownership was just kept solely in the hands of the host country, it may have larger implication on missing out on global market access, know-how and diversified business integration. In recent years, tour operators and travel agencies have intensive experience of mergers and acquisition (Čavlek, 2000). Therefore, making arrangements for transnational ownership will facilitate tourism marketing at the global arena.

Tourism board representation in foreign countries

The branding of the tourism sector of any country follows certain courses of action in the national and international arena. When it comes to global tourism marketing, host countries should build tourism stations in different parts of the world from where the interested tourists may easily get information about the diversified tourism services that the host country is offering to the global citizen. Establishing tourism board representation in foreign countries is an effective initiative to uphold the country's tourism sector in a trustworthy manner. Tourism has the nature to flow internationally, therefore it is essential to establish the brand value of any country's tourism and get global recognition in the international arena. With this view, the host country can trigger efforts to market the products and tourism services jointly with other countries to reap maximum benefits (Mitchell and Orwig, 2002). Joint venture marketing can be facilitated through the tourism representation in foreign countries.

Integration of digital connectivity and data security

The tourism sector of any country can be effectively promoted through digital connectivity. The recent advancement of technology offers a great opportunity for the tourism industry. Lately, people around the world have placed their trust in technology and consider digital online sources like smartphones, online digital apps, wearables, digital payment methods and so on as potential avenues to be motivated for planning their trips to the different parts of the world (PATA and Oxford Economics, 2018). The report further mentions that 62% to 92% of travel is researched and booked through online applications. The countries included in this range of online digital travel are Vietnam, Australia, China, India, Indonesia, Thailand, New Zealand, Hong Kong SAR, Republic of Korea, Chinese Taipei, the Philippines and Japan (PATA and Oxford Economics, 2018). In the meantime, since the internet-based digital connectivity highway is opened for all, smaller tourism enterprises and business can take advantage of these benefits and compete potentially with the big tourism market players. Now, tourists can share their experiences in social media, which has a magnificent impact on tourism promotion for any destination. However, digital connectivity also poses challenges like data management skills, online scams, online payment problems and data protection of the tourists. Bangladesh has been focusing much on the digitalization of services in diversified sectors in recent years, which opens potential avenues for the tourism industry to promote this growing industry throughout the world. The tour and travel industry can greatly be transformed by the digital platforms since recently travellers have been booking airlines tickets and hotels online. Hence, tourism businesses in Bangladesh should integrate digital services in their regular marketing approaches where tourists can receive services from the far end and they can share their experiences in blogs and social media recommending and promoting tourism for prospective tourists who may intend to pay a visit to Bangladesh.

Integrated transport network

Tourism marketing in the globalization process demands an easy and integrated transportation network where tourists find it convenient and comfortable moving from one destination to another. An organized and integrated transport network promotes the tourism industry to grow quickly and exponentially. It is important to combine air, rail, roads and water transportation systems in an effective network where tourists can be attracted and offered convenient travelling from the beginning until the end of the trip. The tourism sector has been growing rapidly throughout the globe and countries are developing local and international transportation networks to ensure quick congregation and evacuation of travellers through easing access to travel documents and transportation. The globalization process has quickened the movement of the people from one country to another, consequently transportation networks have been widened. The development of integrated transportation networks will facilitate the accessibility of tourists in different destinations. For example, the direct link of the tourists' origin to the desired destination will improve connectivity and minimize the cost of travelling significantly (Van Truong and Shimizu, 2017). The authors further argue that transportation is not merely the main factor that links the demand (origin) and supply (destination) sides of the tourism industry but it is one of the important factors that determine the attractiveness of the destination and thereby influences the demand side (Van Truong and Shimizu, 2017). However, transportation, attractions, services, information and promotions available at the destination influence tourism marketing significantly at the global stage.

Challenges of global tourism marketing

The globalization process not only encourages growth, expansion and benefits, it also triggers many challenges for this growing sector; especially when a small tourism and travel operator strives to attract tourists from the global domain, they face numerous challenges (e.g. lack of brand awareness, the decline of high street travel agents, social media and reputation management, trust, lack of popularity of sharing economies and peer-to-peer travel) in global tourism marketing. The tourism industry is challenged with intense global competition in designing tourism packages that offer the average tourists or travellers something that they have never experienced before. Besides, the tourism sector is challenged due to having high taxation. For example, a traveller has to pay taxes for air tickets and hotel bookings. However, in Bangladesh the airport tax is relatively higher, consequently, the air tickets become expensive for tourists which often works as an influencing factor in destination selection by tourists from different parts of the world (Elite Asia Marketing Team, 2017). Infrastructural development is another challenge faced by the host country. However, when it comes to improving infrastructure including technological advancement by a developing nation like Bangladesh (Rahman, 2016), the country needs to search for international funding sources which subsequently make it costlier for the travellers to afford services during their visit. Global tourism marketing faces another challenge of harmonizing cultural tourism aspects with the key enabling technologies. For example, the harmonization between the long-vested historical and cultural heritage with internet of things, big data, blockchain, artificial intelligence, virtual reality and augmented reality (Peceny et al., 2019). The branding of tourism Bangladesh has proved to be another challenge since it does not wholly reflect the tourist attractions of the country (Howladar, 2012). For nation branding, like "Incredible India" and "Truly Malaysia", Bangladesh should rethink the effectiveness and reflection of the nation in tourism branding. Extreme events such as terrorism or political vulnerability have a negative impact on demand and supply of tourism (Richter, and Waugh, 1986; Ryan, 1993). Feeling unsafe in Bangladesh is a great barrier that discourages not only foreign but also local tourists (Quader, 2008). Political and terrorism violence in Bangladesh have created a bad country image internationally. In the meantime, diversification is a big challenge for Bangladeshi tourism. Bangladesh is only relying on traditional mode of tourism, but our competitors like India, Thailand, Singapore and Malaysia build health tourism, education tourism, religious tourism, halal tourism, business tourism and sports tourism effectively. To promote Bangladesh as a tourist destination, every window of possibility needs consideration.

Methodology

Literature search and interview method have been used for the identification and investigation of the globalization effects of tourism marketing in this study. This study is qualitative in nature through the search information from Emerald and Google Scholar database and the interview was conducted with tourism operators including the ministry of tourism in Bangladesh. Open-ended questionnaires were used to gather information from the respondents. The information was gathered from the three interviews with a representative from tourism operators at Dhaka city in Bangladesh. In this context, since this study is related to finding the globalization effects of tourism marketing in Bangladesh, the study attempted to obtain the appropriate information from the respective experts from the tourism board authority who could give policy-related insights in tourism marketing initiative in the globalization process. The researchers also collected documents from these authorities related to the study and analysed accordingly. Data were gathered about global tourism, historical background of tourism in Bangladesh, global factors affecting tourism in Bangladesh, tourism marketing plans at the global stages, cross-border marketing collaboration, outsourcing, tourism representation, digital connectivity, tourism information security, transport networking and global tourism challenges and further analysed thematically.

Discussion and policy guidance for tourism marketing in Bangladesh

This chapter discusses the globalization effect of tourism marketing in Bangladesh. With this view to understand how globalization matters to promoting the tourism sector of Bangladesh on the global stage, this chapter meticulously analyses the present context of global tourism, tourism sector requirements to enter the global arena, challenges for global tourism marketing and so on. From the experts' opinions and documents collected from the tourism authorities in Bangladesh, this chapter briefly discusses some policies that will help promote the tourism sector of Bangladesh globally. Bangladesh is geographically a small country but lately it is one of the fastest-growing economies in the world. Being a growing economy, Bangladesh is testing every possibility to reap economic benefits from prospective sectors including tourism. As a result, the government of Bangladesh has been taking long-term planning for the sustainability of the tourism sector. However, a huge amount of effort needs to be put in place to promote Bangladesh tourism at the global stage. As one of the tourism experts (Chief Executive Officer, Bangladesh Tourism Board) has opined during the interview,

> We are way behind other countries in promoting our tourism products and services for not having proper policy framework at this very stage. On the other hand, the government is trying to develop required infrastructure to make this sector a real contributor to economic progression. Nevertheless, we are trying our level best to take the tourism sector of Bangladesh at the global stage and with this motto we all are working together and signing new agreements with other countries of the world so that our tourism sector get easy access to the prospective tourists.

However, this chapter as comes up with a set of strategic actions that will play a vital role in developing a marketing strategy with regards to meet the overall objectives of the tourism industry. Figure 10.2 explains the process of developing this strategy.

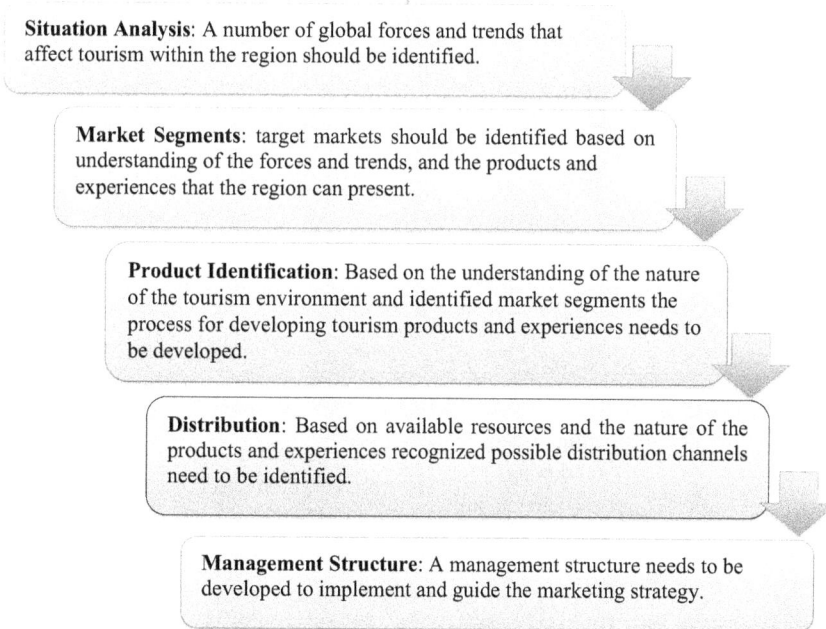

Situation Analysis: A number of global forces and trends that affect tourism within the region should be identified.

Market Segments: target markets should be identified based on understanding of the forces and trends, and the products and experiences that the region can present.

Product Identification: Based on the understanding of the nature of the tourism environment and identified market segments the process for developing tourism products and experiences needs to be developed.

Distribution: Based on available resources and the nature of the products and experiences recognized possible distribution channels need to be identified.

Management Structure: A management structure needs to be developed to implement and guide the marketing strategy.

Figure 10.2 Strategy development process
Source: Developed by the authors, 2019

Figure 10.2 summarizes the entire situation that the tourism of Bangladesh needs to take into account to promote this sector globally. Moreover, this sector needs to acquire a certain level of competence to absorb the global heat of competition. The country should also devise strategies to mitigate challenges which are the blockade for tourism promotion. Meanwhile, the government should take initiatives to ensure a pool of skilled human resources in this particular sector. Moreover, the government should give incentives to invest in tourism business. Foreign investment in tourism businesses should be welcomed providing full-fledged national and local administrative support. Taxation policies (airlines taxes, hotel taxes, transportation and services taxes etc.) related to tourism businesses must be revised to attract more new tourists in the country. It is important for the tourism industry in Bangladesh to organize and participate in more international tourism fairs where the country's tourism brand can strongly be upheld and represented. The tourism board of Bangladesh must send representatives to the foreign countries and simultaneously accomplish cross-border agreements which will ensure the flow of tourists in the country. However, at the same time, there are a number of major trends and forces that will have an impact on tourism in the global over the next five years, which recognizes three key areas for research and analysis that are illustrated in Figure 10.3.

Therefore, Bangladesh should get ready and prepare advance strategies to deal with these emerging trends that will have a big impact on future marketing, advertising and distribution which will further influence visitors' motivation and behaviour.

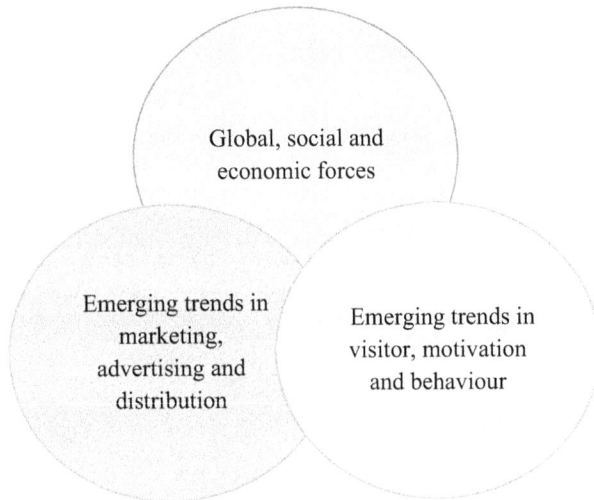

Figure 10.3 Global emerging trends and forces

Source: Developed by the authors, 2019

Conclusion

Tourism is not only a potential economic generator but also promotes a country's image, culture and supremacy on the global stage. The tourism sector is widely known as a "hyper-globalizer", which has created dominance in the major political, economic and cultural spectrums and established a substantial people-to-people connection. Bangladesh has been emerging as a developing nation which has attracted many people around the world as a surprising land in the South Asian region which has opened wide avenues to promote its historical, traditional and cultural heritage throughout the world to make a long-lasting impression about beautiful Bangladesh. Therefore, this chapter aimed to look for strategies and policy frameworks to promote tourism in the globalization process. Through the extensive literature review, document analysis and expert opinion from the tourism board authority, it is apparent that Bangladesh has been challenged along many dimensions in promoting tourism marketing in the globalization process. However, the government of Bangladesh has taken long-term policies to develop the tourism sector to gain economic benefits in the coming years. The predominant factors identified for the promotion of global tourism marketing include transportation, sophisticated technologies, digital transaction development, development of skilful human resources, ease of foreign investment, providing incentives to local and foreign tourism businesses, reduction of exorbitant taxation in tourism activities, social and economic security of tourists and tourism-related firms, representatives in the foreign countries and simultaneously accomplish cross-border agreements and so on. Considering all these predominant factors of tourism marketing in the globalization process, this chapter offers to overcome challenges and accomplishment of these predominant factors that will help promote Bangladesh tourism at the global stage through the framework of a strategy development process. This chapter further has pointed out future global emerging trends and forces that may have a significant impact on tourists' motivation and behaviour. Therefore, a country like Bangladesh should get ready and equipped to face emerging challenges in this dynamic sector.

References

Alam, G. M., Hoque, K. E., Khalifa, M. T. B., Siraj, S. B. and Ghani, M. F. B. A. (2009). The role of agriculture education and training on agriculture economics and national development of Bangladesh. *African Journal of Agricultural Research*, 4(12), pp. 1334–1350.

ASEAN (2019). *ASEAN tourism marketing strategy*. Retrieved from: https://bit.ly/37GC0Kj (accessed: the 24th November 2019).

Bangladesh Tourism (2009). *History of Bangladesh*. Retrieved from: https://bit.ly/2DgzYma (accessed: the 24th November 2019).

Brida, J. G. and Risso, W. A. (2009). Tourism as a factor of long-run economic growth: An empirical analysis for Chile. *European Journal of Tourism Research*, 2(2), pp. 178–185.

Čavlek, N. (2000). The role of tour operators in the travel distribution system. In W. Gartner and D. W. Lime (eds.), *Trends in outdoor recreation, leisure and tourism.* Oxfordshire: CABI, pp. 325–334.

Daily Sun (2017). *Bangladesh tourism: A short historical perception.* Retrieved from: https://bit.ly/2qCuP5n (accessed: the 11th November 2019).

Dhaka Tribune. (2018). *Expert: Bangladesh can be Asia's prime tourist destination.* Retrieved from: https://bit.ly/2QTvI41 (accessed: the 24th November 2019).

Digital Marketing Institute (2019). *11 digital marketing campaign tips for the tourism sector (part i).* Retrieved from: https://bit.ly/33lVssr (accessed: the 24th November 2019).

Elite Asia Marketing Team (2017). Retrieved from: https://resources.elitetranslations.asia/2017/08/29/challenges-confronting-travel-industry/

Held, D., McGrew, A., Goldblatt, D. and Perraton, J. (1999). *Global transformations: Politics, economics and culture.* Stanford, MA: Stanford University Press.

Hjalager, A. M. (2007). Stages in the economic globalization of tourism. *Annals of Tourism Research*, 34(2), pp. 437–457.

Howladar, Z. H. (2012). *Tourism in Bangladesh: Problems and prospects.* Retrieved from: https://bit.ly/34p8arM (accessed: the 24th November 2019).

Ivanov, S. H. (2005). Measurement of the macroeconomic impacts of tourism. *Unpublished PhD Thesis.* Varna: University of Economics.

Ivanov, S. H. and Webster, C. (2013). Tourism's impact on growth: The role of globalisation. *Annals of Tourism Research*, 41, pp. 231–236.

Johanson, J. and Vahlne, J. E. (1977). The internationalization process of the firm – a model of knowledge development and increasing foreign market commitments. *Journal of International Business Studies*, 8(1), pp. 23–32.

Middleton, V. T. C. (1988). *Marketing in travel and tourism.* Oxfordshire: Butterworth-Heinemann.

Mitchell, M. A. and Orwig, R. A. (2002). Consumer experience tourism and brand bonding. *Journal of Product and Brand Management*, 11(1), pp. 30–41.

PATA and Oxford Economics (2018). *Data and digital platforms: Driving tourism growth in Asia Pacific.* Retrieved from: https://s3.amazonaws.com › craftTE_APAC-Data-Digital-Platforms-2018 (accessed: the 24th November 2019).

Peceny, U. S., Urbančič, J., Mokorel, S., Kuralt, V. and Ilijaš, T. (2019). *Tourism 4.0: Challenges in marketing a paradigm shift.* Retrieved from: file:///C:/Users/USER/Downloads/65836.pdf (accessed: the 11th November 2019).

Quader, S. B. (2008). *A land with potential in tourism.* Retrieved from: www.thedailystar.net/story.php?nid=21277 (accessed: the 11th November 2019).

Rahman, M. T. (2016). Impacts of social branding on tourism business: The case of Bangladesh. *World Review of Business Research*, 6(2), pp. 58–65.

Richter, L. K. and Waugh, Jr. W. L. (1986). Tourism politics and political science: A case of not so benign neglect. *Annals of Tourism Research*, 10, pp. 313–315.

Roy, S. C. and Roy, M. (2015). Tourism in Bangladesh: Present status and future prospects. *International Journal of Management Science and Business Administration*, 1(8), pp. 53–61.

Ryan, C. (1993). Crime, violence, terrorism and tourism: An accidental or intrinsic relationship? *Tourism Management*, 14, pp. 173–192.

Scheyvens, R. (2011). *Tourism and poverty.* New York, NY: Routledge.

Seetanah, B. (2011). Assessing the dynamic economic impact of tourism for island economies. *Annals of Tourism Research*, 38(1), pp. 291–308.

Swarbrooke, J. and Horner, S. (2001). *Business travel and tourism.* Oxfordshire: Butterworth-Heinemann.

The Financial Express (2019). *Tapping tourism potential crucial to BD economy: USAID.* Retrieved from: https://bit.ly/35zfscH (accessed: the 11th November 2019).

UNWTO (2018). *Tourism towards 2030: Global overview.* Retrieved from: https://bit.ly/33iwV7C (accessed: the 11th November 2019).

UNWTO (2019). *International tourist arrivals reach 1.4 billion two years ahead of forecasts.* Retrieved from: https://bit.ly/33lT6tB (accessed: the 24th November 2019).

Van Truong, N. and Shimizu, T. (2017). The effect of transportation on tourism promotion: Literature review on application of the Computable General Equilibrium (CGE) Model. *Transportation Research Procedia*, 25, pp. 3096–3115.

Part 6
Tourism marketing
Risk perception

11 Perceived risks of tourism in Bangladesh

Nazmoon Akhter and Azizul Hassan

Introduction

Tourism is one of the major contributing sectors in the economic growth of a country, creating revenues and employment and supporting cultural value and entertainment. The World Tourism Organization's "Tourism 2020 Vision" (UNWTO, 2003) anticipates that international tourist arrivals will be more than 1.56 billion during year 2020. However, planning and anticipation time before the trip is not a simple task as a tourist must make decisions regarding timing, transportation mode, budget, secondary destinations and activities along with the primary destination. Besides, tourism is connected with risk and fear for both physical integrity and belongings of the travellers. For this reason, tourists' perceived risk influences tourists' choice of destinations (Fuchs and Reichel, 2006). Tourists try to avoid high-risk destinations and select ones that they consider safe (Sonmez and Graefe, 1998a). Thus, successful tourism development relies on the risk reduction that is related with a destination and an event along with infrastructural improvements and heightened awareness on the world stage.

Perceived risk is reviewed across various disciplines, namely geology (Burton et al., 1978), sociology (Douglas and Wildavsky, 1982), psychology (Kahn and Sarin, 1988), marketing (Bauer, 1960; Dholakia, 2001) and tourism (Carter, 1998; Lepp and Gibson, 2008). In the field of tourism, Roehl and Fesenmaier (1992) define three groups of tourists on the basis of their risk perception: risk neutral, considering no risk involved in tourism or their destination; functional risk, considering the probability of mechanical, equipment or organizational problems associated with tourism; and place risk, perceiving fairly risky vacations and very risky destinations that have been recently visited by tourists during their vacations. Some scholars have reported that perceived risk generates from various potential losses (Dholakia, 2001; Jacoby and Kaplan, 1972; Roselius, 1971). Performance risk arises from purchases that will not perform according to buyers' desire or expectation (Horton, 1976). Financial risk indicates potential net financial loss that is having the possibility of repairing, replacing or price refunding of purchased product (Laroche et al., 2004). Psychological risk means the possibility of anxiety or psychological discomfort like worry and regret, generated from post-purchase affective reactions (Roehl and Fesenmaier, 1992). Social risk is the probability that purchaser motive to purchase a product may be influenced by other people's opinions (Murray and Schlacter, 1990). Physical risk is concerned about the potential threat of a person's health or appearance due to the consumed product (Mitchell, 1998). Time risk has the potentiality of consuming time or wasting time due to the products purchased (Roselius, 1971). Physical risks (Sonmez and Graefe, 1998a; 1998b), financial risks (Um et al., 2006), health risks (Larsen et al., 2007) and social risks (Carter, 1998) are imperatively related to travel. Roehl and Fesenmaier (1992) found seven different types of risk involved in travel decisions, namely equipment, financial, physical, psychological, satisfaction, social and time risks. They found different risk perceptions among tourists, where some tourists are more risk averse than others. A strongly influential "generalization effect" of perceived risk is exited which can result in serious economic losses. Besides, when the consumer has faced certain types of risks, his/her behaviour has changed from delaying the purchase to following strategies designed to reduce risks to a "tolerable" level. For this reason, it is much more challenging to investigate tourists' risk perception about destination and risk reduction strategies that encourage them to return to the risky destination.

Literature review

Perceived risk

Perceived risk is the uncertainty that a consumer has when buying items. Mitchell et al. (1999) reports that researchers, i.e. practitioners and academicians, have focused on perceived risk that is related to various fields, including food technology, banking services, dental services, apparel catalogue shopping, intercultural comparisons and so on. Mowen and Minor (1998) defined perceived risk as a consumer's perception regarding risk based upon assessing the possibility of negative outcomes as well as the likelihood of occurring such outcomes. According to several literatures regarding consumer behaviour (Engel et al., 1995; Assael, 1995; Mowen and Minor, 1998; Schiffman and Kanuk, 2007), consumer perceived risk is classified in different ways such as physical risk, indicating probability of physical harm to the consumer arisen from product malfunction; performance risk, representing the probability of not operating the purchased product as expected; financial risk, defining probability of losing investment in the product; social risk, meaning the fear that the purchased product will not conform to the standards of the reference group; psychological risk, indicating the fear that the purchased product will not suit the consumer's self-image; time risk, explaining the risk of taking time excessively for product consumption; and opportunity loss risk, defining the possibility that by taking an action, the consumer will miss out on alternative preferred activities. According to Quintal, Lee and Soutar (2010) and Dowling and Staelin (1994), perceived risk is the perception of probable loss because of uncertainty involved in purchasing product or service. Additionally, Reichel et al. (2007) classified consumers' risk perception into seven categories: physical risk (i.e. potential harm result from consuming products and services), monetary risk (i.e. potential loss of the money invested in a product/service purchase), performance risk (i.e. the possibility of the product/service not reaching consumers' expectations), social risk (i.e. anxiety that the product will not be approved by the reference group), psychological risk (i.e. the possibility that the product will not be consistent with self-image), time risk (i.e. concern that consuming product/service will be time consuming), and opportunity loss risk (i.e. fear that the consumer will forgo other better consumption options due to making the decision of purchasing a particular product/service).

Tourists' perceived risk

Travel intention is influenced mainly by destination image regarding its risk level. Roehl and Fesenmaier (1992), who primarily research on risk perception in tourism, determined three basic factors of perceived risk, namely physical-equipment risk, vacation risk and destination risk. Promsivapallop and Kannaovakun (2017) report that risk perception regarding travel has negative effect on tourists' intention to travel only in risky destinations but no effect on their intention to visit low-risk or no-risk destinations. Further, Promsivapallop and Kannaovakun (2018) examine travel risk determinants and their relation with young educated adults who lived in Germany and visited Thailand through considering their role, past experience, types of gender and their intention to travel. They found six determinants of tourists' risk perception regarding travel, namely crime and false practice risk, health risk, hazard risk, communication risk, over-commercialization risk and political risk, which have relationships with tourists' roles and their past travel experience. They stated that over-commercialization risk is a more crucial risk which has relationships with tourist roles and their travel experience than other types of risks, where political risk is a less concerned risk having a negative impact on tourists' travel intention.

Garg (2013) stated that tourists perceive terrorist activities, SARS, swine flu, earthquakes and tsunamis as key travel risk concerns at the time of choosing destinations and they try to avoid destinations having high risk from these factors. Low-risk perception about a destination has positive effect on that destination. In this regard, Tavitiyaman and Qu (2013) argued that travellers with low-risk perceptions about a destination will normally establish a more positive image about that destination, and improve travellers' visit intention to that destination which results in gaining greater overall satisfaction than travellers with high-risk perceptions. Rittichainuwat et al. (2018) examine tourists' risk perception and travel decisions using variables like demographics, knowledge about safety and country of residence where samples were gathered in Thailand, Japan, Australia and Indonesia. They explained that tourist risk perception was influenced by tsunami occurrence probability and was destination specific. Additionally, Kozak et al. (2007) found that the majority of travellers have changed their decisions to travel to a destination due to elevated risk. They also observed that different continents perceive travellers' risk differently and concluded that travellers from varied nations may face different degrees of risk.

Parrey et al. (2018) investigated both government initiatives and media influence as mediating roles between perceived risks and destination image regarding conflict zones. They found that unrest (terrorist) and political risk is mostly followed by both psychological risk and socio-cultural risk which is against the assumption that unrest (terrorist) risk is the crucial source of risk that domestic tourists perceived during visiting any conflict zone. Further, they found that the media's role improves the risk perception and government initiatives decrease the destination's risk image as well as its competitiveness. Moreover, they affirmed that performing best under the dimension of control as government initiatives plays very weak role as compared to performing under the dimension of concern as media for the destination image in the conflict zone. Mitchell and Vassos (1997) and Mitchell et al. (1999) found a list of 43 risk factors about a holiday package, which ranged from serious occurrences like natural disasters to trivial matters like when a tour representative did not participate in activities. Moreira (2007) investigated stealth and catastrophic risks to the development of tourism destinations. He explains that stealth risks are more significant than catastrophic risks to residents and tourists alike. Stealth risks are the gradual degradation of neutral or positive present conditions diffused over time such as air quality, while catastrophic risks include the frequency of sudden negative impacts on present reality by serious accidents or natural disasters such as earthquakes and typhoons. According to both domestic and international markets, Dolnicar (2005) analysed various risks or "fears" such as political risk, composing of terrorism and political instability; environmental risk, involving natural disasters and landslides; health risk, indicating lack of access to healthcare and life-threatening diseases; planning risk, representing unreliable airlines and inexperienced operators; and property risk, including theft and loss of luggage. Fuchs and Reichel (2006) investigated destination risk perception among foreign tourists visiting Israel and identified six destination risk perception factors: human-induced; financial; service quality; social–psychological; natural disasters and car accidents; and food safety problems and weather. In this regard, Reisinger and Mavondo (2005,2006) established an integrative theory of risk perception and anxiety as determinants of international travel intentions. They found significant differences in perceptions of international tourists about travel risk and safety, anxiety and intention to travel. They explained that tourists from the United States, Hong Kong and Australia perceived more travel risk, felt less safe, were more anxious and reluctant to travel as compared to tourists from the United Kingdom, Canada and Greece. In another study (2005), they utilized path analysis to show travel risk perception as a function of cultural orientation and psychographic factors and found that terrorism and socio-cultural risk emerged as the most significant predictors of travel anxiety where intentions to travel internationally were determined by both travel anxiety levels and level of perceived safety.

Reichel et al. (2007) reported that perceived risk of backpackers' experience is composed of multi-dimensional phenomenon which varies according to an individual's characteristics such as gender, past backpacking experience and preference for fellow travellers. In addition, Dayour et al. (2019) explored backpackers' risk perceptions towards smartphone usage and consequent risk reduction strategies. They explained that backpackers' innovativeness, trust and familiarity with a smartphone are established as inhibitors of their perceived risk where levels of trust had a significant positive impact on their intentions to reuse the device, as did their satisfaction levels with the device and travel. They found that backpackers used a mix of both cognitive and non-cognitive measures to manage their risk perceptions. Adam (2015) investigated backpackers' risk perceptions and risk reduction strategies in Ghana and explored backpackers' perceived risks, determinants of perceived risk and risk reduction strategies. He showed that there are six dimensions of backpackers' perceived risks in Ghana, namely expectation, physical, health, financial, political and socio-psychological risks. He used a binary logistic regression model to determine backpackers' perceived risk in Ghana where he found that religion, continent of origin, sex, repeat visits and travel arrangements were significant determinants. He also focused on risk reduction strategies by backpackers where risk reduction strategies were found to vary by type of perceived risk.

A more comprehensive explanation to the concept of travel risk is given by Cui et al. (2016) who summarized three views of tourism risk perception including tourist's subjective feelings of adverse effects which may occur during the trip, tourist's objective evaluation of the adverse effects, and tourist's cognitive of exceeding the threshold part of the expected adverse effects during travel. Further, Fennell (2017) developed a comprehensive model about travel perceived risk and fear where he found six components: characteristics of tourists, fear-inducing factors of a trip, strategies to reduce fear, travel stage, fear intensity and fear responses.

Risk reduction strategies

The significant role of information as a means for risk reduction is highlighted by several other scholars. According to Mitchell et al. (1999), if the tolerance level of the consumer about risk is crossed, it will either result in abandonment of the purchasing process or the consumer's engaging in risk reduction. They state that consumers seek to reduce the uncertainty or consequences of an unsatisfactory decision by following risk reduction, or "risk handling" strategies. They explain that uncertainty can be reduced by collecting additional information and by "the importance of a name that can be trusted". They also argue that "risk tolerance" directly affects the risk threshold at which consumers begin to engage in risk reduction strategies, as "risk tolerance" not only represents a level of risk that the consumer cannot tolerate, but also represents the ability of the consumer to absorb the risks involved in the decision. Byzalov and Shachar (2004) argued that advertising increases consumers' tendency to purchase the promoted product as the informative content of advertising helps to resolve some of the uncertainty that risk-averse consumers face and thus reduces the risk associated with the product. According to Boshoff (2002), tourism can follow several strategies to reduce risk perceptions and hence to directly or indirectly enhance the purchase intentions of prospective buyers. He highlighted that such risk strategies include providing potential buyers with general information about the service, providing potential buyers with price information and providing a service guarantee prior to actual purchase. In this regard, Tideswell and Faulkner (1999) report that familiarity with a destination and information search behaviour can also be useful tools for risk reduction. Law (2006) suggested several risk reduction strategies such as free insurance coverage, local government guarantees of tourists' personal safety, an increase in transparency of information related to risk incidents and the introduction of surveillance or protection measures to deal with the aforementioned risks of pandemics, terrorist attacks and natural disasters.

Mitchell (1993) identified factors such as age, socio-economic group and education that can influence the use of risk reduction strategies. He explained that increased age of people lowers the propensity of the search and processing of information. Higher educational levels tend to increase levels of searching, but not in all product categories. He also explained that consumers with high self-confidence and high risk perception tend to use more risk reduction strategies.

Tan (1999) investigated Internet shopping risk reduction strategies and found that reference groups appeal as the most preferred risk reliever for Internet shopping, particularly product endorsements by expert users. In addition, he also finds that the marketer's reputation, brand image and specific warranty strategies effectively reduce risk for potential Internet shoppers.

Due to the significance of tourism on the one hand and the considerable effects of perceived risks on tourism on the other hand, it is important to investigate the factors related to perceived risk and how this risk can be minimized which is beneficial to both travellers and marketers.

Travel and tourism in Bangladesh

Travel and tourism provides a great contribution to a nation's economy through creating jobs, driving exports and generating prosperity. In Bangladesh, both domestic and international tourists play crucial roles in its economic sector. The direct contribution of travel and tourism is 2.2% of total GDP in 2017 and is forecast to rise by 6.1% in 2018. The total contribution of travel and tourism to GDP is 4.3% of GDP in 2017, and is forecast to rise by 6.4% in 2018. Additionally, travel and tourism directly supported 1,178,500 jobs (1.8% of total employment) during 2017 which is expected to rise by 3.0% in 2018. The total contribution of travel and tourism to employment, including jobs indirectly supported by the industry, was 3.8% of total employment (2,432,000 jobs) in 2017 that is expected to rise by 2.5% in 2018. In Bangladesh, domestic travel spending generated 97.4% of direct travel and tourism GDP in 2017 compared with 2.6% for visitor exports (i.e. foreign visitor spending or international tourism receipts). Domestic travel spending is expected to grow by 6.3% in 2018 to BDT725.9 billion and visitor exports are expected to grow by 6.3% in 2018 to BDT19.5 billion (World Travel & Tourism Council [WTTC], 2018).

In addition, Bangladesh's visitor arrivals have dropped 73.9% in December 2018 as compared to an increase of 23.6% in 2017 (CEIC Data, 2018). On the other hand, foreign tourist arrivals in Bangladesh have increased over the last five years. In 2014, the number of foreign tourists is about 0.16 million which slightly reduced in 2015 reaching 0.14 million, before increasing again to 0.20 million in 2016, about 0.26 million in 2017, around 0.27 million in 2018 and about 0.20 million up to July in 2019 (Imam, 2019).

Research methods

Sample design

The sample for this study is the international tourists and domestic tourists, who stayed in Chattogram in Bangladesh and visited Cox's Bazar, Saint Martin, Dhaka and Sylhet from July to December 2019.

Data collection

In the study, apart from using primary data to conduct the research, a number of articles and textbooks have been reviewed to find out related variables and existing models on the perceived risk of tourism and various information have been supplemented through the browsing of related web pages on the internet. Then a printed survey questionnaire was prepared on the basis of reviewed literature to collect primary data where such data were collected by direct personal visit to the respondents.

Survey instrument

Questionnaire survey method is administered among the respondents by providing an 83-item questionnaire developed by the researchers on the basis of reviewed literatures to gather primary data where 12 questions are about demographic characteristics of the respondents, 54 questions are about the perceived risks of tourism in Bangladesh and the remaining 17 questions were about the strategies to reduce risks associated with tourism. The respondents were asked to rank each of 71 items on a 5-point Likert scale (5 = Strongly Agree . . . 1 = Strongly Disagree).

Mode of data analysis

The study used Statistical Product and Service Solutions (SPSS) version 22.0 and AMOS 23 software to analyse the data. Five main statistical tools were employed in the analysis, namely exploratory factor analysis (principal component analysis), confirmatory factor analysis, binary logistic regression, one-way analysis of variance (ANOVA) and ordinal logistic regression.

A sophisticated method of statistics, factor analysis (principal component analysis) by varimax rotation method, is used in this present study to obtain interpretable dimensions of perceived risk of tourism in Bangladesh. Here, researchers have followed the initial factor matrices to varimax rotation procedures to provide orthogonal common factors (Kaiser, 1974). Finally, dimensions regarding the perceived risk of tourists are made on the basis of factors scores. After that, the first-order confirmatory factor analysis (CFA) is performed in the overall dataset through the maximum likelihood (ML) estimate to measure the validity of a theoretical construct (Byrne and Gavin, 1996). Then, CFA results are evaluated to confirm the undimensionality and reliability of each contract. The model fit indicators evaluated are CMIN/DF, RMSEA, SRMR, GFI, AGFI, CFI, PCLOSE and HOELTER.

To discuss the model fit of SEM, the study considers the criteria of the various model fit indices as follows:

Further, to assess the determinants of tourists' perceived risk, a binary logistic regression function is used as such tool has the ability to accept independent variables of varying measurement levels (Pallant, 2005). Additionally, binary logistic regression is an appropriate tool for categorical dependent variables in a binary

Table 11.1 Criteria of model fit indices

Model Fit Indices	Description	Criteria	Source
CMIN/DF	Relative Chi-square value	< 3	Hu and Bentler, 1999
GFI	Goodness-of-Fit	≥ 0.90 (Depends on the sample size) ≥ 0.80 (Marginal)	Mulaik et al., 1989 Chandra et al., 2018
AGFI	Adjusted Goodness-of-Fit	≥ 0.90 (Depends on the sample size) ≥ 0.80 (Marginal)	Mulaik et al., 1989 Chandra et al., 2018
CFI	Comparative Fit Index	≥ 0.90 (Very Good Fit) ≥ 0.80 (Satisfactory) ≥ 0.75 (Fair Fitting Model)	Konovsky and Pugh, 1994, p. 662; Du Plessis, 2010; Moolla and Bisschoff, 2013, p. 9
RMSEA	Root Mean Square Error of Approximation	< 0.08 (Good Fit) 0.08–0.10 (Mediocre Fit)	MacCallum et al., 1996
SRMR	Standardized Root Mean Square Residual	< 0.05 (Well Fit Model) < 0.08 (Deemed Acceptable)	Byrne, 1998; Diamantopoulos and Siguaw, 2000; Hu and Bentler, 1999
HOELTER	Hoelter's Index	Critical Sample Size > 75 at p value 0.05 and 0.01	Arbuckle, 2012; Newsom, 2005

Source: Data compiled by the authors, 2020

format (Pallant, 2005). In using binary regression, the tourists' perceived risk dimensions are turned into one (by summing the scores and dividing them by the number of items making up the dimension). Then, the response to the dimension is recorded into a binary function where "strongly agree" and "somewhat agree", indicating the presence of perceived risk, are coded as one (1) and "strongly disagree" and "somewhat disagree", indicating the absence of perceived risk, are coded as zero (0) (Adam, 2015). The risk dimension is regressed against a set of independent variables to determine which factors influence tourists' perceived risks. Further, ANOVA is used to assess the differences in risk reduction strategies across tourists' perceived risks dimensions. Lastly, ordinal logistic regression is used in the study as the dependent variables i.e. risk reduction strategies are categorical (Likert scale) and the value of each category has a significant chronological order and each value is certainly higher than the previous one.

Identifications of variables that influence tourists' perceived risks in Bangladesh

The variables influencing tourist's risk perception in Bangladesh are presented in Table 11.2.

Table 11.2 List of variables that represent tourists' perceived risks in Bangladesh

Tourists' Perceived Risks	Variables
It will result in physical danger or injury.	X1
I may experience or witness violence.	X2
It is not safe for me.	X3
I may become sick from food or water.	X4
There is a possibility of contracting infectious diseases.	X5
There is possibility of risk of unhygienic surroundings at public places.	X6
There is possibility of risk of ill hygiene and cleanliness at hotels.	X7
There is possibility of risk of wrong medication.	X8
Potential health problems are a concern at the tourist spots.	X9
It will not provide value for the money spent.	X10
There is a possibility of charging additionally for visiting attractions.	X11
I worried that the trip would involve unexpected extra expenses (such as extra costs in hotels).	X12
I worried that the trip would involve more incidental expenses than I had anticipated, such as clothing, maps, sports equipment.	X13
I worried that the trip would have an impact on my financial situation.	X14
I would rather spend money on purchases at home.	X15
It will negatively affect others' opinion of me.	X16
Friends and relatives will disapprove my vacation in different tourist spots of Bangladesh.	X17
I want a vacation in different tourist spots of Bangladesh because that is where everyone goes.	X18
It is too time consuming.	X19

(*Continued*)

Table 11.2 (Continued)

Tourists' Perceived Risks	Variables
It will be a waste of time.	X20
Your transport for travel may be delayed.	X21
It will not reflect my personality.	X22
I'll experience inconvenience of telecommunication facilities.	X23
My baggage may be misplaced or delayed (by the transport or hotel).	X24
It may be a disappointment considering everything that can go wrong during the vacation.	X25
It is likely to enhance my feeling of well-being.	X26
I feel the government is committed in promoting destination's positive image.	X27
I feel the policies/regulation of the government at the vacation spot are favourable for tourists.	X28
It may result in mechanical or equipment problems.	X29
The thought of vacationing at tourist spots of Bangladesh will give me a feeling of unwanted anxiety.	X30
The thought of vacationing at tourist spots of Bangladesh will make me feel uncomfortable.	X31
The thought of vacationing at tourist spots of Bangladesh will cause me to experience unnecessary tension.	X32
Tourist spots of Bangladesh are avoided by tourists because of its political instability.	X33
I would not let political instability keep me from vacationing at tourist spots of Bangladesh.	X34
There are risks of strikes at the tourist spots.	X35
There are risks of curfews at the tourist spots.	X36
There are risks of local violence at the tourist spots.	X37
There is a risk of sexual harassment/rape at the tourist spots of Bangladesh.	X38
There is a risk of kidnapping at the tourist spots of Bangladesh.	X39
There is a risk of murder at the tourist spots of Bangladesh.	X40
I would like to vacation at the tourist spots of Bangladesh but negative news about various tourist spots discourages me from it.	X41
Travellers at the tourist spots of Bangladesh have a high probability of being targeted by terrorists.	X42
I'll not be intimidated by terrorism when vacationing at the tourist spots of Bangladesh.	X43
Terrorism will not influence my vacation at the tourist spots of Bangladesh.	X44
It is important that people who I meet while vacationing at the tourist spots of Bangladesh speak both Bangla and English.	X45
I have concerns about having possible communication problems due to local language when vacationing at the tourist spots of Bangladesh.	X46
I will not have problems in communication with others whom I meet during my vacation at the tourist spots of Bangladesh.	X47
At the tourist spots, there is a possibility of:	
Being sold under quality, pirated and duplicate products.	X48
Being misled by false advertisements.	X49
Being compelled by the guide to shop at local market.	X50
Entering the restricted sites by anyone without permission.	X51
Road/rail/air accidents.	X52
Earthquakes.	X53
Landslides.	X54

Source: Data compiled by the authors, 2020

Table 11.3 Variables that represent tourists' risk reduction strategies in Bangladesh

Risk Reduction Strategy	Variables
Travel in the company of friends	X56
Travel in the company of other nationals	X57
Make use of travel intermediaries	X58
Avoid places crowded by locals	X59
Make use of local guides	X60
Dress like locals	X61
Avoid public transport when alone	X62
Seek advice from police	X63
Seek advice from local tourist board	X64
Gather information from travel agencies	X65
Search for information from friends and relatives and make decisions in cooperation with relatives and friends	X66
Search for information on the Internet	X67
Watch television programmes about tourist places where I want to visit	X68
Rely on information from theWorld Tourism Organization (WTO)	X69
Read articles about tourist places where I want to visit	X70
Consult with people who have previously visited the destination	X71
Choose a popular destination	X72

Source: Data compiled by the authors, 2020

Identifications of variables that represent tourists' risk reduction strategies in Bangladesh

The variables that represent tourists' risk reduction strategies in Bangladesh are presented in Table 11.3.

Data analysis and findings

This section is discussed under the following heads.

Analysis and findings of sample respondents

The total number of respondents involved in the interviews is 320, of whom 60.9% are males and the remaining 39.1% are females. In terms of age, 64.4% respondents lie within the age group of 18–30 years. This is followed by 31–40 years (30.0%), 41–50 years (3.4%), less than 18 (1.3%) and above 50 years (0.9 %) respectively. Most of the respondents are Muslims (82.5%) which is followed by Hindus (13.1%), Christians (2.2%), Buddhists (1.6%) and atheists (0.6%), respectively. Among respondents, 45.9% have bachelor degrees, 45.0% have post-graduate degrees, 6.6% have completed high school and the remaining 2% have master's degrees. Regarding relationships, 60.9% respondents are single and the remaining 39.1% are married. In terms of occupation, 45.9% respondents are students, 40.9% are private jobholders, 6.9% are businessmen, 4.7% are house-wives and the remaining 1.6% are private jobholders. Among the sample, 6.3% are international tourists and 93.7% are domestic tourists. Most of the respondents

(48.4%) are visiting the tourist place at the time of visiting their friends and relatives while about 45.9% are visiting tourist spots during holiday and the remaining 5.6% are visiting tourist spots for other purposes such as honeymoon, study
tour and so on. Most of the respondents (44.7%) stay the tourist places for two
days. Most of the respondents (60%) have visited tourist places in Bangladesh
before while about 40% are visiting for the first time. Additionally, 47.5% are
visiting with friends and have mostly organized their trip by themselves (69.1%)
without using travel intermediaries. Most of the respondents (43.3%) have trav-
elled with a budget of less than or equal to Tk. 10,000 (Table 11.4).

Table 11.4 Demographic analysis

Demographic Analysis		
Background Variable	*Frequency*	*Percent*
Sex		
Male	195	60.9
Female	125	39.1
Age (Years)		
<18	4	1.3
18–30	206	64.4
31–40	96	30.0
41–50	11	3.4
>50	3	.9
Religion		
Christianity	7	2.2
Islam	264	82.5
Buddhism	5	1.6
Hinduism	42	13.1
Atheistic	2	.6
Marital status		
Single	195	60.9
Married	125	39.1
Level of education		
High School	21	6.6
University/College	147	45.9
Post Graduate	144	45.0
Others	8	2.5
Occupation		
Student	147	45.9
Businessman	22	6.9
Public Servant	5	1.6
Private Jobholder	131	40.9
Housewife	15	4.7
Tourist type		
International	20	6.3
Domestic	300	93.7

Demographic Analysis

Background Variable	Frequency	Percent
Purpose of visit		
Holiday	147	45.9
Visit to Friends and Relatives	155	48.4
Others	18	5.6
Length of stay		
One Day	98	30.6
Two Days	143	44.7
More Than Two Days	79	24.7
Repeat visit to the same place		
First Time Visitor	128	40.0
Visited Before	192	60.0
Travel party to tourist spots of Bangladesh		
Alone	23	7.2
Friends	152	47.5
Family	137	42.8
Relatives	8	2.5
Travel arrangements		
Self-arranged	221	69.1
Use Intermediaries	99	30.9
Travel budget (Tk.)		
<= 10000	139	43.4
10001–20000	73	22.8
20001–30000	36	11.3
30001–40000	10	3.1
40001–50000	20	6.3
>50000	42	13.1

Source: Data compiled by the authors, 2020

Appropriateness of data for factor analysis

Kaiser-Meyer-Olkin (KMO) is a method of representing the appropriateness of data for factor analysis to show sampling adequacy. The value of KMO statistics varies between 0 and 1. Kaiser (1974) argued that values greater than 0.5 are acceptable. Again, Bartlett's test of sphericity (Bartlett, 1950) is another statistical tool used in the study to verify its appropriateness. This test will be significant with a value less than 0.5. The values of KMO and Bartlett's test for adequacy of the sample of the study have been presented in Table 11.5.

Table 11.5 shows that in this study, the value of KMO is 0.770 which is greater than 0.5, indicating that the sample taken to process the factor analysis is acceptable. Besides, the significance value of Bartlett's test of sphericity is also less than 0.5, indicating that the data is appropriate for the factor analysis.

After testing appropriateness of data, the next step has been carried out to show factor analysis to simplify the diverse relationship that exists among a set of observed variables.

Table 11.5 KMO and Bartlett's test

KMO and Bartlett's Test		
Kaiser-Meyer-Olkin Measure of Sampling Adequacy		0.770
Bartlett's Test of Sphericity	Approx. Chi-Square	10711.095
	df	1326
	Sig.	0.000

Source: Data compiled by the authors, 2020

Factor analysis

For factor analysis, principal component analysis (PCA) followed by varimax rotation is performed. It is required to mention here that factors loading greater than 0.3 are considered significant, 0.4 are considered more important and 0.5 or greater are considered very significant (Hair et al., 2003). For parsimony, only those factors with loading above 0.5 are considered significant (Hair et al., 2003; Pal and Bagai, 1987; and Hair et al., 1998). Communality values indicate the proportion of the variance in the response to the variables that are explained by the identified factors. Moreover, a factor's eigenvalue, which is the sum of the squares of its factor loading, has also been computed hereafter to indicate how well each factor fits the data from all of the respondents on all of the variables.

Determinants of perceived risks of tourism in Bangladesh

In order to find out the determinants of tourists' perceived risk in Bangladesh, 54 reasons are taken into consideration. In this case, the Bartlett's test of sphericity and KMO of sampling adequacy are applied to evaluate the adequacy of the survey data for factor analysis. Then, a minimum 0.30 value of communality of each item has been considered for further analysis in principle components analysis. As a result, two items, X10 and X43, have been eliminated from the analysis. Therefore, a principal component factor analysis with varimax rotation is performed for the remaining 52 items/reasons of which four items, X24, X31, X50 and X54, have been deleted due to have factor loading less than 0.40 and 48 items with factor loading greater than 0.40 have produced a clear structure of factor. Here, six-factor results are emerged from the output with eigenvalues greater than 1 (Table 11.6). The total variance of 51.520% found afterwards indicates that the six-factor solutions explain 51.520% of the total variance where the first factor explains the most and is about 12.454%; the second, third, fourth, fifth and sixth factors explain 11.725%, 10.659%, 6.044%, 5.322% and 5.315%, respectively. These six factors of tourists' perceived risks along with their ordered variables have been named and their results are shown in Table 11.6.

Table 11.6 shows that the name of six factors of tourists' perceived risks are financial and communication risk, political instability and natural risk, health and time risk, physical risk, social risk and psychological risk where the results of Cronbach's alpha coefficients of the six dimensions are 0.883, 0.870, 0.853, 0.663, 0.607 and 0.604, respectively.

Table 11.6 Principle component analysis with rotated component matrix and communalities

Dimensions	Variables	Component						Communalities	Cronbach's Alpha
		1	2	3	4	5	6		
Financial and Communi- cation Risk	X15	.721						.639	0.883
	X27	.710						.54	
	X18	.676						.552	
	X28	.668						.605	
	X13	.645						.529	
	X11	.640						.545	
	X46	.622						.599	
	X47	.621						.545	
	X34	.468						.466	
	X44	.462						.468	
	X45	.441						.582	
	X49	.439						.574	
	X32	.427						.414	
	X36	.400						.576	
Political Instability and Natural Risk	X39		.822					.737	0.87
	X40		.783					.635	
	X38		.758					.672	
	X37		.746					.665	
	X41		.620					.513	
	X42		.607					.386	
	X33		.513					.394	
	X52		.512					.574	
	X35		.474					.569	
	X53		.442					.550	
	X48		.402					.403	
Health and Time Risk	X6			.715				.585	0.853
	X8			.698				.578	
	X4			.635				.567	
	X19			.602				.416	
	X7			.597				.556	
	X9			.583				.471	
	X14			.570				.542	
	X21			.562				.543	
	X25			.478				.459	
	X12			.449				.346	
Physical Risk	X2				.703			.649	0.663
	X3				.667			.466	
	X1				.513			.349	
	X5				.477			.532	
Social Risk	X26					.678		.625	0.607
	X30					.569		.533	
	X16					.444		.432	
	X20					.439		.428	
Psychological Risk	X22						.631	.460	0.604
	X17						.538	.421	
	X23						.445	.451	
	X29						.437	.446	
	X51						.434	.477	
Eigenvalues		10.585	6.175	3.220	2.827	2.087	1.898		
% of Variance		12.454	11.725	10.659	6.044	5.322	5.315		
Cumulative %		12.454	24.179	34.838	40.882	46.204	51.520		

Source: Data compiled by the authors, 2020

Result of various dimensions of perceived risk

The relationship between each latent variable of perceived risk is displayed in Figure 11.1.

From Table 11.6, the main dimension of financial and communication risk is government committed in promoting the destination's positive image (0.715), while the main dimension of political instability and natural risk is risk of kidnapping at the tourist spots of Bangladesh (0.862). Furthermore, the main dimension of health and time risk is risk of ill hygiene and cleanliness at hotels (0.753), the main dimension of physical risk is experience or witnessing of violence (0.697) and the main dimension of social risk is the thought of vacationing at tourist spots of Bangladesh giving a feeling of unwanted anxiety (0.628). Finally, the main dimension of psychological risk is not reflecting personality (0.668). All

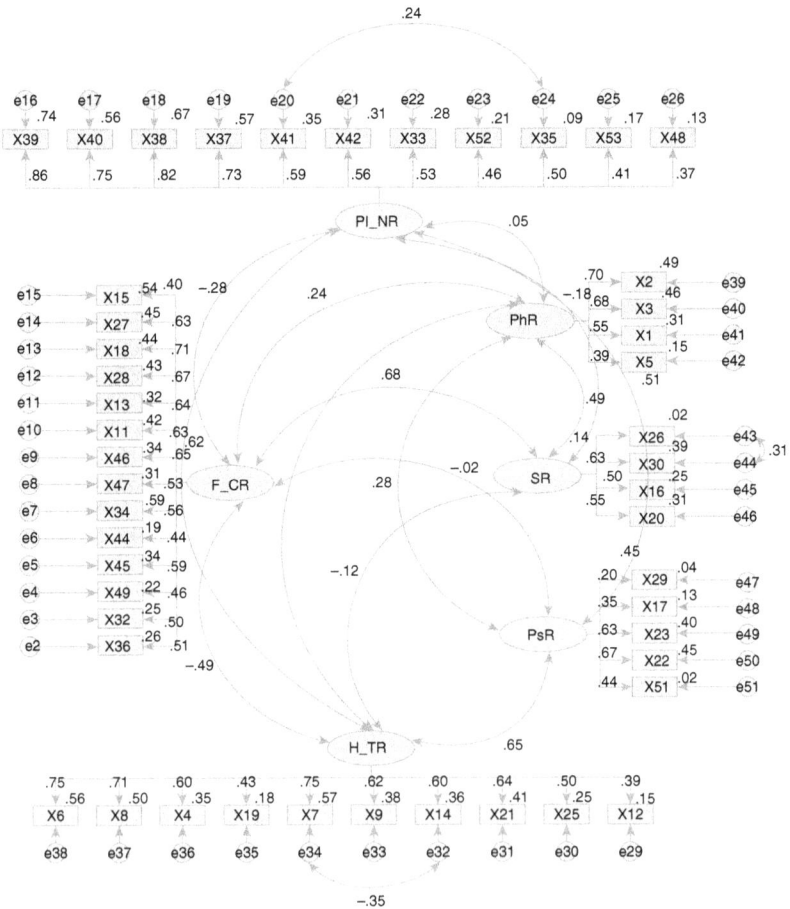

Figure 11.1 Path diagram of tourists' perceived risk

Source: Data compiled by the authors, 2020

dimensions affect each variable significantly (Table 11.3). The results of model fit indices in the study reports that CMIN/DF = 2.267, CFI= 0.868, GFI=0.853, AGFI=0.81, RMSEA = 0.075, SRMR = 0.0631, Hoelter's N returns value at 5% significant level = 119 and Hoelter's N returns value at 1% significant level = 127 are found in the model. Based on overall indices, this sample has an acceptable fit to the model as chi-square, CMIN/DF, CFI, RMSEA, RMR and SRMR lie in the acceptable ranges. Although there are some indicators that do not meet the criteria goodness of fit, overall the model has met the criteria of goodness of fit (Meesala and Paul, 2016) as CMIN/DF, CFI, RMSEA, SRMR and Hoelter's N return values lie in the acceptable ranges (Table 11.7).

Table 11.7 Result of regression weight

	Unstandardized Estimate	Standardized Estimate	S.E.	C.R.	P
X36 <--- F_CR (Financial and Communication Risk)	1	0.512			
X32 <--- F_CR (Financial and Communication Risk)	0.887	0.504	0.126	7.032	***
X49 <--- F_CR (Financial and Communication Risk)	0.765	0.464	0.115	6.641	***
X45 <--- F_CR (Financial and Communication Risk)	1.19	0.586	0.154	7.75	***
X44 <--- F_CR (Financial and Communication Risk)	0.873	0.438	0.137	6.372	***
X34 <--- F_CR (Financial and Communication Risk)	1.036	0.558	0.138	7.522	***
X47 <--- F_CR (Financial and Communication Risk)	1.164	0.587	0.15	7.76	***
X46 <--- F_CR (Financial and Communication Risk)	1.213	0.65	0.147	8.236	***
X11 <--- F_CR (Financial and Communication Risk)	1.269	0.623	0.158	8.039	***
X13 <--- F_CR (Financial and Communication Risk)	1.187	0.635	0.146	8.124	***
X28 <--- F_CR (Financial and Communication Risk)	1.327	0.641	0.162	8.169	***
X18 <--- F_CR (Financial and Communication Risk)	1.415	0.67	0.169	8.37	***
X27 <--- F_CR (Financial and Communication Risk)	1.479	0.715	0.171	8.66	***
X15 <--- F_CR (Financial and Communication Risk)	1.125	0.635	0.138	8.124	***
X39 <--- PI_NR (Political Instability and Natural Risk)	1	0.862			
X40 <--- PI_NR (Political Instability and Natural Risk)	0.837	0.747	0.054	15.548	***
X38 <--- PI_NR (Political Instability and Natural Risk)	0.932	0.821	0.052	17.941	***

(*Continued*)

Table 11.7 (Continued)

	Unstandardized Estimate	Standardized Estimate	S.E.	C.R.	P
X37 <--- PI_NR (Political Instability and Natural Risk)	0.84	0.755	0.053	15.787	***
X41 <--- PI_NR (Political Instability and Natural Risk)	0.646	0.588	0.057	11.251	***
X42 <--- PI_NR (Political Instability and Natural Risk)	0.602	0.556	0.057	10.495	***
X33 <--- PI_NR (Political Instability and Natural Risk)	0.695	0.526	0.071	9.812	***
X52 <--- PI_NR (Political Instability and Natural Risk)	0.499	0.459	0.06	8.36	***
X35 <--- PI_NR (Political Instability and Natural Risk)	0.337	0.303	0.064	5.301	***
X53 <--- PI_NR (Political Instability and Natural Risk)	0.401	0.414	0.054	7.447	***
X48 <--- PI_NR (Political Instability and Natural Risk)	0.368	0.365	0.057	6.489	***
X12 <--- H_TR (Health and Time Risk)	1	0.39			
X25 <--- H_TR (Health and Time Risk)	1.386	0.505	0.239	5.797	***
X21 <--- H_TR (Health and Time Risk)	1.719	0.643	0.269	6.388	***
X14 <--- H_TR (Health and Time Risk)	1.648	0.603	0.266	6.205	***
X9 <--- H_TR (Health and Time Risk)	1.741	0.618	0.276	6.3	***
X7 <--- H_TR (Health and Time Risk)	2.301	0.753	0.344	6.684	***
X19 <--- H_TR (Health and Time Risk)	1.186	0.425	0.223	5.317	***
X4 <--- H_TR (Health and Time Risk)	1.619	0.596	0.261	6.214	***
X8 <--- H_TR (Health and Time Risk)	1.996	0.706	0.303	6.583	***
X6 <--- H_TR (Health and Time Risk)	2.07	0.746	0.31	6.689	***
X2 <--- PhR (Physical Risk)	1	0.697			
X3 <--- PhR (Physical Risk)	0.904	0.68	0.11	8.225	***
X1 <--- PhR (Physical Risk)	0.714	0.554	0.096	7.456	***
X5 <--- PhR (Physical Risk)	0.463	0.389	0.083	5.598	***
X26 <--- SR (Social Risk)	1	0.137			
X30 <--- SR (Social Risk)	5.356	0.628	2.469	2.17	**
X16 <--- SR (Social Risk)	3.948	0.495	1.954	2.021	**
X20 <--- SR (Social Risk)	4.468	0.554	2.196	2.035	**
X29 <--- PsR (Psychological Risk)	1	0.197			
X17 <--- PsR (Psychological Risk)	1.606	0.355	0.582	2.758	***
X23 <--- PsR (Psychological Risk)	2.684	0.635	0.887	3.024	***
X22 <--- PsR (Psychological Risk)	2.912	0.668	0.959	3.036	***
X51 <--- PsR (Psychological Risk)	0.767	0.142	0.422	1.817	*

Source: Data compiled by the authors, 2020

Notes: *, **, *** Significant at alpha 10 %, 5 %, and 1 % respectively, S.E: Standard Error, CR: Critical Ratio, CP: Constant Parameter

Results of binary logistic regression

The binary logistic regression model is a good predictor of tourists' perceived risk as indicated by the omnibus tests of model coefficients and the Hosmer and Lemeshow test. To be a good predictor, the alpha values of the omnibus tests of model coefficients need to be less than 0.05 and the Hosmer and Lemeshow test has to be greater than 0.05 (Pallant, 2005). The study is reliable as the alpha values of both tests fulfilled the requirement as shown in Table 11.8.

The model predicted 47.7% (Nagelkerke R square) and 35.7% (Cox and Snell R square) of tourists' perceived risk in Bangladesh as shown by the pseudo R2 values. However, despite the significance of the model in predicting tourists' perceived risk in Bangladesh, not all the predictor variables are significant in predicting tourists' perceived risk. Six out of the 13 predictor variables are found to be significant to the model. Age, type of tourist, purpose of visit and travel budget emerge as the most significant predictors of tourists' perceived risk in Bangladesh (Table 11.9).

Under age, tourists within the 31–40 age group are more likely to associate Bangladesh with risk compared to the age group greater than 50. This is

Table 11.8 Results of model fit

		Chi-square	df	Sig.
Omnibus Tests of Model Coefficients	Step 1	141.551	32	.000
	Block	141.551	32	.000
	Model	141.551	32	.000
Hosmer and Lemeshow Test	Step 1	5.259	8	.730

Source: Data compiled by the authors, 2020

Table 11.9 Results of binary logistic regression

Variables	B	S.E.	Sig.	Exp(B)
Sex				
Male	–.431	.423	0.024	.650
Female (RC)				1
Age (Years)				
<18	–17.160	19557.676	.067*	1.323
18–30	–17.089	19557.676	0.016**	1.474
31–40	–19.679	19557.676	0.044**	1.609
41–50	–15.472	19557.676	0.075*	0.054
>50 (RC)				1
Religion				
Christianity	–3.282	29430.374	0.355	.038
Islam	.158	29430.374	0.010**	1.171
Buddhism	–42.755	33289.279	0.217	.000
Hinduism	.137	29430.374	0.023**	1.147
Atheistic (RC)				1
Marital Status				
Single	–1.063	.749	.156	.345
Married (RC)				1

(*Continued*)

Table 11.9 (Continued)

Variables	B	S.E.	Sig.	Exp(B)
Level of Education				
High School	–.676	2.284	.767	.509
University/College	–.307	1.974	.877	.736
Post Graduate	–.600	1.968	.760	.549
Others (RC)				1
Occupation				
Student	–1.225	1.611	0.012**	.294
Businessman	–.254	1.772	.886	.776
Public Servant	17.311	17974.843	.999	3.772
Private Jobholder	–1.483	1.563	0.005*	.227
Housewife (RC)				1
Tourist Type				
International	–2.446	2.025	.045**	.087
Domestic (RC)				1
Purpose of Visit				
Holiday	.500	.841	0.003***	1.649
Visit to Friends and Relatives	.783	.836	0.001***	2.187
Others (RC)				1
Length of Stay				
One Day	–.333	.509	.513	.717
Less Than Three Days	.033	.480	.945	1.034
More Than Three Days (RC)				1
Repeat Visit to the Same Place				
First Time Visitor	–.027	.326	.935	.974
Visited Before (RC)				1
Travel Party to Tourist Spots of Bangladesh				
Alone	21.980	9003.413	.998	9.676
Friends	.029	1.138	.980	1.029
Family	1.016	1.180	.389	2.762
Relatives (RC)				1
Travel Arrangements				
Self-arranged	.078	.374	.834	1.082
Use Intermediaries (RC)				1
Travel Budget (Tk.)				
<= 10000	3.307	1.218	.007***	27.300
10001–20000	3.894	1.217	.001***	49.114
20001–30000	4.084	1.265	.001***	59.353
30001–40000	2.983	1.423	.036**	19.745
40001–50000	4.677	1.373	.001***	107.454
>50000 (RC)				1
Constant	16.135	35336.535	1.000	15.686

Source: Data compiled by the authors, 2020

Notes: *, **, *** Significant at alpha 10%, 5%, and 1% respectively, RC: Reference Category

followed by tourists' age groups within 18–30, less than 18 and 41–50. Under religion, Muslim tourists are more likely to associate risk with Bangladesh than atheist counterparts. This is followed by Hindu tourists. Under occupation, students are 0.294 times less likely to risk as compare to housewives, which is followed by private jobholders. Under tourist type, international tourists are 0.087 times less likely to associate risk as compared to domestic travellers in Bangladesh. Tourists who are travelling during holidays are 1.649 times more likely to associate risk than those who are travelling for other purposes such as honeymoon, physical treatment and so on. This is followed by tourists who are travelling to visit to meet their friends and relatives. Tourists who have tourist budget Tk. 40,001–50,000 are 107.454 times more likely to associate Bangladesh with risk as compared those have tourist budget greater than Tk. 50,000. This is followed by tourists having budget Tk. 20,001–30,000, Tk. 10,001–20,000, less than or equal to Tk. 10,000 and Tk. 30,001–40,000, respectively (Table 11.9).

Descriptive statistics of risk reduction strategies

On the whole, 17 risk reduction strategies are considered (Table 11.10). The most used risk reduction strategy is to consult with people who had previously visited the destination as indicated by a mean score of 4.2000. This was followed by

Table 11.10 Descriptive statistics of risk reduction strategies

	Descriptive Statistics		
Variables	*Risk Reduction Strategy*	*Mean*	*Std. Deviation*
RRS1	Travel in the company of friends	3.8188	1.19230
RRS2	Travel in the company of other nationals	2.9063	1.26092
RRS3	Make use of travel intermediaries	3.2688	1.12382
RRS4	Avoid places crowded by locals	3.6219	1.17343
RRS5	Make use of local guides	3.6063	1.19349
RRS6	Dress like locals	2.5906	1.40238
RRS7	Avoid public transport when alone	3.4938	1.46857
RRS8	Seek advice from police	4.0969	1.16371
RRS9	Seek advice from local tourist boards	4.0281	1.20442
RRS10	Gather information from travel agencies	3.6563	1.12284
RRS11	Search for information from friends and relatives and make decisions in cooperation with relatives and friends	3.4406	1.26302
RRS12	Search for information on the Internet	3.6281	1.21478
RRS13	Watch television programmes about tourist places where I want to visit	2.7750	1.28873
RRS14	Rely on information referred by WTO	3.9031	1.14743
RRS15	Read articles about tourist places where I want to visit	3.1406	1.33517
RRS16	Consult with people who had previously visited the destination	4.2000	1.00344
RRS17	Choose a popular destination	3.8500	1.26788

Source: Data compiled by the authors, 2020

seek advice from police (4.0969), seek advice from local tourist boards (4.0281), rely on information referred by WTO (3.9031), choose a popular destination (3.8500), travel in the company of friends (3.8188), gather information from travel agencies (3.6563), search for information on the Internet (3.6281), avoid places crowded by locals (3.6219), make use of local guides (3.6063), avoid public transport when alone (3.4938), search for information from friends and relatives and make decisions in cooperation with relatives and friends (3.4406), make use of travel intermediaries (3.2688)and read articles about tourist places where I want to visit (3.1406). The strategies that are not preferred by the sample tourists are travel in the company of other nationals (2.9063), watch television programmes about tourist places where I want to visit (2.7750) and dress like locals (2.5906).

Perceived risk by risk reduction strategies

Now the central tendency as to mean of the data and their dispersion as to standard deviation have been calculated along with one-way ANOVA and are shown in Table 11.11. The table reveals that the average results of financial and communication risk, political instability and natural risk, health and time risk, physical risk, social risk and psychological risk are 3.1049, 3.4688, 3.2875, 3.0406, 2.718 and 2.9438, respectively, with standard deviation of 0.79253, 0.786, 0.82529, 0.55028, 0.66155 and 0.72056, respectively. This shows that most of the respondents of the commercial banks in Bangladesh are satisfied with the first four dimensions and unsatisfied with the last two dimensions. However, the results of one-way ANOVA explain the variation in risk reduction strategies across the various perceived risk dimensions at different significant levels (Table 11.11). The results report that statistically significant differences exist between the various dimensions of perceived risk and type of risk reduction strategy except the variation between physical risk dimension and RRS12, i.e. search for information on the Internet. Tourists with the perceived risks dimensions want to use almost all types of risk reduction strategy discussed in Table 11.11 to keep safe and enjoy their tour.

Results of ordinal regression analysis

The results of regression analysis help us to understand how dependent variables are affected by various independent variables included in the 17 types of risk reduction strategy. For this independent variable ordinal regression is performed separately. Regression results presented six tourists' perceived risks dimensions and 17 types of risk reduction strategy variables where tourists' perceived risk dimensions have both a positive and negative relationship with the types of risk reduction strategies but only positive relationships are statistically significant (Table 11.12). The model also shows global significance according to the likelihood ratio and pseudo R2 measures that provide indication of model explanatory power.

Table 11.11 Risk reduction strategies by perceived risk

Dimensions	Mean	Std. Deviation	ANOVA Analysis																
			RRS1	RRS2	RRS3	RRS4	RRS5	RRS6	RRS7	RRS8	RRS9	RRS10	RRS11	RRS12	RRS13	RRS14	RRS15	RRS16	RRS17
Financial and Communication Risk	3.1049	0.79253	11.573***	6.325***	7.852***	6.886***	5.896***	6.821***	5.701***	4.189***	5.179***	8.342***	8.061***	5.879***	6.529***	3.776***	3.597***	5.419***	4.572***
Political Instability and Natural Risk	3.4688	0.786	7.323***	5.529***	3.847***	6.362***	4.521***	4.1***	3.429***	6.103***	5.995***	6.034***	4.885***	2.064***	9.779***	4.528***	6.287***	5.781***	2.967***
Health and Time Risk	3.2875	0.82529	9.501***	5.469***	6.23***	5.283***	6.374***	5.598***	4.842***	6.119***	5.476***	6.402***	7.53***	3.675***	7.567***	4.08***	4.09***	4.032***	4.819***
Physical Risk	3.0406	0.55028	2.762***	3.899***	4.152***	4.583***	3.89***	3.887***	3.527***	4.245***	2.914***	4.591***	9.405***	0.906	2.064***	3.41***	5.007***	2.094***	3.708***
Social Risk	2.718	0.66155	6.559***	6.622***	10.455***	4.975***	9.152***	8.582***	4.104***	7.41***	11.111***	3.329***	7.295***	1.576*	4.912***	7.527***	3.472***	7.405***	2.08**
Psychological Risk	2.9438	0.72056	7.846***	10.907***	8.65***	5.379***	4.075***	5.754***	3.755***	4.282***	4.204***	4.057***	15.381***	2.153***	4.412***	3.838***	3.93***	2.311***	4.111***

Source: Data compiled by the authors, 2020

Notes: * , ** , *** Significant at alpha 10%, 5% and 1%, respectively

Table 11.12 Results of ordinal regression model

Threshold	Parameter Estimates of Dependent Variable	RRS1 (OR1)	RRS2 (OR2)	RRS3 (OR3)	RRS4 (OR4)	RRS5 (OR5)	RRS6 (OR6)	RRS7 (OR7)
1	Estimate	-.364	1.928**	.989	-3.141***	-5.555***	2.459***	-3.209*⋯
	Std. Error	.886	.835	.858	.884	.927	.855	.8⋯
	Wald	.169	5.331	1.329	12.612	35.921	8.279	12.7⋯
2	Estimate	1.704**	3.346***	3.275***	-.915	-4.618***	3.359***	-2.171⋯
	Std. Error	.867	.849	.851	.847	.914	.863	.8⋯
	Wald	3.864	15.543	14.815	1.166	25.536	15.161	5.9⋯
3	Estimate	3.076***	4.590***	4.677***	.251	-3.545***	4.448***	-1.65⋯
	Std. Error	.882	.866	.871	.851	.905	.875	.8⋯
	Wald	12.169	28.120	28.807	.087	15.357	25.855	3.4⋯
4	Estimate	4.901***	6.554***	6.520***	1.709**	-.970	5.681***	-.4⋯
	Std. Error	.907	.897	.903	.856	.888	.894	.8⋯
	Wald	29.233	53.358	52.186	3.984	1.192	40.420	.2⋯
Dimensions	Parameter Estimates of Independent Variable							
Financial and Communication Risk	Estimate	1.807***	.773***	.737***	1.037***	.788***	.580***	1.082*
	Std. Error	.201	.174	.174	.184	.196	.186	.2⋯
	Wald	80.941	19.737	17.938	31.641	16.220	9.738	24.0⋯
	Exponential Estimate Values	6.094	2.167	2.089	.355	.455	1.786	.3⋯
	Odd Ratio (%)	509.384	116.690	108.930	64.547	54.517	78.617	66.1⋯
Political Instability and Natural Risk	Estimate	.289*	-.146	.229	.970***	.792***	-.079	.460*
	Std. Error	.158	.148	.149	.158	.163	.150	.1⋯
	Wald	3.334	.969	2.376	37.856	23.520	.279	9.4⋯
	Exponential Estimate Values	1.335	.864	1.258	2.639	2.207	.924	1.5⋯
	Odd Ratio (%)	33.513	-13.582	25.759	163.891	120.690	-7.599	58.4⋯
Health and Time Risk	Estimate	.386**	-.166	.539***	-.100	.603***	.404**	.618⋯
	Std. Error	.176	.163	.165	.166	.177	.167	.1⋯
	Wald	4.826	1.037	10.629	.361	11.604	5.842	12.2⋯
	Exponential Estimate Values	.680	.847	.583	.905	1.828	.668	1.8⋯
	Odd Ratio (%)	32.034	-15.308	41.661	-9.515	82.845	33.246	85.4⋯
Physical Risk	Estimate	-.051	-.318	.681***	.185	-.330	-.148	-.3⋯
	Std. Error	.213	.207	.210	.209	.218	.210	.2⋯
	Wald	.056	2.355	10.563	.785	2.296	.498	2.0⋯
	Exponential Estimate Values	.951	.727	1.977	1.204	.719	.862	.7⋯
	Odd Ratio (%)	-4.928	-27.255	97.653	20.376	-28.115	-13.792	-26.3⋯
Social Risk	Estimate	.646***	.413**	.026	.250	-.276	1.105***	.36⋯
	Std. Error	.204	.192	.191	.191	.197	.198	.1⋯
	Wald	10.019	4.656	.019	1.721	1.960	31.163	3.7⋯
	Exponential Estimate Values	.524	1.512	1.026	1.284	.759	3.019	.6⋯
	Odd Ratio (%)	47.567	51.170	2.644	28.434	-24.087	201.860	30.6⋯
Psychological Risk	Estimate	.177	.844**	.340**	-.128	1.059***	.202	.2⋯
	Std. Error	.179	.172	.168	.167	.181	.165	.1⋯
	Wald	.980	23.917	4.086	.588	34.352	1.501	1.7⋯
	Exponential Estimate Values	1.1934608	2.3246518	1.4051936	0.8798185	0.3469037	1.2241651	1.24461⋯
	Odd Ratio (%)	19.346082	132.46518	40.51936	-12.018152	65.309629	22.416509	24.4612⋯
Model Fitting Information	Likelihood Ratio (χ^2)	142.320***	97.936***	82.099***	86.465***	154.354***	97.207***	101.346
Pseudo R-Square	R^2 Cox and Snell	.359	.264	.226	.237	.383	.262	.2⋯
	R^2 Nagelkerke	.382	.276	.238	.250	.405	.274	.2⋯
	R^2 McFadden	.157	.098	.086	.092	.167	.098	.1⋯

Source: Data compiled by the authors, 2020

Notes: *, **, *** Significant at alpha 10%, 5%, and 1%, respectively, OR: ordinal regression result

RRS8 (OR8)	RRS9 (OR9)	RRS10 (OR10)	RRS11 (OR11)	RRS12 (OR12)	RRS13 (OR13)	RRS14 (OR14)	RRS15 (OR15)	RRS16 (OR16)	RRS17 (OR17)
−1.418	−.980	−3.041***	−14.306***	−2.976***	−13.259***	1.398***	−7.964***	−3.367***	1.771***
.910	.904	.888	1.189	.852	1.133	.880	.964	.936	.870
2.431	1.174	11.719	144.703	12.200	136.908	2.523	68.186	12.939	4.139
−.928	.033	−1.768	−12.285***	−1.974*	−11.343***	2.394***	−6.480***	−3.150***	2.404***
.903	.893	.869	1.122	.838	1.073	.870	.933	.930	.868
1.05	.001	4.143	119.953	5.552	111.811	7.578	48.218	11.477	7.670
.181	.760	−.224	−10.443***	−1.038	−10.038***	3.097***	−5.509***	−1.589*	3.552***
.902	.894	.871	1.068	.832	1.033	.872	.914	.912	.877
.040	.722	.066	95.698	1.555	94.479	12.605	36.339	3.036	16.408
1.562	2.222	1.596*	−8.528***	.639	−7.694***	5.041***	−3.605***	.214	4.840***
.910	.906	.880	1.012	.831	.978	.902	.886	.914	.896
2.946	6.016	3.288	71.018	.591	61.934	31.227	16.538	.055	29.211
−.220	−.065	1.223***	1.303***	.731***	1.678***	.925***	.433***	−.231	.855***
.205	.200	.202	.212	.173	.196	.181	.170	.206	.182
1.147	.105	36.493	37.933	17.800	73.521	26.142	6.490	1.251	21.954
.803	.937	.294	.272	2.077	.187	2.522	.649	.794	2.350
−19.741	−6.295	70.569	72.833	107.709	81.329	152.184	35.149	−20.617	135.027
.345**	.368**	.475***	.157	−.234	1.785***	.232	1.857***	.631***	−.206
.153	.153	.151	.169	.149	.182	.152	.191	.164	.150
5.105	5.806	9.915	.858	2.458	96.536	2.322	94.570	14.899	1.879
1.412	1.444	1.608	1.170	.792	.168	1.261	.156	1.880	.814
41.248	44.439	60.819	16.990	−20.827	83.219	26.052	84.384	88.017	−18.638
.955***	.912***	.431***	.560***	.492***	−.102	.617***	.949***	.527***	.937***
.181	.177	.170	.188	.165	.170	.170	.179	.183	.175
27.939	26.609	6.386	8.907	8.909	.361	13.184	28.171	8.302	28.611
2.599	2.489	1.539	1.751	1.635	.903	1.853	2.583	1.693	2.552
159.875	148.867	53.860	75.067	63.498	−9.697	85.256	158.316	69.334	155.208
.615***	.002	.431**	.724***	.359***	.519***	.089	−.138	−.236	.018
.225	.219	.215	.223	.207	.217	.212	.209	.222	.214
7.475	.000	4.014	10.552	3.011	5.695	.176	.436	1.135	.007
.540	1.002	1.539	.485	.698	1.680	1.093	.871	.790	1.018
45.962	.220	53.867	51.521	30.160	68.006	9.303	−12.898	−21.045	1.840
−.271	.551***	−.313	.012	.671***	−.150	.459**	.351*	.930***	.829***
.197	.197	.192	.201	.194	.200	.198	.195	.204	.202
1.890	7.810	2.658	.004	12.030	.560	5.351	3.231	20.807	16.802
.763	.576	.732	1.012	.511	.861	.632	.704	.394	.437
−23.712	42.369	−26.843	1.226	48.888	−13.891	36.813	29.595	60.556	56.350
.185	−.148	.236	−2.191	−.202	−.173	−.123	.100	.057	.545***
.175	.172	.169	.213	.167	.176	.172	.173	.176	.176
1.119	.739	1.964	105.394	1.473	.975	.508	.330	.103	9.552
1.202929	0.8623785	1.2665176	0.111826	0.8167281	0.8408451	0.8843713	1.1046498	1.0582934	1.7245144
0.292902	−13.76215	26.651764	−88.817405	−18.327193	−15.91549	−11.562866	10.464978	5.8293442	72.45144
8.417***	69.789***	119.801***	275.634***	27.987***	208.430***	39.246***	130.920***	71.937***	57.026***
.192	.196	.312	.577	.084	.479	.115	.336	.201	.163
.209	.211	.331	.605	.088	.501	.124	.351	.223	.174
.085	.083	.129	.280	.030	.209	.046	.130	.096	.063

The table shows that the tourists who perceived financial and communication risk want to follow almost all types of risk reduction strategies except RRS8, RRS9 and RRS16. Tourists with political instability and natural risk dimension resort to the use of RRS1, RRS4, RRS5, RRS7, RRS8, RRS9, RRS10, RRS13, RRS15 and RRS16. Tourists who associated Bangladesh with health and time risk use 14 risk reduction strategies: RRS1, RRS3, RRS5, RRS6, RRS7, RRS8, RRS9, RRS10, RRS11, RRS12, RRS14, RRS15, RRS16 and RRS17. Those who perceived Bangladesh to be associated with physical risk use RRS3, RRS8, RRS10, RRS11, RRS12 and RRS13 risk reduction strategies. Tourists who perceived social risk in Bangladesh use RRS1, RRS2, RRS6, RRS7, RRS9, RRS12, RRS14, RRS15, RRS16 and RRS17 as risk reduction strategies. Tourists who associated Bangladesh with psychological risk use RRS2, RRS3, RRS5 and RRS17 risk reduction strategies. In the model, the exponentiated coefficient offers an idea of magnitude marginal effects on the probability of observing higher categories for risk reduction strategies against lower categories. For instance, a one-unit increase in the financial and communication risk would increase the odds 509.384% that tourists follow RRS1, while all other variables in the model are held constant under ordinal regression result one (OR1) (Table 11.12).

Conclusion and implications

This chapter examined the perceived risk of tourists in Bangladesh as well as the determinants of the perceived risk. The chapter also determined and assessed the risk reduction strategies used by tourists in Bangladesh. To conduct the study, primary data was collected from 320 tourists of whom 6.3% are international tourists and the remaining 93.7% are domestic tourists who stayed in Chattogram in Bangladesh and visited Cox's Bazar, Saint Martin, Dhaka and Sylhet from July to December 2019.

The study found six dimensions of tourists' perceived risk on Bangladesh, namely financial and communication risk, political instability and natural risk, health and time risk, physical risk, social risk and psychological risk where these six factor explain 12.454%, 11.725%, 10.659%, 6.044%, 5.322% and 5.315%, respectively and 51.520% in total of the total variance. The main dimension of financial and communication risk is government committed in promoting destination's positive image, while the main dimension of political instability and natural risk is risk of kidnapping at the tourist spots of Bangladesh. Furthermore, the main dimension of health and time risk is risk of ill hygiene and cleanliness at hotels, the main dimension of physical risk is experience or witnessing of violence and the main dimension of social risk is the thought of vacationing at tourist spots of Bangladesh giving a feeling of unwanted anxiety. Finally, the main dimension of psychological risk is not reflecting personality. In addition, the study results report that a good number of individual factors affect perception of travel risk. Age, type of tourist, purpose of visit and travel budget emerge as the most significant predictors of tourists' perceived risk on Bangladesh, suggesting that demographic and travel characteristics of tourists are relevant in understanding tourists' risk perceptions. Additionally, the study also shows that tourists follow both consumption behaviour modification and information search as risk reduction strategies. The study also revealed that the usage of risk reduction strategies of tourists relied on the type of risk perceived. Thus, tourism planners should pay attention to which risks might cause stress among tourists, an awareness that should also inform marketing strategies.

The study has indicated that financial and communication risk is a highly perceived risk among tourists, implying that improvement in service provision as well as standardization of services at reasonable costs and good communication facilities will help to allay such risk. Subsequently, the Ministry of Tourism, Bangladesh Parjatan Corporation and hotel industry in Bangladesh should educate and train employees in tourists' facilities on quality service delivery. Additionally, the study revealed that tourists are a heterogeneous group since certain demographic and travel characteristics such as age, type of tourist, purpose of visit and travel budget influenced their perceptions of risk on Bangladesh. This means that further research is required on the dynamic features of tourists especially in relation to risk perceptions. Although this study has developed the risk reduction strategies used by tourists, issues such as socio-demographic characteristics, past travel experience and so on that influence their risk reduction strategies largely remain unexplored. In this regard, future studies on tourists should delve into the risk reduction strategies. As another limitation, qualitative data such the top travel concerns is not considered in the study. In this regard, open-ended questions might be helpful to report which are the top travel concerns.

References

Adam, I. (2015). Backpackers' risk perceptions and risk reduction strategies in Ghana. *Tourism Management*, 49(2015), pp. 99–108.

Arbuckle, J. L. (2012). *Users guide. IBM® SPSS® Amos™ 21*. Retrieved from: ftp://public.dhe.ibm.com/software/analytics/spss/documentation/amos/21.0/en/Manuals/IBM_SPSS_Amos_Users_Guide.pdf (accessed: the 27th December 2019).

Assael, H. (1995). *Consumer behaviour and marketing action*. Chicago, IL: South-Western College Publishing.

Bartlett, M. S. (1950). Tests of significance in factor analysis. *British Journal of Statistical Psychology*, 3, pp. 77–85.

Bauer, R. (1960). Consumer behaviour as risk-taking. In R. S. Hancock (ed.), *Dynamic marketing for a changing world*. Chicago, IL: American Marketing Association, pp. 389–398.

Boshoff, C. (2002). Service advertising: An exploratory study of risk perceptions. *Journal of Service Research*, 4, pp. 290–298.

Burton, I., Kates, R. and White, G. (1978). *The environment at hazard*. Oxford: Oxford University Press.

Byrne, B. M. (1998). *Structural equation modeling with LISREL, PRELIS and SIMPLIS: Basic concepts. Applications and programming*. Mahwah, NJ: Lawrence Erlbaum Associates.

Byrne, B. M. and Gavin, D. A. (1996). The Shavelson Model revisited: Testing for the structure of academic self-concept across pre-, early, and late adolescents. *Journal of Educational Psychology*, 88(2), pp. 215–228.

Byzalov, D. and Shachar, R. (2004). The risk reduction role of advertising. *Quantitative Marketing and Economics*, 2(4), pp. 283–289.

Carter, S. (1998). Tourists' and travellers' social construction of Africa and Asia as risky locations. *Tourism Management*, 19(4), pp. 349–358.

Chandra, T., Ng, M., Chandra, S. and Priyono. (2018). The effect of service quality on student satisfaction and student loyalty: An empirical study. *Journal of Social Studies Education Research*, 9(3), pp. 109–131.

Cui, F., Liu, Y., Chang, Y., Duan, J. and Li, J. (2016). An overview of tourism risk perception. *Natural Hazards*, 82(1), pp. 643–658.

Dayour, F., Park, S. and Kimbu, A. N. (2019). Backpackers' perceived risks towards smartphone usage and risk reduction strategies: A mixed methods study. *Tourism Management*, 72, pp. 52–68.

Dholakia, U. (2001). A motivational process model of product involvement and consumer risk perception. *European Journal of Marketing*, 35(11/12), pp. 1340–1360.

Diamantopoulos, A. and Siguaw, J. A. (2000). *Introducing LISREL*. London: Sage Publications.

Dolnicar, S. (2005). Understanding barriers to leisure travel: Tourist fears as a marketing basis. *Journal of Vacation Marketing*, 11(3), pp. 197–208.

Douglas, M. and Wildavsky, A. (1982). *Risk and culture*. Berkeley, CA: University of California Press.

Dowling, G. R. and Staelin, R. (1994). A model of perceived risk and intended risk-handling activity. *Journal of Consumer Research*, 21, pp. 119–134.

Du Plessis, J. (2010). *Statistical consultation services*. Potchefstroom: North-West University.

Engel, J. F., Blackwell, R. D. and Miniard, P. W. (1995). *Consumer behaviour*. Chicago, IL: Dryden Press.

Fennell, D. A. (2017). Towards a model of travel fear. *Annals of Tourism Research*, 66, pp. 140–150.

Fuchs, G. and Reichel, A. (2006). Tourist Destination risk perception: The case of Israel. *Journal of Hospitality & Leisure Marketing*, 14(2), pp. 83–108.

Garg, A. (2013). A study of tourist perception towards travel risk factors in tourist decision making. *Asian Journal of Tourism and Hospitality Research*, 7(1), pp. 47–57.

Hair, J. F., Anderson, R. E., Tatham, R. L. and Black, W. C. (1998). *Multivariate data analysis*. Upper Saddle River, NJ: Prentice Hall.

Hair, J. F., Anderson, R. E., Tatham, R. L. and Black, W. C. (2003). *Multivariate data analysis*. New Delhi: Pearson Education.

Horton, R. L. (1976). The structure of perceived risk. *Journal of the Academy of Marketing Science*, 4, pp. 694–706.

Hu, L. T. and Bentler, P. M. (1999). Cutoff criteria for fit indexes in covariance structure analysis: Conventional criteria versus new alternatives. *Structural Equation Modeling*, 6(1), pp. 1–55.

Imam, S. H. (2019). Tourists arrivals rise in five years. *Financial Express Report*. Dhaka: the 26th September 2019.

Jacoby, J. and Kaplan, L. (1972). The components of perceived risk. *Proceedings of the Third Annual Conference of the Association for Consumer Research*, pp. 382–393.

Kahn, B. and Sarin, R. (1988). Modeling ambiguity in decisions under uncertainty. *Journal of Consumer Research*, 15(2), pp. 265–271.

Kaiser, H.F. (1974). An index of factorial simplicity. *Psycometrica*, 39, pp. 31–36.

Konovsky, M. A. and Pugh, S. D. (1994). Citizen behaviour and social change. *Academy of Management Journal*, 37(3), pp. 656–669.

Kozak, M., Crotts, J. C. and Law, R. (2007). The impact of the perception of risk on international travellers. *International Journal of Tourism Research*, 9(4), pp. 233–242.

Laroche, M., McDougall, J., Bergeron, J. and Yang, Z. (2004). Exploring how intangibility affects perceived risk. *Journal of Service Research*, 6(4), pp. 373–389.

Larsen, S., Brun, W., Torvald, O. and Selstad, L. (2007). Subjective food-risk judgments in tourists. *Tourism Management*, 28(6), pp. 1555–1559.

Law, R. (2006). The perceived impact of risks on travel decisions. *International Journal of Tourism Research*, 8(4), pp. 289–300.

Lepp, A. and Gibson, H. (2008). Sensation seeking and tourism: Tourist role, perception of risk and destination choice. *Tourism Management*, 29(4), pp. 740–750.

MacCallum, R. C., Browne, M. W. and Sugawara, H., M. (1996). Power analysis and determination of sample size for covariance structure modeling. *Psychological Methods*, 1(2), pp. 130–149.

Meesala, A. and Paul, J. (2016). Service quality, consumer satisfaction and loyalty in hospitals: Thinking for the future. *Journal of Retailing and Consumer Services*, 40, pp. 261–269.

Mitchell, V. W. (1993). Factors affecting consumer risk reduction: A review of current evidence. *Management Research News*, 16(9), pp. 6–21.

Mitchell, V. W. (1998). A role for consumer risk perceptions in grocery retailing. *British Food Journal*, 100(4), pp. 171–183.

Mitchell, V. W., Davies, F., Moutinho, L. and Vassos, V. (1999). Using neural networks to understand service risk in the holiday product. *Journal of Business Research*, 46(2), pp. 167–181.

Mitchell, V. W. and Vassos, V. (1997). Perceived risk and risk reduction in holiday purchases: A cross-cultural and gender analysis. *Journal of Euromarketing*, 6(3), pp. 47–97.

Moolla, A. I. and Bisschoff, C. A. (2013). An empirical model that measures brand loyalty of fast-moving consumer goods. *Journal of Economics*, 4(1), pp. 1–9.

Moreira, P. (2007). Stealth risks and catastrophic risks: On risk perception and crisis recovery strategies. *Journal of Travel & Tourism Marketing*, 23(2/3/4), pp. 15–27.

Mowen, J. and Minor, M. (1998). *Consumer behaviour*. Englewood Cliffs, NJ: Prentice-Hall.

Mulaik, S. A., James, L. R., Van Alstine, J., Bennet, N., Lind, S. and Stilwell, C.D. (1989). Evaluation of goodness-of-fit indices for structural equation models. *Psychological Bulletin*, 105(3), pp. 430–445.

Murray, K. and Schlacter, J. (1990). The impact of services versus goods on consumers' assessment of perceived risk and variability. *Journal of the Academy of Marketing Science*, 18(1), pp. 51–65.

Newsom, M. (2005). *Some clarifications and recommendations on fit indices*. Retrieved from: www.google.co.za/#q=ecvi+CFA+model+fit+interpretation (accessed: the 27th December 2019).

Pal, Y. and Bagai, O. P. (1987). A common factor better reliability approach to determine the number of interpretable. Paper presented at the *IX Annual Conference of the Indian Society for Probability and Statistics*. New Delhi: University of Delhi.

Pallant, J. (2005). *SPSS survival manual: A step by step guide to using SPSS for windows (version 12)*. New South Wales: Allen & Unwin.

Parrey, S. H., Hakim, I. A. and Rather, R. A. (2018). Mediating role of government initiatives and media influence between perceived risks and destination image: A study of conflict zone. *International Journal of Tourism Cities*, 5(1), pp. 90–106.

Promsivapallop, P. and Kannaovakun, P. (2017). A comparative assessment of destination image, travel risk perceptions and travel intention by young travellers across three ASEAN countries: A study of German students. *Asia Pacific Journal of Tourism Research*, 22(6), pp. 634–650.

Promsivapallop, P. and Kannaovakun, P. (2018). Travel risk dimensions, personal-related factors, and intention to visit a destination: A study of young educated German adults. *Asia Pacific Journal of Tourism Research*, 23(7), pp. 639–655.

Quintal, V. A., Lee, J. A. and Soutar, G. N. (2010). Risk, uncertainty and the theory of planned behaviour: A tourism example. *Tourism Management*, 31(6), pp. 797–805.

Reichel, A., Fuchs, G. and Uriely, N. (2007). Perceived risk and the noninstitutionalized tourist role: The case of Israeli student ex-backpackers. *Journal of Travel Research*, 46(2), pp. 217–226.

Reisinger, Y. and Mavondo, F. (2005). Travel anxiety and intentions to travel internationally: Implications of travel risk perception. *Journal of Travel Research*, 43(3), pp. 212–225.

Reisinger, Y. and Mavondo, F. (2006). Cultural differences in travel risk perception. *Journal of Travel & Tourism Marketing*, 20(1), pp. 13–31.

Rittichainuwat, B., Nelson, R. and Rahmafitria, F. (2018). Applying the perceived probability of risk and bias toward optimism: Implications for travel decisions in the face of natural disasters. *Tourism Management*, 66, pp. 221–232.

Roehl, W. S. and Fesenmaier, D. R. (1992). Risk perception and pleasure travel: An exploratory analysis. *Journal of Travel Research*, 30(4), pp. 17–26.

Roselius, T. (1971). Consumer rankings of risk reduction methods. *Journal of Marketing*, 35(1), pp. 56–61.

Schiffman, L. G. and Kanuk, L. L. (2007). *Consumer behaviour*. Englewood Cliffs, NJ: Prentice-Hall.

Sonmez, S. and Graefe, A. (1998a). Determining future travel behaviour from past travel experience and perceptions of risk and safety. *Journal of Travel Research*, 37(2), pp. 171–177.

Sonmez, S. and Graefe, A. (1998b). Influence of terrorism risk on foreign tourism decisions. *Annals of Tourism Research*, 25(1), pp. 112–144.

Tan, S. J. (1999). Strategies for reducing consumers' risk aversion in Internet shopping. *The Journal of Consumer Marketing*, 16(2), pp. 163–180.

Tavitiyaman, P. and Qu, H. (2013). Destination image and behaviour intention of travellers to Thailand: The moderating effect of perceived risk. *Journal of Travel & Tourism Marketing*, 30(3), pp. 169–185.

Tideswell, C. and Faulkner, B. (1999). Multidestination travel patterns of international visitors to Queensland. *Journal of Travel Research*, 37(4), pp. 364–374.

Um, S., Chon, K. and Ro, Y. (2006). Antecedents of revisit intention. *Annals of Tourism Research*, 33(4), pp. 1141–1158.

UNWTO (2003). *Tourism highlights: Edition 2003*. Madrid: UNWTO.

World Travel &Tourism Council (2018). *Annual report, 2018: The economic impact of travel& tourism, 2018*. London: WTTC.

Part 7

Tourism marketing and human resource management

12 A conceptual study on human resource compensation practices in Bangladesh

Md Yusuf Hossein Khan, Md Zahid Al Mamun and Azizul Hassan

Introduction

The potential use of skills, knowledge and competencies of employees in an organization would help to improve organizational performance. Certainly, the necessity of strategic human resource management in a business cannot be undermined as human resource management practices and policies influence work, attitudes and performance of employees. Having said that, human resource policies are focused on many essential practices, and in turn can positively impact organizational performance, such as human resource planning, recruitment, selection, training and development, compensation, performance management and employee relations (Khan, 2018). Since gaining independence in 1971 Bangladesh has been progressing gradually towards her dream of a hunger- and poverty-free society. Until 1990, the government of Bangladesh adopted a socialist economic model. Therefore, the government and policymakers did not prioritize issues such as private sector development, industrialization, globalization, competitiveness and human resource management (HRM).

It is noteworthy that Bangladesh shifted its economic policies from socialism to a free market economy in the early 1990s and achieved commendable economic growth which has an industry contribution to GDP of about 28.5% (ILO, 2013; PwC, 2015). If this growth continues without interruption Bangladesh is predicted to become the world's 23rd largest economy in terms of Purchasing Power Parity (PPT) by 2050 and has been included by Goldman Sachs in the N-11 countries (Chowdhury and Mahmood, 2012; PwC, 2015). To become a middle-income country by 2021, given the country's limited natural resources and abundance of human resources, the efficiency and efficacy of HRM practices could be pivotal and driving forces for Bangladesh's economic development (Absar et al., 2014). The main aspect of the study is to provide an overview of the stature of research conducted so far on HRM and compensation practices in Bangladesh. The present research has four distinct objectives: first, to study the current compensation systems and HRM practices in Bangladesh; second, to accumulate the studies conducted so far on HR compensation practices in context of Bangladesh; third, to identify the research gaps with respect to compensation practices in Bangladesh; and fourth, to recommend some directions for conducting future research on HR compensation practices in context of Bangladesh.

Importance of compensation in achieving organizational competitive advantages

According to McDonald and Smith (1995), companies that manage reward systems successfully outperform companies that do not, with higher profits, better cash flows, stronger stock market preference, productivity gains, higher sales growth per employee and overall better financial performance. Competitiveness has become the norm in these challenging and turbulent times. Noe et al. (2003, p. 4) defines competitiveness as "a company's ability to maintain and gain market share in its industry". Achieving competitive advantage has become the rule in a market saturated with players. With competition being a "key element in any analysis of the specific or task environment of the organization," competitive advantage "refers to something that an organization does extremely well, a core competency that clearly sets it apart from competitors and gives it an advantage over them in the marketplace" (Schermerhorn, 2010, p. 66). How companies are able to attain a certain level of performance and organizational effectiveness becomes a measure of organizational competitiveness. Organizational effectiveness encompasses all measures indicating satisfaction of shareholders' interests: acceptable returns for stockholders; products or services of value for the customers; equitably compensated human and motivating work for the employees; and environment-friendly and ethical business practices for the society (Noe et al., 2010a). We assume the extremely crucial role of human resources in achieving organizational competitiveness; human resource management has evolved to become one of the strategic means in bringing competitive advantage to a firm (Huselid, 1995 ; Huselid et al., 1997; Noe et al., 2010b) from merely regarding it as an administrative function. Compensation is defined by Mondy (2010, pp. 268–269) as the "total of all rewards provided to employees in return for their services", the overall purpose of which is to attract, retain and motivate employees. When organizational behaviour modification interventions have been systematically applied over the years using both financial and non-financial rewards, it was found that performance increased by an average of 17% (Luthans and Stajkovic, 1999). It must be noted that employees have different perceptions on various types of rewards in terms of ability to motivate. For instance, non-monetary rewards significantly influence an employee's willingness to engage in extra-task performance (Chiang and Birtch, 2008).

A statistically significant and positive relationship was found to govern rewards and motivation, implying that if rewards being offered to employees were to be altered, then there would be a corresponding change in satisfaction and work motivation while the period salary increments, allowances, bonuses, fringe benefits and other compensations on regular and specific periods keep their morale high and make them more motivated (Danish and Usman, 2010).

Another total compensation model being used by more managers and academics alike is total rewards strategy. This model was developed by Milkovich, Newman, and Gerhart (2011) in order to integrate all the elements of rewards that has monetary value, employee learning and development opportunities, quality of the work environment and other employee benefits and privileges.

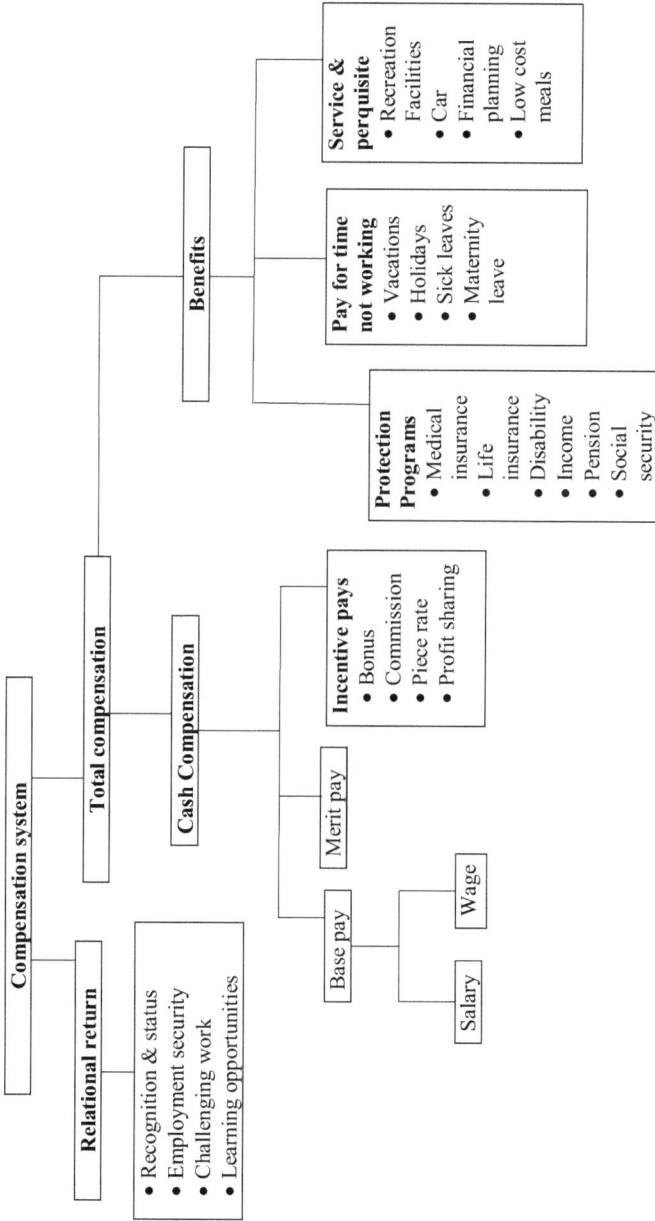

Figure 12.1 Theoretical compensation system
Source: Milkovich, Newman and Gerhart, 2011.

Organizations would attain significant returns with the appropriate use of the integrated total rewards strategy.

Methodology

This chapter applies the research design methodology based on the review of publications aiming at its analysis from the field of HRM with the use of computer search. Because there is no database or digital warehouse for information, the researchers collected different studies on HRM practices in Bangladesh physically from different libraries, universities, bookstores and corporate offices, and from the web. The authors also used information published by national and international organizations to assess the current situation and future challenges with regard to HRM in Bangladesh. Finally, to collect relevant information of HRM practices in Bangladesh, researchers used different websites of public and private sector organizations. After reviewing the literature, both theoretical and empirical studies were taken into consideration. In total, 37 studies were collected for the review. The following keywords were applied: reward, reward strategy, reward policy, compensation strategy, pay policy, incentives and motivators. By analysing and comparing certain information included in the selected articles from the HR field with the results of an analysis of reward system conducted from the perspective of HRM, significant conclusions were formulated. Due to the wide scope of the reward issues the current chapter is devoted to one main problem: analysis of the selected HRM publications available to the authors referring to the reward problems.

Emerging trends of HRM practices in Bangladesh: a critical assessment

Nowadays, it is proven that the human resource is the most important resource out of the four basic resources of an organization: human capital, physical structure, financial and information resources. Most organizations are family owned and controlled by family members, and HRM activities tend to be viewed as just a company owner's wish. In reality, with few exceptions most of the private organizations have no proper HR department, hence there is no recognized structure with regards to base pay, rewards, termination policy, recruitment and selection and so on. Researchers have indicated that improved working conditions and better wage rates (rewards) could improve the productivity and profitability of organizations in Bangladesh (Ahmed and Peerlings, 2009).

With the help from the World Bank, the Bangladesh government formed an agency called Bangladesh Institute of Management (BIM), which offers specialized degrees and diplomas in HRM and related areas. Following in its footsteps, most of the higher educational institutes such as public and private universities in Bangladesh now also offer bachelor of business administration (BBA) and master of business administration (MBA) degrees, with specialization in HRM. In addition to the educational institutions, two professional associations, the Institute of Personnel Management (IPM) and the Bangladesh Society for Human Resource

Management (BSHRM), have been formed by recognized HRM specialists to promote the HRM profession and development of HRM practices in Bangladesh. The contribution of BSHRM is widely recognized by the global HR community, and it has granted membership of Asia Pacific Federation of Human Resource Management (APFHRM) and the World Federation of People Management Association (WFPMA). Therefore, further developments in terms of HRM practices and recognition of HRM professions are expected in the near future.

The lack of comprehensive studies on HRM practices in Bangladesh makes it very difficult to detect overall scenarios, and therefore, recently, as a rare initiative, the Ernst and Young LLP and the Bangladesh Society for Human Resource Management (BSHRM) jointly conducted a survey on HRM practices in Bangladesh with a sample of 1,000 HRM managers from different sectors of the economy. The findings of this survey showed that more than 55% of the organizations do not have any defined employee reward programme. Also 66% of the organizations that represented fast-selling consumer goods, pharmaceutical sectors and telecommunications sectors reported short-term incentive programmes. Nonetheless, about 95% of the organizations do not provide any long-term incentives to retain employees which would be shocking in developed organizations in present time.

The highlights of the survey are presented in Figure 12.2 which could help to understand the present scenario of HRM in Bangladesh.

- About 33% of organizations used internal referral and word of mouth in the employee recruitment process. Use of social media and mobile applications are very limited (i.e. less than 10%) in the recruitment process.
- Around 40% of organizations use formal background or reference checks in the selection process.
- About one-third of the organizations indicated outsourcing recruitment processes sometimes in the past to recruitment managerial employees.
- More than 25% of organizations do not conduct training need analysis and do not provide training to employees on a regular basis.
- Performance appraisals indicators include mostly functional achievements rather than behavioural aspects of the jobs. However, IT and telecommunications sectors are leading to introduce structured performance appraisal systems and employee training need identification processes.
- More than 55% of the organizations do not have any defined employee reward programme. Also 66% of the organizations that represented fast-selling consumer goods, pharmaceutical sectors and telecommunications sectors reported short-term incentive programmes.
- About 95% of organizations do not provide long-term incentives to retain employees.
- About 75% of organizations have structured mechanisms to deal with employee grievances.
- About 30% of organizations reported to have preventive mechanisms to avoid employee conflicts.

Figure 12.2 Features of HRM in Bangladesh

Source: The Daily Star, 2014

HRM compensation perspective and practices in Bangladesh

In the field of HRM, the idea that pay policies have a strategic impact has become a major theme within the compensation literature since the mid-1980s, although a limited number of studies have addressed the idea of "strategic" reward systems (Boyd and Salamin, 2001). The strategic impact of pay policies has received little empirical attention (Montemayor, 1996). This strategic perspective on compensation is based on the fact that organizations differ in pay policies and the belief that matching pay policies to business strategy results in higher organizational performance. Effectiveness in strategy implementation at the corporate level depends significantly on the existence of a match between reward strategies and firms' strategies (Shaw, 2002). Pay policy is frequently mentioned in strategic compensation theory and includes first, compensation philosophy; second, external competitiveness; third, incentive-base mix; fourth, individual (merit) pay increases; and fifth, pay administration (Montemayor, 1996). Additionally, compensation decisions can be classified into four distinct areas of compensation policy (Gerhart and Milkovich, 1990). The first is the pay-level policy, which determines whether a firm will pay above, meet or pay below the market wage level (Milkovich and Newman, 1990). The second area is how firms make pay differentiation decisions at the individual level; how, for example, pay is related to performance or to organizational tenure. The third is the pay-structure policy, which governs the relationships between pay rendered at the various levels of an organization. Finally, the fourth area, the benefits policy, is the basis for how a firm provides employees with indirect financial compensation (Werner and Tosi, 1995).

Previous studies (see Figure 12.3) of compensation have typically emphasized only one component of pay. Yet, because pay level and pay structure are both essential characteristics of a compensation/reward system, it is important to consider them simultaneously in order to relate pay policy to organizational

POLICIES (Foundation on which pay system are built)	TECHNIQUES (Make-up of the system)	OBJECTIVES
INTERNAL ALIGNMENT	INTERNAL STRUCTURE	• EFFICIENCY • FAIRNESS • COMPLIANCE • ETHICS
COMPETITIVENESS	PAY STRUCTURE	
CONTRIBUTIONS	PAY FOR PERFORMANCE	
MANAGEMENT	EVALUATIONS	

Figure 12.3 Compensation model

Source: Milkovich, Newman and Gerhart, 2011

outcomes. Both theory and empirical research suggest that pay level and pay structure are each important for understanding the organization-level implications of pay policy (Bloom and Michel, 2002; Huselid et al., 1997).

The introduction of performance-related pay (PRP) is explicit in private sector organizations, mainly in multinationals and leading local organizations in Bangladesh (Absar et al., 2013). However, public sector organizations are still lagging behind in implementing such practices because of legal complexities. Most public sector organizations still consider seniority as the main criteria for pay and promotion. According to civil service employment regulation, entry-level seniority needs to be maintained throughout the service period. Employees who receive a minimum pass mark in the annual performance evaluation automatically qualify for a pre-specified annual salary increment, as determined in the national pay scale. However, there is no scope to reward any employee by more than this pre-specified increment, irrespective of their level of performance. In contrast, private sector organizations are concerned about productivity and employee motivation (Khan, 2013; Mia and Hossain, 2014).

Pay structure in public sector organizations

The Bangladeshi government created the National Pay Commission (NPC) to design the pay structure of public sector organizations and, in 1997, the NPC recommended a 20-grade pay structure for public sector employees, excluding workers in manufacturing organizations. Although the basis of this pay structure in not explicitly clear, the NPC adopted the historical categorization of employees used by the British government (Islington Commission, 1917, cited in Obaidullah, 1995). The Islington Commission proposed that there would be four broad categories of employees: Class I (officers/executives), Class II (junior officers), Class III (clerical/secretarial) and Class IV (custodial). For employees of the public sector organizations, the government formed the National Pay Commission (NPC) and the National Wages and Productivity Commission (NWPC) to formulate two separate pay structures. The NPC recommends the pay structure, salary and benefits for employees in the non-manufacturing sectors, while the NWPC recommends the wage structure for those in the manufacturing sector. Both commissions consider four main parameters when recommending wage and salary structures and benefits for employees: (1) the minimum wage should be adequate to provide for the basic needs of a worker's family, comprising three adult consumption units; (2) industrial wages should be higher than agricultural or rural wages, and the wage of official employees; (3) wages should be linked to productivity; and (4) the ability of the enterprises to pay should be considered. The formation of a separate wage structure for manufacturing workers was justified on the grounds that manufacturing workers at the lower grades perform more demanding manual/physical work than do lower-grade non-manufacturing employees. In addition, manufacturing requires more grades to provide promotional opportunities to the workers because of their limited upward mobility to become supervisors or managers (Sarker, 2006). The latest pay commission, that is, the Eighth National Pay and Service Commission (NPSC) was formed in June 2014 and the commission submitted its recommendations to the finance ministry in December 2014 (The Dhaka Tribune, 2014). This commission suggested pay raise up to 112.5% for different categories of public sector employees, compared to 94% of 2005 NSPC and 96% of 2009 NSPC. The suggestions for comparatively higher salary increase aim to recruit better qualified talents into public services and to minimize the pay gap between public and private sector employees. It is expected that the recommendations of Eighth NSPC, if implemented properly, will bring significant positive changes in the pay and performance of the employees of public sector organizations.

Pay structures in private sector organizations

The pay structure in large private sector organizations has some similarities with that in public sector organizations. Generally, there is one pay structure for managerial employees and another for non-managerial employees. In the absence of any legal obligations and very low state regulations, management unilaterally designs the pay structure for managerial employees. Almost no local private sector organizations follow any established pay structure for managerial employees. Usually, the head of the organization decides the salary of the managerial employees after discussing the jobs with departmental heads. Non-managerial employees' pay structure is determined through consultation and bargaining with collective bargaining agents (CBA), depending on the organizational HR policy and CBA management bargaining outcomes (Mia and Hossain, 2014). Therefore, formal private sector pays structures follow two main procedures. Sectors in which collective bargaining is absent, owing to the non-existence of effective trade unions, use minimum wage provisions, while unionized organizations use collective bargaining arrangements at the enterprise level. Owing to the level of private sector development, most local private sector organizations have not yet developed any formal or institutionalized pay structure for their employees. Except for a few large organizations, employees' pay or salary is determined arbitrarily (e.g. through personalized pay contracts), rather than any formal structure or grade (Chowdhury and Mahmood, 2012). In the absence of any legal or recognized grading system for private sector organizations, management usually divides jobs into job clusters for grading purposes. Local private sector organizations place entry-level jobs in different departments within the same grade (comparable educational background and subsequent practical experience), as well as their corresponding pay scales, if they already have these (Absar and Mahmood, 2010; Khan, 2013). On the other hand, pay and job grades in the private sector depend on individual employee competencies and bargaining with management, but there is no link to employee pay and fringe benefits. Sometimes, the identification of managerial or non-managerial posts depends on the opportunistic behaviour of management. For example, a number of private sector banks and manufacturing organizations have categorized non-managerial jobs as managerial posts to avoid unionization and to control different terms and conditions (Khan, 2013).

Future challenges of HR compensation practices in Bangladesh

As Bangladesh is moving from being a developing economy to an emerging economy, people's expectations and values are also changing. In the last 20 years, because of globalization and the IT revolution, people are better informed about the world, work and civic facilities of modern life. While previous generations worried about savings and job security, the new generation of employees are more concerned about relationships, work-life balance and meaning in their work. Women are joining the labour market in increasing numbers, and the number of working couples is increasing too. While women professionals are moving towards higher-level employment positions, they face a daunting challenge to maintain work-life balance and career development in traditional Bangladesh society (ILO, 2013). HR managers are facing a big challenge in adjusting to diversity management and inclusion issues, as these require changes in employment policies, especially compensation systems, such as those dealing with working hours, health and safety measures, statutory maternity leave, work-life balance, childcare facilities and so on (Bowden, 2014). As a signatory of International Labour Organization conventions, Bangladesh may need to implement laws and regulations to deal with these emerging trends, which is a daunting task in a traditional, conservative, male-dominated and hierarchical society. The legal context of Bangladesh is viewed by some as a barrier to introducing the latest HRM practices. For example, organizations cannot initiate any pay restructuring or retrenchment without the approval of trade unions. Most public sector organizations are overstaffed and incurring huge operating losses year after year. However, labour laws in Bangladesh make it very complicated for organizations to hire part-time employees or to change an employment contract from a full-time permanent position to a temporary or adjunct position. Such a rigid regulatory environment handicaps HRM managers who need to implement appropriate and effective practices in their organizations (Mahmood, 2008; Sarker, 2006). Many multinationals have been operating in Bangladesh for a long period of time, and their financial performance is better than that of the local competitors. Multinational corporations are role models in developing systematic HRM practices and leading local companies are trying to imitate their practices (Chowdhury and Mahmood, 2012). Therefore, without a proper compensation system organizations would not be able to acquire or retain high-skilled and talented employees to combat competitive challenges for different professions. Though organizations are introducing formal performance appraisal systems, their links with employee compensation systems and promotions are not widely visible. In addition, the global consulting firm's 2018 Bangladesh Total Remuneration Survey suggests the rise in salary will be 10% in the next year. The study also concludes that a strong and consistent GDP growth of 7% per year over the last three years has led to expanded employment opportunities and consequently competition for talent, which has further led to double digit increments, a positive hiring outlook and the need to manage increasing levels of attrition.

Kwak and Lee (2009) proposed a new strategic compensation model (see Figure 12.4) for the future benefits of the organizations to achieve sustainable

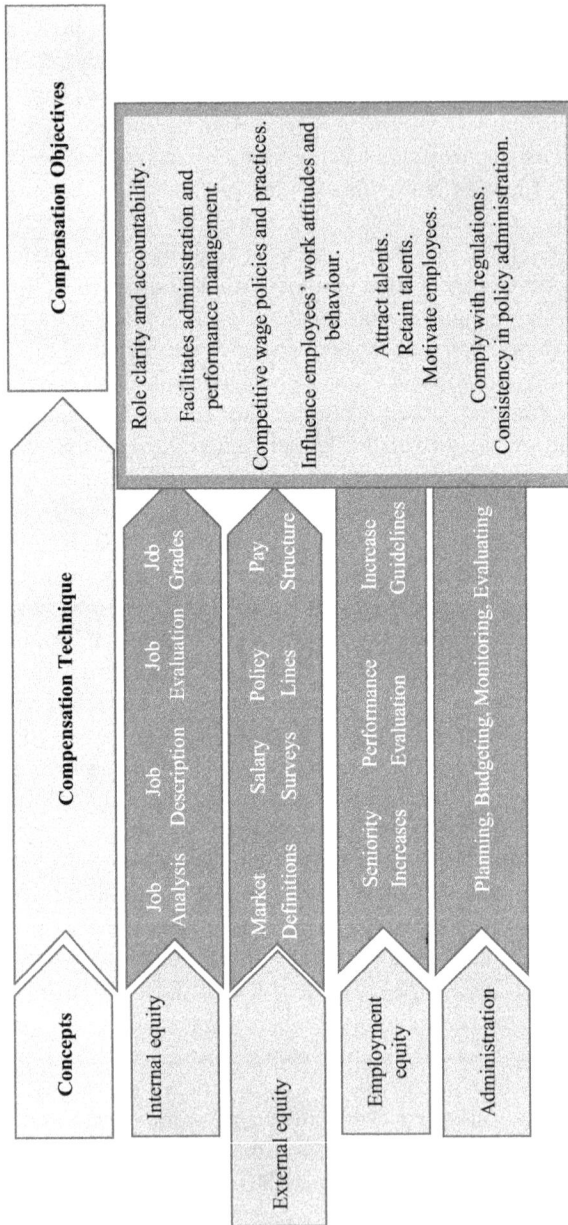

Figure 12.4 Strategic compensation model

Source: Kwak and Lee, 2009

competitive advantages. In this model he mentions internal equity, external equity and employee equity to facilitate administration and performance management, competitive wage policies and practices, influence employees' work attitudes and behaviour, attract and retain talents, motivate employees and so on.

In studies on rewards and job satisfaction (Galanou et al., 2011; Ghazi et al., 2010) a common premise is that when employees are satisfied, they feel a sense of fulfilment, achievement and joy in their jobs which are considered to be positive factors to employee productivity and creativity as well as organizational profitability. Empirical studies further support the positive relationship between employee benefits and performance which serves as proxy for organizational competitiveness.

In public sector organizations, HRM practices are still centralized and all practices respond to government directives. Private sector organizations are moving towards more strategic HRM practices, which is a positive sign for the future economic development of the country. In both cases, base pay or fixed salary as employee compensation practices are considered to be main motivational factors for employee performance in organizations, which is very old fashioned in modern HRM practices. There is no other way to obtain sustained competitive advantages in HRM without establishing satisfactory compensation systems for employees. With increased openness of the global economy and transnational labour movement, business organizations in Bangladesh soon will face acute shortages of high-skilled talents for different professional areas. Therefore, HR managers need to be proactive to combat this anticipated problem, as the study indicated. Both authors agreed that the findings of the review study could have some implications for HR managers and policy makers in Bangladeshi organizations and other organizations in developing countries.

Conclusion

As a review study, this chapter tried to assess current trends and scenarios and future prospects of HRM compensation practices in Bangladesh. This research found that in Bangladesh, base pay or fixed salary is still the only HRM compensation practice. This is also an important factor of employee motivation. Though both review type and empirical type studies were carried out on all HRM practices, strategic compensation and benefit has been found to be an attractive area of research. Findings show that the relationship between HRM compensation practices and organizational performance have not been carried out in the Bangladesh context. As HRM compensation practices have not been studied extensively in Bangladesh, there are a number of areas where future studies can be directed: First, in-depth studies may be conducted to evaluate the impact of compensation practices on organizational performance through using descriptive and difference inferential statistics. It will help organizations to plan and produce strategic compensation systems to enhance employee's performance as well as organizational competitiveness. Second, research may be carried out on the effectiveness of strategic compensation practices in service sector organizations. These outcomes may accelerate service sector organization employee performance to achieve ultimate organizational goals. Third, studies may be undertaken to compare HRM compensation practices with respect to public and private sector organizations in Bangladesh. The results of this study would genuinely help public sector organizations to update their existing compensation policies to cope with current compensation practices around the world. Fourth, case studies may be undertaken on compensation practices of different organizations for thorough analysis. This would certainly trigger the competition level among different organizations in terms of compensation strategy in a positive direction. Fifth, studies can be undertaken to portray compensation and employee motivation in the context of Bangladeshi organizations. This study would be very relevant for both employers and employees because most of the organizations ignored employee motivation, which is really important for employee performance. Sixth and lastly, more research may be carried out on the future trends of compensation strategy of HRM practices extensively in Bangladesh. The world is changing in the blink of an eye with the blessing of ultra-modern technology. Employers and organizations need to collaborate with the dynamic business world to gain sustainable competitive advantages.

References

Absar, N., Amran, A. and Nejati, M. (2014). Human capital reporting: Evidences from the banking sector of Bangladesh. *International Journal of Learning and Intellectual Capital*, 11(3), pp. 244–258.

Absar, N. and Mahmood, M. (2010). New HRM practices in the public and private sectors enterprises in Bangladesh: A comparative assessment. *International Review of Business Research Papers*, 7(2), pp. 118–136.

Absar, N., Nimalathasan, N. and Mahmood, M. (2013). HRM-market performance relationship: Evidence from Bangladeshi organizations. *South Asian Journal of Global Business Research*, 1(2), pp. 238–255.

Ahmed, N. and Peerlings, J. (2009). Addressing workers' rights in textile and apparel industries: Consequences for Bangladesh economy. *World Development*, 37(3), pp. 661–675.

Bloom, M. and Michel, J. G. (2002). The relationships among organizational context, pay dispersion, and among managerial turnover. *Academy of Management Journal*, 45(1), pp. 33–42.

Bowden, B. (2014). Commentary – Bangladesh clothing factory fires: The way forward. *South Asian Journal of Human Resource Management*, 1(2), pp. 283–288.

Boyd, B. and Salamin, A. (2001). Strategic reward systems: A contingency model of pay system design. *Strategic Management Journal*, 22(8), pp. 777–792.

Chiang, F. F. and Birtch, T. A. (2008). Achieving task and extra-task-related behaviours: A case of gender and position differences in the perceived role of rewards in the hotel industry. *International Journal of Hospitality Management*, 27, pp. 491–503.

Chowdhury, S. and Mahmood, M. (2012). Societal institutions and HRM practices: An analysis of four multinational subsidiaries in Bangladesh. *The International Journal of Human Resource Management*, 23(9), pp. 1808–1831.

Danish, R. Q. and Usman, A. (2010). Impact of reward and recognition on job satisfaction and motivation: An empirical study from Pakistan. *International Journal of Business and Management*, 5, pp. 159–167.

Galanou, E., Sotiropoulos, I. Georgakopoulos, G. and Vasilopoulos, D. (2011). Job satisfaction in an organizational chart of four hierarchical levels: A qualitative study. *International Journal of Human Sciences*, 8, pp. 485–520.

Gerhart, B. A. and Milkovich, G. T. (1990). Organizational differences in managerial compensation and financial performance. *The Academy of Management Journal*, 33, pp. 663–691.

Ghazi, S. R., Ali, R., Shahzada, G. and Israr, M. (2010). University teachers' job satisfaction in the North West Frontier Province of Pakistan. *Asian Social Science*, 6, pp. 188–192.

Huselid, M. A. (1995). The impact of human resource management practices on turnover, productivity, and corporate financial performance. *The Academy of Management Journal*, 38, pp. 635–672.

Huselid, M. A., Jackson, S. E. and Schuler, R. S. (1997). Technical and strategic human resource management effectiveness as determinants of firm performance. *The Academy of Management Journal*, 40, pp. 171–188.

International Labour Organization (ILO) (2013). Bangladesh: Seeking better employment condition for better socio-economic outcomes. In *Studies on growth with equity report*. Dhaka: ILO Dhaka Office.

Khan, M. Y. H. (2018). Strategic human resource practices and its impact on performance towards achieving organizational goals. *Business Ethics and Leadership*, 2(2), pp. 66–73.

Khan, S. (2013). High performance work systems in the context of the banking sector in Bangladesh. *Unpublished PhD Thesis*. Melbourne: La Trobe University.

Kwak, J. and Lee, E. (2009). An empirical study of "Fringe Benefits" and performance of the Korean firms. *International Journal of Business and Management*, 4, pp. 3–9.

Luthans, F. and Stajkovic, A. D. (1999). Reinforce for performance: The need to go beyond pay and even rewards. *The Academy of Management Executive*, 13, pp. 49–57.

Mahmood, M. (2008). Sharing the Pie: Trade unionism and industrial relations in multinationals in Bangladesh. *Sri Lankan Journal of Human Resource Management*, 1(2), pp. 28–45.

McDonald, D. and Smith, A. (1995). A proven connection: Performance management and business results. *Compensation and Benefits Review*. 27(1), pp. 59–64.

Mia, K. and Hossain, M. (2014). A comparative study of HRM practices between foreign and local garment companies in Bangladesh. *South Asian Journal of Human Resource Management*, 1(1), pp. 67–89.

Milkovich, G. T. and Newman, J. M. (1990). *Compensation*. Homewood, IL: BPI/ lrwin.

Milkovich, G. T., Newman, J. M. and Gerhart, B. A. (2011). *Compensation* (10th ed.). London: McGraw Hill Higher Education.

Mondy, R. Q. (2010). Job satisfaction and organizational commitment of university teachers in public sector of Pakistan. *International Journal of Business and Management*, 5, pp. 17–26.

Montemayor, E. F. (1996). Congruence between pay policy and competitive strategy in high-performing firms. *Journal of Management*, 22(6), pp. 889–908.

Noe, R., Hollenbeck, J., Gerhart, B. and Wright, P. (2003). *Human resource management: Gaining a competitive advantage* (4th ed.). Boston, MA: McGraw Hill, p. 4.

Noe, R. A., Hollenbeck, J. R., Gerhart, B. and Wright, P. M. (2010a). *Fundamentals of human resource management*. New York, NY: McGraw-Hill/Irwin.

Noe, R. A., Hollenbeck, J. R., Gerhart, B. and Wright, P. M. (2010b). *Human resource management: Gaining a competitive advantage*. New York, NY: McGraw-Hill/Irwin.

Obaidullah, A. (1995). Reorganization of pay policy and structure in Bangladesh: Quest for living wage. *Journal of Asiatic Society of Bangladesh*, 40(1), pp. 135–155.

PwC (2015). *The world in 2050: Will the shift in global economic power continue?* London: PricewaterhouseCoopers

Sarker, A. (2006). New public management in developing countries: An analysis of success and failure with particular reference to Singapore and Bangladesh. *International Journal of Public Sector Management*, 19(2), pp. 180–203.

Schermerhorn, J. R. (2010). *Introduction to management*. Hoboken, NJ: John Wiley and Sons, Inc.

Shaw, J. D., Gupta, N. and Delery, J. E. (2002). Pay dispersion and workforce performance: Moderating effects of incentives and interdependence. *Sri Lankan Journal of Human Resource Management*, 2(1), pp. 28–45.

The Daily Star (2014). *HR practices pick up: A study report*. Retrieved from: www.thedailystar.net/hr-practices-pick-up-study-9507 (accessed: the 7th January 2019).

The Dhaka Tribune (2014). *Public servants to get up to double salaries*. Retrieved from: www.dhakatribune.com/bangladesh/2014/dec/22/public-servants-get-double-salaries (accessed: the 7th January 2019).

Werner, S. and Tosi, H. L. (1995). Other people's money: The effects of ownership on compensation strategy and managerial pay. *Academy of Management Journal*, 38(6), pp. 1672–1691.

Part 8

Tourism marketing and capital investment

13 Economic contribution of tourism in Bangladesh

Capital investment perspective

Mallika Roy, FAN Yajing and Bablo Biswas

Introduction

Investment in the tourism industry implies the making of capital or products fit for producing different goods or services in the tourism industry for maximizing profits in the private sector or local revitalization and growth of the economy for public purposes. Tourism, as one of the most promising indicators of growth and development for the world economy, can play an important role in driving the transition to a high-income economy, and contributing to more GDP. Investment and financing are inevitable parts of this. The possibilities are diversified and include both public and private investment. Strategic investors (SI), construction investors (CI) and financial investors (FI) are all tourism investors.

Thoroughly investigating all aspects of tourism development and economic growth is extremely important for all concerns (Leea and Chang, 2008). Though some qualitative analyses have been done in this context, there is no statistical or econometric analysis on the economic contribution of tourism in Bangladesh. In this context, this study is an attempt by the researchers to evaluate present trends and growth of the tourism industry which can contribute a lot to the economy of Bangladesh. By the end of this chapter, readers will be able to first, understand and find additional information on tourism trends in Bangladesh; second, become familiar with tourism investment trends and opportunities; third, become familiar with basic investment concepts and learn about their relation with GDP; fourth, understand the benefits of tourism investment. Investment trends influence the viability of a plan. Understanding this trend is the first step in identifying investment opportunities, quantifying market potential, determining the contribution of investment to GDP and assessing the feasibility of investment. To expand the tourism sector, tourism investment plays an important role. In this research, the trend of tourism investment and its impact on tourism revenue are discussed. The specific research objectives of this study are first, to know about the policies for tourism sector development in Bangladesh; second, to examine the impact of capital investment and revenue on tourism in Bangladesh; third and finally, to find out limitations and suggest some policy implications for the development of the tourism sector.

Literature review

A review of the existing literature shows that there have been no published studies in academic journals concerning the forecasting of tourism of Bangladesh. One of the reasons for this is that Bangladesh has not been regarded as a traditional and popular destination by Western or Asian tourists. Indeed many European and North Americans do not even know where Bangladesh is, let alone consider choosing Bangladesh as a holiday destination. Bangladesh is located in South Asia with a coastline of 580 km (360 miles) on the northern littoral of the Bay of Bengal. The delta plain of the Ganges (Padma), Brahmaputra (Jamuna) and Meghna Rivers and their tributaries occupy 79% of the country. Four uplifted blocks (including the Madhupur and Barind Tracts in the centre and northwest) occupy 9%, and steep hill ranges up to approximately 1,000 metres (3,300 ft) high occupy 12% in the southeast (the Chittagong Hill Tracts) and in the northeast. Bangladesh is the world's eighth most populous country with a population exceeding 167 million (United Nations, 2017).

Type of investors in the tourism industry

Strategic investors (SI)

Individuals or firms that invest resources into tourism industry improvement projects to gain the privilege to oversee them and take liability for running an organization. For example, tourism businesses and real estate developers specializing in planning and financing, brand possessors and so on.

Construction investors (CI)

Individuals or firm that invests into tourism industry advancement projects to take responsibility for the development.

Financial investors (FI)

Individuals or firms that invest capital into tourism industry improvement projects to get a dividend. For example, banks, insurance companies, private asset management companies, REITs (real estate investment trusts), PEF (private equity fund) and so on.

Patterns of economic growth, public and private gross capital formation in Bangladesh in terms of GDP growth rate

In the wake of enduring significant difficulties during the Liberation War and a log jam in growth, Bangladesh's economy has quickened since the finish of the 1980s. The economy of Bangladesh saw an average of 4% growth or more per annum all through the 1990s. GDP growth rate in Bangladesh averaged 5.69% from 1994 until 2016, reaching an all-time high of 7.11% in 2016 and a record low of 4.08% in 1994 (Trading Economics, 2019).

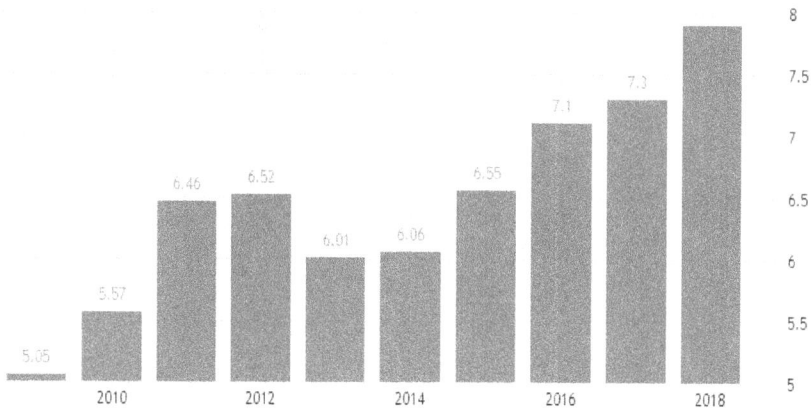

Figure 13.1 Trend of GDP growth rate
Source: Trading Economics, 2019

The growth of GDP has been accelerating in each successive period since the early 1990s. Three components go into increasing growth. The two that have played roles since 1990 are physical capital and human capital, the latter estimated as far as the nature of the workforce and their aptitudes and skills level. In spite of the fact that labour expansion is significant as a growth source, it doesn't clarify the growth variety crosswise over nations; driving the distinction crosswise over nations are capital aggregation and profitability upgrades.

Investment-to-GDP ratio

The expansion in the investment-to-GDP ratio has been for the most part because of the dynamism in private investment, with investment in the public sector remaining practically unaltered as an extension of GDP. Both investment and saving rates have relentlessly improved, in this manner preparing for predominant growth performance. In the interim, because of powerful and continued growth in export income and the accompanying increment in imports, there has been a rapid growth in the trade openness of the economy (that is, the combined proportion of imports and exports to GDP).

Overall investment crossed 31% of GDP for the first time in Bangladesh's history in fiscal 2017–2018, which was 30.51% the previous year, according to data from the Bangladesh Bureau of Statistics. The Bangladesh Bureau of Statistics (BBS) (2019) data showed the slow pace of growth in investment is mainly due to public investment: in fiscal 2017–2018, public investment to GDP was 7.97%. On the other hand, private investment was 23.26% of GDP, up from 23.10% in fiscal 2016–2017. For the last decade, the private investment-to-GDP ratio has been stuck at 21 to 23%.

Investment to GDP ratio (%)

■ Public Investment ■ Private Investment

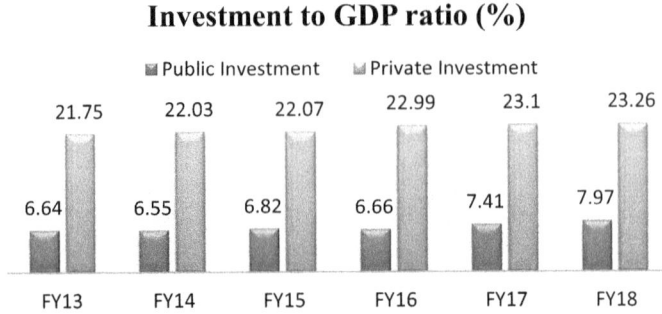

	21.75	22.03	22.07	22.99	23.1	23.26
	6.64	6.55	6.82	6.66	7.41	7.97
FY13	FY14	FY15	FY16	FY17	FY18

Figure 13.2 Public/private investment-to-GDP ratio (in percentage)
Source: Bangladesh Bureau of Statistics, 2019

Growth experience of South Asian countries

The last two decades have seen reasonable economic growth in most South Asian countries. Toward the millennium's end in the Development Goals period, Bangladesh outranked numerous other developing nations of a similar income level in poverty and human capital advancement statistics including its economically propelled neighbours, for example, India and Pakistan.

Figure 13.3 plots data on GDP growth rates of the South Asian countries over 1990–2017. Bangladesh has been growing around the South Asian average in terms of GDP growth rate. Barring a few years, it has been consistently above other regional competitors such as Pakistan. Figure 13.3 also shows that during 1990–2017, 1990–2003 and 2004–2017, Bangladesh's average GDP growth rate was 5.49, 4.7 and 6.16% while Pakistan's figures over the respective periods were 4.14, 3.86 and 4.45% and South Asian averages were 6.27, 5.28 and 7.27%.

Investment assists in stimulating and restructuring economic activities for accomplishing higher economic growth rates in all economies. Investments, being a part of aggregate demand as well as a source of capital accumulation, have been given a lot of significance in previous research pertaining to sector-wise growth, while the tourism sector has been given less consideration. Governments, as well as international development agencies, in developing or less developed countries (LDCs) have, for many years, regarded the tourism sector as a main wellspring of employment and income generation. Holzner (2011) and Brau et al. (2007) found that countries dependent on tourism face higher than average economic growth rates. Solow (1956) states that poorer countries have higher than average economic growth rates. These two studies could be connected to my results, since I discovered that countries that are dependent on tourism are relatively poor.

As contended by Baum and Szivas (2008), the motivation behind legislative backing to the tourism industry is the capacity of this area to create employment and add to social and economic development. Investment activities can be made either by the public or private segment; furthermore, the results are generally

GDP growth rate (%)

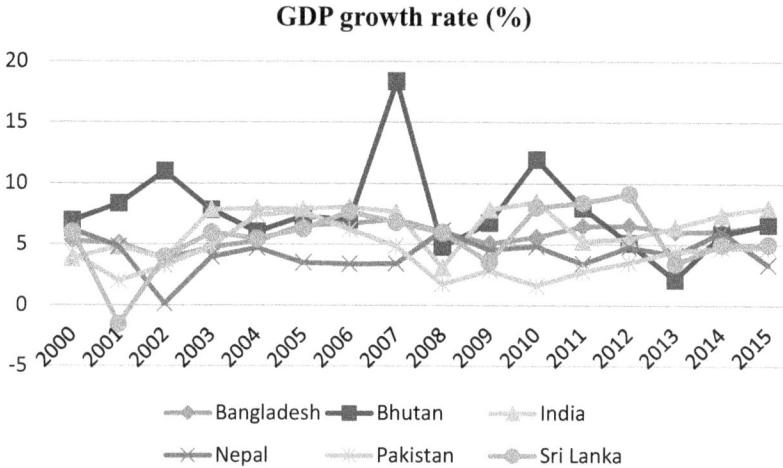

Figure 13.3 GDP growth rate of South Asian countries, 1990–2017
Source: World Bank, 2017

determined by the domestic social, political and economic structure. From the economic viewpoint, public investment is defended when the private segment fails to create a sufficient amount. Investment by the public sector decreases the risks to the private arena and guarantees profitability (Rosentraub and Joo, 2009).

In general, public investment improves both sectoral growth (for example, the tourism industry) and economic growth. Munnell (1992) declared that the profitable limit of a part or a territory can be extended by public capital investment, by improving the efficiency of current resources and including more resources also. In many developing countries, the role of public sector has been significant in the advancement of the tourism industry (Akama, 2002).

Other than the tourism industry strategy formulation and developing a national tourism plan, governments have been effectively occupied with the arrangement of the travel industry and cordiality offices and administrations. Given that the tourism industry is a profoundly divided sector that includes numerous stakeholders in the arrangement of different administrations, the prime role of governments in encouraging and advancing the industry through the provision of ideal socio-political and legitimate condition is of most extreme significance (Akama, 1997, 2002; Gunn, 1988; Hughes, 1994; Jenkins and Henry, 1982). Thus, in developing nations, the public sector is required to contribute effectively to tourism industry improvement, not just in the foundation of legislative systems and strategies, but also in the investment and management of the tourism sector.

Brau et al. (2007) explain the effects of tourism on countries through underlying mechanisms in their work. They expressed that allocation of labour is one of these mechanisms, hence when a country faces higher relative endowment of natural resources, it allocates more labor into the tourism sector and comparative

228 Mallika Roy, FAN Yajing and Bablo Biswas

advantage in tourism is obtained (Brau et al., 2007). According to them, the relative value of tourism services grows over time; hence average economy grows at a lower rate. They concluded that the underlying mechanism indicates tourism-led high steady-state growth of a sustainable nature. Instead of physical expansion, the growth is driven by increasing appreciation of tourism services (Brau et al., 2007).

Balalia and Petrescu (2011) asserted that the role of the state is focal in regulating and controlling the tourism activities and in certain circumstances, even encouraging it. In addition, he proclaimed that the public segment helps tourism industry development by improving infrastructure advancement, empowering private investment in hotel development, keeping up the standard of quality and securing tourists against any sort of instabilities or insecurities. The activities taken by the state to make the best conditions to animate the development of overall production has an immediate impact on the tourist industry too and government intervention is genuinely necessary in the tourism industry (Ribarić and Ribarić, 2013). In such manner, the state needs to deliberately focus on investment to make favourable conditions for the betterment of tourism industry.

Bakan and Bosnic (2012), in a study on public-private partnerships in sustainable tourism development in Croatia, claimed that the low size of investment in tourist infrastructure is an important factor responsible for the slow pace of tourism development. Empirical evidence also proposes that small islands, for example, Zanzibar in Tanzania, where the legislature has not been effectively engaged with direct investment, other than policy formulation and monitoring. The government has not developed any policies related to investment (Sharpley and Telfer, 2014).

Methodology

Overview

This chapter uses both qualitative and quantitative methods to fulfil the objective. The chapter firstly introduces the current situation and existing problems of tourism investment in Bangladesh. Second, we employ time series models to predict the trend of tourism investment in the future. Third, in order to promote the growth of tourism revenue in Bangladesh, it is necessary to analyse the relationship between tourism investment and tourism revenue. This research selects capital investment as the tourism investment index and international tourism receipts as the tourism revenue growth index. Through the correlation analysis method, the correlation degree between each input index and total tourism revenue in Bangladesh is analysed. Based on the correlational analysis, this chapter also analyses the impact of tourism investment on tourism revenue by constructing vector autoregressive (VAR) model.

Model

Since our research focuses on the total contribution of travel and tourism to GDP, capital investment on travel and tourism, and international tourism receipts, and all those variables are included in the model. To account for the direct and indirect impact of capital investment on travel and tourism and international tourism receipts on the total contribution of travel and tourism to GDP, we use the VAR model, an equation system in which all variables are treated as endogenous and variables at the current time point are regressed against lagged values of all the variables in the system. It enables us to consider the complex relationship among all variables, with particular emphasis on the total contribution of travel and tourism to GDP (Seetanah et al., 2011). The VAR model can be written as follows:

$$\Upsilon_t = \Pi_1 \Upsilon_{t-1} + \Pi_2 \Upsilon_{t-2} + \ldots + \Pi_p \Upsilon_{t-p} + U_t$$
$$U_t \sim IID(0, \Sigma) \tag{1}$$

Υ_t is a vector including all interested variables in the study (i.e. the total contribution to GDP of travel and tourism investment and international tourism receipts in our research). U_t is a vector of regression errors that are assumed to be contemporaneously correlated but not autocorrelated. If a VAR model has m equations, there will be $m + mp^2$ coefficients that need to be estimated where p is lag length of variables in each equation. Over-parameterizing may occur when p is too large but the information is not enough because degrees of freedom run out. However, if p is too small, the model cannot represent correctly the data-generating process (Song and Witt, 2006). In the current research, we have 2 variables and 23 observations so the maximum value of p is 4. Then we use Akaike information criterion (AIC) to determine the lag length of the VAR model.

Impulse response analysis

One advantage of this approach is that an impulse response analysis can be carried out, which can provide useful information for policymaking purposes (Song and Witt, 2006). By repeatedly substituting the lagged variables into themselves based on equation (1), the vector moving average (VMA) process is obtained:

$$\Upsilon_t = \sum_0^\infty \Pi_1^i U_{t-i} \tag{2}$$

The elements of Π_1^i represent the effects of unit shocks in the variables of the system after i periods. They are called impulse responses or dynamic multipliers. Introducing forecast error term, we have the impulse-responds function:

$$\Upsilon_{t+n} = \sum_0^\infty \Pi_1^i U_{t+n-i} \tag{3}$$

It represents the response of $\Upsilon_{i,t+n}$ to a one-time impulse from $\Upsilon_{j,t}$ with all other variables dated to t or earlier held constant. The response of variable i to a unit shock (forecast error) in variable j will be depicted graphically to get a visual impression of the dynamic inter-relationships within the system.

Through empirical results and recommendations, the main objective of this study is to improve the understanding of the impact of investment on economic growth and point to policy measures aimed at further strengthening economic growth in Bangladesh. In this regard, the study analysed the impact of investment on economic growth in Bangladesh. The methodology adopted is ARIMA model selection, VAR model and orthogonal impulse response analysis.

Empirical results

In an attempt to study the trend of tourism investment and its impact on tourism development, our study makes use of data from Bangladesh over a period of 23 years (1995–2017). We collected the data of the total contribution of travel and tourism to GDP and travel and tourism investment from world travel and tourism councils instead of the direct contribution of travel and tourism. The total contribution of travel and tourism includes its "wider impacts" (i.e. the indirect and induced impacts on the economy) (World Travel & Tourism Council, 2018). In addition, we collected complementary data of international tourism receipts from World Bank.

From Figure 13.4, we can see that the total contribution of travel and tourism to GDP kept increasing during 1995–2017. Figure 13.4 also shows that the total contribution of travel and tourism to GDP grew to over US$10.62bn in 2017, an increase of US$0.81bn (8.23%) over the previous year. International tourism receipts experienced a year-on-year average growth rate of 6.15% from 1995 to 2017. Moreover, in recent years, international tourism receipts show an

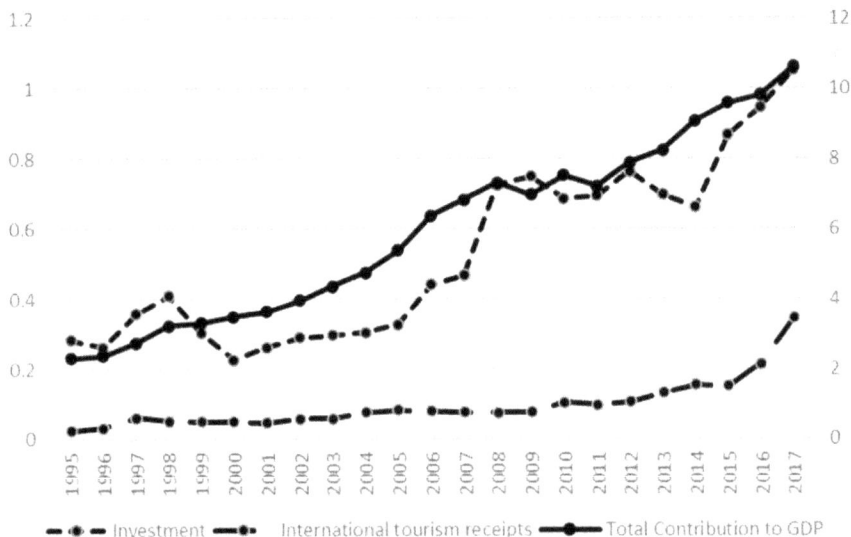

Figure 13.4 Trend of tourism investment, international tourism receipts and contribution to GDP (US$ in billions)

Source: World Travel & Tourism Council (2018)

increasing growth rate, around 10.83% from 2012 to 2017. Travel and tourism investment decreased from 2012 to 2014 and then grew rapidly with growth rate of 30.72%.

Modelling trends using univariate time series models

It is straightforward to generate the forecasts for all three indicators by the univariate time series model first. We select the optimal model for each indicator based on AIC as presented in Table 13.1. The forecast results are presented in Table 11.3 and Figures 13.5–13.7.

Table 13.1 ARIMA model selection

Indicator	Model	AIC
Total contribution to GDP	ARIMA(0,2,0)	31.60
	ARIMA(1,2,0)	**23.05**
	ARIMA(0,2,1)	24.33
	ARIMA(1,2,1)	24.55
Investment	ARIMA(0,3,0)	−7.69
	ARIMA(1,3,0)	−11.89
	ARIMA(0,3,1)	−23.12
	ARIMA(1,3,1)	**−23.33**
International tourism receipts	ARIMA(0,3,0)	−75.8
	ARIMA(1,3,0)	−78.67
	ARIMA(0,3,1)	**−82.28**
	ARIMA(1,3,1)	−80.23

Table 13.2 Forecast results (US$ in billions)

Indicator	Year	Point Forecast	Lo 80	Hi 80	Lo 95	Hi 95
Total contribution to GDP	2018	11.01	10.57	11.45	10.34	11.69
	2019	11.70	10.98	12.42	10.59	12.80
	2020	12.18	11.01	13.35	10.39	13.97
	2021	12.81	11.19	14.43	10.33	15.28
	2022	13.33	11.18	15.49	10.04	16.63
Travel and tourism investment	2018	1.15	1.01	1.30	0.93	1.38
	2019	1.26	0.97	1.56	0.81	1.71
	2020	1.38	0.90	1.86	0.64	2.11
	2021	1.50	0.80	2.19	0.43	2.56
	2022	1.62	0.67	2.57	0.17	3.07
International tourism receipts	2018	0.67	0.58	0.75	0.54	0.79
	2019	0.86	0.71	1.01	0.63	1.09
	2020	1.07	0.84	1.30	0.71	1.42
	2021	1.30	0.97	1.63	0.79	1.81
	2022	1.55	1.10	2.01	0.86	2.25

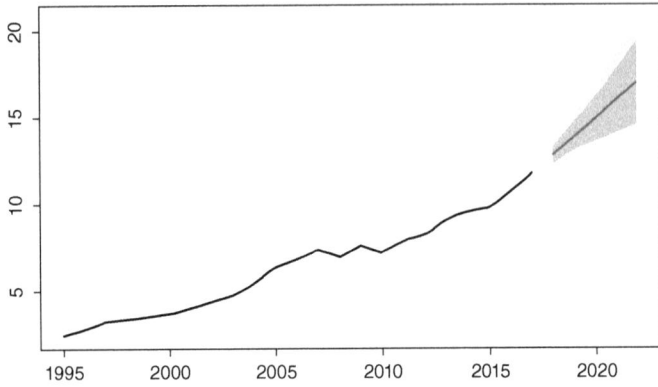

Figure 13.5 Forecast of contribution to GDP

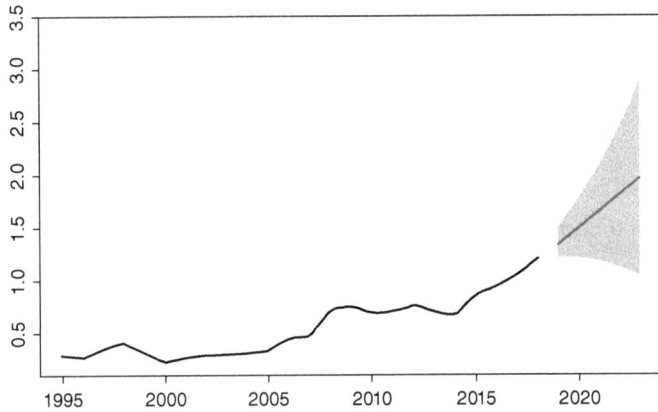

Figure 13.6 Forecast of travel and tourism investment

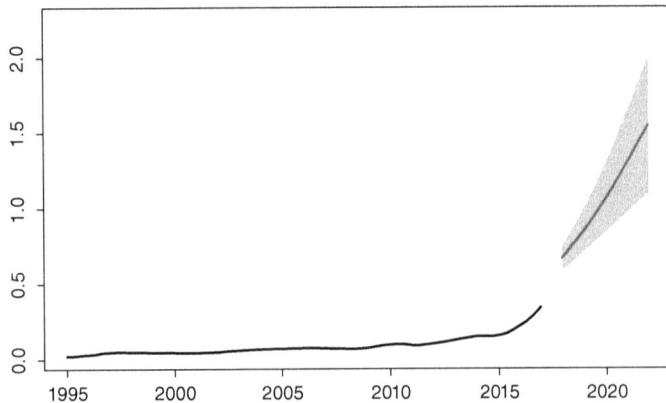

Figure 13.7 Forecast of international tourism receipts

The total contribution of travel and tourism to GDP, including employment by hotels, travel agents, airlines and other passenger transportation services also includes, for example, the activities of the restaurant and leisure industries directly supported by tourists. This value was US$10.61bn in 2017 (4.3% of GDP) and is expected to grow by 25.54% to US$13.33bn in 2022, with travel and tourism investment reaching US$1.62bn and international tourism receipts US$1.55bn. Columns 4–5 show the 80% confidence intervals and columns 6–7 show the 95% confidence intervals. For example, 80% of the time, when we calculate a confidence interval in this way, the true value of the total contribution of travel and tourism to GDP will be between US$11.18bn and US$15.49bn.

Modelling trends using VAR

In this section, the VAR model is used to study the relationship among contribution of tourism to GDP, travel and tourism investment and international tourism receipts. The estimation of the VAR models generated a large number of parameters, as three equations were estimated in total. As the limitation of our sample size, maximum order of lag is four, so we select the optimal model based on AIC as shown in Table 13.3. The estimation results of the VAR models are omitted as the focus of this chapter is the contribution of investment to tourism development. We just present the graph of the responses of the total contribution of travel and tourism to GDP, international tourism receipts and tourism investment here.

These figures demonstrate the impulse response relationships among all three variables. The X-axis represents the number of lag or years and the Y-axis is the magnitude of the shock. The second graph in first column presents the impulse response of travel and tourism investment to the total contribution of travel and tourism to GDP. At the initial period, the shock will be negative and not obvious. It means that an investment shock leads to a drop in contribution of tourism to GDP in the short term. The plot also illustrates that, as time passes, the effects of a positive shock on the contribution to GDP increases. Similarly, international tourism receipts have a positive impact on the total contribution of travel and tourism to GDP after around three years.

Table 13.3 VAR model selection

Lag	AIC
1	–140.3993
2	–128.2794
3	–146.3329
4	–169.1528

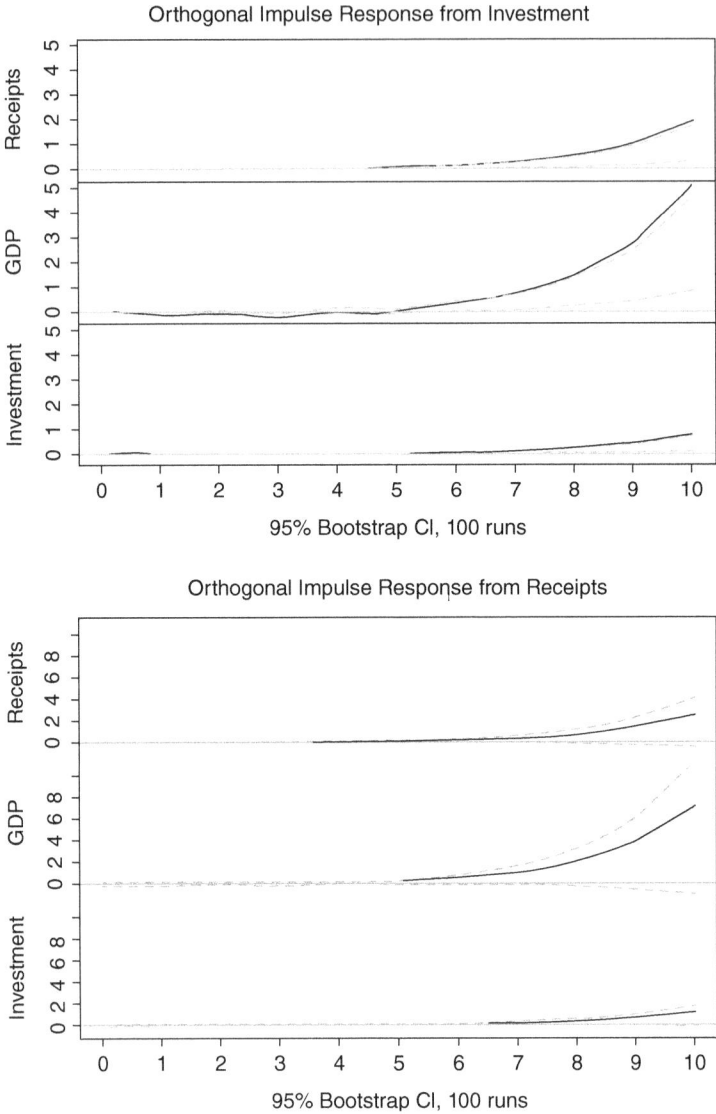

Figure 13.8 Orthogonal impulse response

Recommendations

The fundamental conclusion of this chapter identifies the important role economic policy has in impacting economic growth, principally when developing nations are the concern. In view of forecasting results, contribution to GDP, travel and tourism investment and international tourism receipts are trending upwards. To improve conditions further, Bangladesh should focus on the sectors discussed next.

The aftereffects of the study have helpful ramifications for Bangladesh. One significant proposal to help economic growth in Bangladesh is to put more accentuation on private investment. Consequently, the Bangladesh government must place emphasis on this variable to improve and animate economic growth in Bangladesh. One of the approaches to accomplish this strategy objective is to make more wealth or to create more employment opportunities.

There are certain restricting limitations – land, energy, trade logistics and access to long-term financing, skills and regulatory complexity as well as unpredictability – that make it extremely difficult for existing and potential entrepreneurs to start businesses or go into new zones of business. Bangladesh needs to establish special economic zones, liquefied natural gas terminals, base-load power plants and one-stop shops as well as upgrade ports. Then private investment will surely break out from 21–23% of GDP.

One of the significant drivers of growth in a developing economy like Bangladesh is total factor productivity. Cross-country analyses find that highly developed economies are driven both by growth in their raw materials as well as continued growth in their productivity. The intriguing part of this conclusion isn't that productivity growth must be incredibly high, simply sustained over a long period. In spite of the ongoing expanding pattern in total factor productivity in Bangladesh, the contribution of productivity growth to the overall growth of the economy is low, and that growth has been input-driven as opposed to productivity driven. When taking a gander at the total factor productivity development encounters of different nations, one finds that variables, for example, human capital advancement, physical capital improvement (including infrastructure), financial development, technology absorption and openness (particularly in terms of openness to imports) have a critical effect of growth in total factor productivity. Until the country focuses on these issues, it will be difficult for Bangladesh to accomplish supportable development. Upgrade productivity could be another viable policy target to quicken future economic growth.

One of the basic questions that emerges over all economies is what amount of economic growth is brought about by development in physical and human capital and what amount is brought about by other factors, for example, innovation and institutional change. In spite of the fact that there is discussion about the positive effects of expanded physical and human capital on development, most economists feel that sustainable growth is reliant on continued innovative and institutional development.

Among private sector investments, the manufacturing sector has been a significant driver of GDP growth in Bangladesh. Sustainable dynamism in manufacturing would be significant for Bangladesh's transition to middle-income status. To release the maximum capacity of the manufacturing division and to accomplish more prominent broadening, it would be important that the intensity of Bangladeshi manufacturing segment be reinforced impressively.

The general significance of tradable and non-tradable segments, in spite of changes because of basic movements, are to such an extent that still the non-tradable areas like services, development, small-scale industry and other demand-driven exercises are significant supporters of economic growth in Bangladesh. This proposes future growth policies in the nation, at least in the medium term, ought to at the same time centre around quickening the development of both tradable and non-tradable areas as opposed to concentrating only on tradable areas to drive development.

Conclusion

Tourism is a big sub-sector of the national economy. Without maintaining proper strategies in the sector, it may lag behind in making its potential contribution to the national economy. This chapter focused on the economic contribution of tourism and also shows the way for further improvement. Having reviewed the available literature, both theoretical and empirical and having the results, it is evident that the effect of investment on tourism growth is not negligible. A careful examination of the existing research shows that research on the effect of private and public investment on tourism in Bangladesh is still inadequate and needs more attention. Moreover, the study of the collaborative impact of public and private investment on tourism growth has, so far, received less attention. Therefore, in order to fill the aforementioned gaps in the literature, future research addressing tourism sector growth from the perspective of public and private investment is needed in general.

References

Akama, J. S. (1997). Tourism development in Kenya: Problems and policy alternatives. *Progress in Tourism and Hospitality Research*, 3(2), pp. 95–105.

Akama, J. S. (2002). The role of government in the development of tourism in Kenya. *International Journal of Tourism Research*, 4(1), pp. 1–14.

Bakan, R. and Bosnic, I. (2012). Public-private partnership: A model for sustainable tourism development in Regional park Mura-Drava – the possibility of tourist valorisation of abandoned army barracks. *Economy of Eastern Croatia Yesterday, Today, Tomorrow*, 1, pp. 201–206.

Balalia, A. E. and Petrescu, R. M. (2011). The involvement of the public and private sector – elements with influence on travel & tourism demand during the crisis period. *Tourism and Hospitality Management*, 17(2), pp. 217–230.

Bangladesh Bureau of Statistics (2019). *Public/private investment to GDP ratio*. Dhaka: BBS.

Baum, T. and Szivas, E. (2008). HRD in tourism: A role for government? *Tourism Management*, 29(4), pp. 783–794.

Brau, R., Lanza, A. and Pigliaru, F. (2007). *How fast are small tourism countries growing? The 1980–2003 evidence*. Retrieved from: https://ssrn.com/abstract=951221 (accessed: the 22nd December 2019).

Gunn, C. A. (1988). *Tourism planning*. New York, NY: Taylor & Francis.

Holzner, M. (2011). Tourism and economic development: The beach disease? *Tourism Management*, 32(4), pp. 922–933.

Hughes, H. L. (1994). Tourism multiplier studies: A more judicious approach. *Tourism Management*, 15(6), pp. 403–406.

Jenkins, C. L. and Henry, B. (1982). Government involvement in tourism in developing countries. *Annals of Tourism Research*, 9(4), pp. 499–521.

Leea, C. C. and Chang, C. P. (2008). Tourism development and economic growth: A closer look at panels. *Tourism Management*, 29, pp. 180–192.

Munnell, A. H. (1992). Policy watch: Infrastructure investment and economic growth. *The Journal of Economic Perspectives*, 6(4), pp. 189–198.

Ribarić, H. M. and Ribarić, I. (2013). *Government intervention in driving the development of sustainable tourism*. Retrieved from: https://bib.irb.hr/datoteka/908986. Makarin_Ribari_Helga_Ribari_Ivana.pdf (accessed: the 22nd December 2019).

Rosentraub, M. S. and Joo, M. (2009). Tourism and economic development: Which investments produce gains for regions? *Tourism Management*, 30(5), pp. 759–770.

Seetanah, B., Padachi, K. and Rojid, S. (2011). Tourism and economic growth: African evidence from panel vector autoregressive framework. In A. K. Fosu (ed.), *Growth and institutions in African development*. Oxon: Routledge, pp. 145–158.

Sharpley, R. and Telfer, D. J. (2014). *Tourism and development: Concepts and issues*. Bristol: Channel View Publications.

Solow, R. M. (1956). A contribution to the theory of economic growth. *The Quarterly Journal of Economics*, 70(1), pp. 65–94.

Song, H. and Witt, S. F. (2006). Forecasting international tourist flows to Macau. *Tourism Management*, 27(2), pp. 214–224.

Trading Economics (2019). *Bangladesh GDP growth rate*. Retrieved from: https://tradingeconomics.com/bangladesh/gdp-growth (accessed: the 22nd December 2019).

United Nations (2017). *World population prospects*. Retrieved from: https://population.un.org/wpp/Publications/Files/WPP2017_DataBooklet.pdf (accessed: the 22nd December 2019).

World Bank (2017). *GDP growth rate of South Asian countries, 1990–2017*. Manila: WB.

World Travel & Tourism Council (2018). *World travel & tourism council report-2018*. London: WTTC.

14 The competitive power of capital structure in a tourism destination

Yusuf Babatunde Adeneye, Ei Yet Chu and Fathyah Hashim

Introduction

In the extant literature on strategic management, there is a considerable support for certain factors that determine the competitive forces and advantage of the firm. Among these factors are structural forces (Birkinshaw et al., 1995), information technology and governance (Dehning and Stratopoulos, 2003; Raymond et al., 2019), supply chain capabilities (Yusuf et al., 2004) and high involvement work practices (Guthrie et al., 2002). A critical examination of these studies revealed that much attention is placed largely on business and innovation processes, supply chain, production function, external environmental factors and marketing processes in formulating and implementing competitive strategies. However, one area that has received less attention is the financial strategy/policy of the firm within the context of a tourism destination.

A large number of past studies on tourism strategies have focused on coopetition (Damayanti et al., 2019; Della Corte and Aria, 2016), destination environmental conditions (Lee and King, 2006; Martínez-Román, et al., 2015; Mihalic, 2016), sustainability (Boskov et al., 2018; Mihalic, 2016) and international trade (Fuster et al., 2018; Hong, 2009). While these studies focused largely on external factors, there is a literature gap in relation to corporate financing or financing structure as a contextual factor that influences the level of competitive strategy in the tourism sector. Consequently, the use of corporate finance theories, such as trade-off and pecking, has received less theoretical and practical analyses to explain the level of tourism destination strategies. Therefore, while most of the works on tourism performance placed emphasis on competitive strategies, there is a limited understanding of the level to which tourism financial performance might be influenced following tourism firms' strategy response to a changing capital structure.

The heterogeneity of capital structure in the tourism sector is increasing daily as competition in the tourism sector becomes more dynamic. Eccles (1991) stated that a firm's corporate system depends largely on its financial decisions that may undercut its strategy for competitive advantage. The financial strategy of corporate firms can serve as a sea change strategy that influences the firm's strategic choice and priorities by practising leverage recapitalization and taking additional debt, often in excess of the firm's optimal debt capacity (Eccles, 1991). Thus, a firm's drive for competitive advantage depends on its financing policy.

According to the Phillips et al. (2017), tourism-related investment is locked in fierce competition for investment inflows and access to credit and finance attracts investment in the tourism sector while increasing the level of competition within the industry. Barriers to the tourism investment can also shift the nature of competition between large and small firms. While large firms have the capacity to overcome finance barriers, thereby obtaining monopoly status, similar barriers pose high costs on small firms that have less access to finance, thus impeding their competitiveness and stifling innovation. USAID (2017) submitted that grants, equity financing and debt financing are the three financing approaches that can promote sustainable tourism projects. This implies that there are risks and returns that firms and other players in the tourism industry must consider in designing and changing their competitive strategies. The lower the risk of financing, the shorter the maturity of debt financing and the lower the cost of capital will be.

Bangladesh has some unique tourist centres and destinations that can give better experiences and quality services to tourists. However, the country's tourism competitiveness ranking is declining among other countries of the world and even among their counterparts in the Southeast and Southern Asia regions. Bangladesh is ranked poor in key indicators of travel and tourism competitiveness index (see Table 14.1). Bangladesh is ranked 111th in total travel and tourism government expenditure, ranked 51st in travel and tourism industry GDP, ranked 113th in travel and tourism industry share of employment, ranked 130th in travel and tourism policy and conditions and ranked 135th in travel and tourism sustainable development, despite the fact that the industry ranked high, 15th in its travel and tourism industry employment. Low competition has been related to finance and strategy (Dwyer, 2018; Chen et al., 2018; Zhou, 2019). In addition, evidence has shown that firms that depend largely on sales, profitability and retained earnings face intense tourism competition (Verreynne et al., 2019; Bakker, 2019; Carrillo-Hidalgo and Pulido-Fernández, 2019). Therefore, we conjecture a practical and theoretical link between competitive strategy, financing decisions and performance.

This chapter examines the relationship between capital structure, competitive strategies, and financial performance among tourism firms in Bangladesh. It contributes practically and theoretically. First, we found that internal factors such as corporate financing have significant effects on firms' competitive strategies in tourism destination rather than innovation and information and communication technologies (ICT) as documented by previous studies. We found that firms' level of differentiation strategy is declined by high debt financing while it encourages cost leadership strategy. Second, we establish that trade-off theory and pecking order theory partially support the level of competition in tourism destinations. While trade-off theory supports the cost-leadership strategy, the pecking order supports the differentiation strategy. Third, our findings have practical implications for tourism stakeholders such as government and regulators to consider access to

Table 14.1 Travel and Tourism Competitiveness Index indicators

Countries	T&T GE	T&T IE	T&T iGDP	T&T ISE	T&T isGDP	T&T PC	T&T SD
Bangladesh	111	15	51	113	99	130	135
China	57	2	2	82	109	84	58
Indonesia	67	5	16	79	72	5	57
Malaysia	24	30	30	51	42	21	7
Philippines	48	12	34	70	47	64	56
Singapore	10	67	27	47	36	1	5
South Africa	40	24	39	44	79	97	22
Spain	12	21	9	33	25	7	45
Thailand	14	9	12	22	8	37	61
United Arab Emirates	1	44	23	27	45	54	1
United Kingdom	47	10	5	32	63	73	36
United States	50	3	1	60	88	58	34
Vietnam	79	7	31	34	19	105	105

Source: World Economic Forum and International Finance Corporation, 2017

Note: T&T GE = T&T Government Expenditure, T&T IE = T&T Industry Employment (1,000 jobs), T&T iGDP = T&T Industry GDP (US$ million), T&T ISE = T&T Industry Share of Employment (% of total employment), T&T is GDP = T&T Industry Share of GDP (% of total GDP), T&T PC = T&T Policy and Conditions subindex, 1–7 (best), T&T SD = Sustainability of Travel and Tourism Industry Development, 1–7.

This chapter is organized as follows. The first section presents the research gap, the significance of capital structure as a new strategy paradigm in the tourism sector and the contribution of the study. Next, the chapter presents the literature review and the hypotheses formulated. The third section presents the methodology of the study, while the fourth presents the data analysis and the discussion of results. The last section concludes the study.

Literature review

Theoretical foundation

The moderating effect of capital structure on the relationship between competitive strategy and profitability is closely related to the theory of pecking order and trade-off theory. First, the trade-off theory posits that the benefits of debt equal the costs of debt. The use of debt determines the value of the firm, and the value of the firm is also a function of competitive strategy. Therefore, debt financing would reflect the nature of competitive strategy of firms and organizations. Firms with good credit ratings and financial indicators have more access to debt than their counterparts. Theoretically, firms can invest well and heavily in any competitive strategy in relation to debt finance level. Second, the pecking order theory stipulates that a firm follows a hierarchical financing pattern: taking retained earnings first, followed by external debt and then external equity. In the context of this chapter, firms use internal financing (i.e. retained earnings) first to

Table 14.2 Summary of studies on competitive strategies in the tourism sector/industry

Authors	Competitive Strategies in Tourism	Theory	Context
Mistilis and Daniele (2004)	Public/private sector partnership Destination marketing systems (DMS)	Corporate model of DMS	–
Hong (2009)	Domestic and global environmental conditions	Ricardo's comparative advantages (RCA) theory and international trade theory	Taiwan
Lee and King (2006)	Tourism destination resources and attractor, tourism destination strategies and tourism destination environments	Industry organization theory and resource-based view	Taiwan
Navickas and Malakauskaite (2009)	Market conditions, level of social development, environmental policy, human resources, quality of infrastructure and technological advancement	Theory of competitiveness evaluation	Tourist destinations
Camisón et al. (2016)	Internal resources, capabilities and destination resources	Resource-based view and dynamic capabilities-based view	Spanish tourism firms
Ezeuduji (2015)	Event staging, unique cultural and natural heritage	Generic event-based rural tourism model	Sub-Saharan Africa
Martínez-Román et al. (2015)	Innovative capability, environment and contextual factors	Two-stage interactive model	Andalusian hospitality industry, Spain
Ajmera (2017)	12 strategic indicators	SWOT and TOPSIS models	Indian medical tourism sector
Mihalic (2016)	Environmental awareness, sustainability agenda and environmental action/responsible tourism (Respsustable tourism – a new tourism nomination)	Triple-A Model	European and UNWTO stakeholders
Della Corte and Aria (2016)	Coopetition	Game theory and behavioural theory	Tourism industry in Naples and Sorrento, Italy
Borsekova et al. (2017)	Innovation, spatial development and competitive advantage	Market-oriented theory and resource-based competitive advantage theory	Slovakia
Damayanti, Scott and Ruhanen (2019)	Collaboration, cooperation and coopetition	Institutional analysis and theory	
Fuster, Lillo-Bañuls and Martínez-Mora (2018)	Service offshoring	International trade theory	Spain
Boskov et al. (2018)	Sustainable urban development, international trade and international tourism	International trade theory	Macedonia and Greece

Source: Compiled by the authors, 2019

design their competitive strategy as this level of strategy formulation requires low financing that can be taken care of using the retained earnings of the firm. In the competitive strategy implementation level, firms require more financing to execute the strategy within turbulent tourism destinations and environments. Based on the trade-off theory and the pecking order theory, we formulate hypotheses by discussing the relationship between capital structure, competitive strategies and financial performance.

Hypotheses development

Differentiation strategy and capital structure

Since the postulation of Porter's generic strategies, many studies have examined the financing policy impact of competitive strategies to investigate Porter's theory and find evidence of capital structure adjustment (Barton and Gordon, 1988; Lowe et al., 1994; Jordan et al., 1998; Balakrishnan and Fox, 1993; Frangouli, 2002; Wanzenried, 2003). Using a quantitative study of UK firms, Jordan et al. (1998) did work on the link between strategy and financing policy in small and medium-sized enterprises and found that business-level strategy determines capital structure. Barton and Gordon (1988) also found a weak relationship between differentiation strategy and debt/equity mix. Wanzenried (2003) found a negative relationship between debt usage and the level of product and quantities differentiation under demand uncertainty. Meanwhile, Frangouli (2002) found a positive link between product differentiation as measured by R&D expenses to sales and debt-to-equity ratio. The author argued that internal financing strongly influences product differentiation thereby raising the industry entry barrier. Thus, product differentiation triggers a high cost of production for new entrants and a high cost of differentiation for potential new entrants. Based on this, it is hypothesized that:

Hypothesis 1: Differentiation strategy increases debt and equity usage in firms.

Cost leadership strategy and capital structure

Cost leadership describes the minimum amount of costs and assets required to achieve a high financial performance, asset use and employee productivity (Porter, 1980; Hambrick, 1983). However, Wanzenried (2003) found no support for leadership strategy and debt usage when customers' demands are uncertain. Because firms that engage in cost leadership are likely to spend more on production capabilities and other management functions to configure an operational efficiency model, high financing is required, causing firms to exhaust internal financing and use additional external financing. We hypothesized that:

Hypothesis 2: Cost leadership strategy increases debt and equity usage in firms.

Capital structure, competitive strategies and financial performance

Firms taking generic strategies are likely to have increased financial performance following an optimal capital structure decision. A firm must consolidate its competitive strategies to keep a low level of operational costs and maintain continuous high performance. However, firms must be careful in the choice of differentiation and cost leadership strategies formulated and implemented. For example, operating costs may be both flexible and inflexible. Inflexible operating costs reduce debt payment ability and trigger bankruptcy, thereby reducing a firm's profitability (Chen et al., 2019). Thus, firms may choose a low financial leverage ex ante when they perceived a high ex post distress cost. Consequently, since firms take advantage of unique characteristics of specialized assets, products and services to embark on differentiation strategy (Balakrishnan and Fox, 1993), they achieve increased financial performance. How firms finance operating costs and unique and differentiation strategies to increase and sustain current profitability will result in variances in capital structure. Thus, a firm's adjustment in capital structure decision moderates the link between competitive strategies and financial performance. Matching hypotheses 1 and 2 together and examining their combined effect on financial performance, it is hypothesized that:

> *Hypothesis 3:* Capital structure moderates the positive/negative links between differentiation strategy, cost leadership and financial performance.

Methodology

Sample

The data of this study was collected from the Datastream database for a sample period of 12 years from 2007 to 2018. The analysis of the study focused on the non-financial firms in the tourism destination of Bangladesh. The total number of firms included in this study is 85. We conduct a static panel data estimation in this study. We however excluded firms with high missing values and those registered within the last three years from 2016 to 2018.

Dependent variable – financial performance and competitive strategies

The data on financial performance (measured using return on equity, ROE) was collected from the Datastream database. We used ROE because we wanted to see how a firm's competitive strategies affect its efficient use of the firm's assets to increase shareholders' wealth. The use of ROE to measure the financial performance of tourist firms is common in tourism literature (Chen, 2010; Lee and Manorungrueangrat, 2019). ROE is the ratio of net profits to outstanding shares of the firm. We use ROE over return on assets as there is no empirical model involving the link between financial performance and capital structure since ROE is sensitive to capital structure decisions (Banker et al., 2014). However, we have an equation where capital structure is a moderating variable.

Four measures of competitive strategy are employed in this study. Two competitive strategies each measure differentiation and cost leadership strategies. We use the ratio of research and development expenses to net sales (RDS) and the ratio of net sales to cost of goods sold (SCOGS) to capture differentiation strategies. We use two measures of differentiation strategies because most studies focus on RDS (such as Frangouli, 2002) and submit that it is positively linked to internal finance which supports the pecking order theory. This stand contends that firms do not only practise hierarchical financing (although it is common among small firms); large firms with high profitability do practise a trade-off theory where they take the advantage of debts. While RDS may be financed internally, SCOGS involves large investments in production, processes and distribution that would involve external financing. In relation to cost leadership, we also employed two measures. The ratio of net sales to capital expenditures on items of property, plant and equipment (SCAPEX), and the ratio of net sales to net book value of property, plant and equipment (SPPE) are used as proxies for cost leadership. Firms will achieve higher sales value if they maximize operational efficiency in the use of property, plant and equipment (Kotha and Nair, 1995; Banker et al., 2014).

Independent and moderating variable – capital structure

The moderating variable in this study came from the Datastream database. The measure of capital structure in this study includes both debt financing and equity financing. This contrasts with previous studies that used only leverage to capture capital structure (Bae et al., 2019; Chen et al., 2019). We use four measures of capital structure: debt ratio, long-term debt ratio, short-term debt ratio and equity ratio. Total debt ratio, long-term debt ratio and short-term debt ratio address the debt financing portion of a firm's capital structure while equity ratio address the equity financing portion. Debt ratio is the proportion of total debt to total assets. Long-term debt ratio is the proportion of long-term debt to total assets. Short-term debt ratio is the ratio of short-term debt to total assets while equity ratio is the ratio of total equity to total assets. This study employed four measures of capital structure (total debt ratio, TDR; long-term debt ratio, LTDR; short-term debt ratio, STDR; and equity ratio, ER). TDR is measured as the ratio of total debt to total assets. LTDR is measured as the proportion of long-term debt to total assets. DR is measured as the ratio of short-term debt to total assets. EQR is measured as the ratio of equity to total assets. We employed equity ratio since a firm is financed with either with full debt, full equity or part debt and part equity.

Control variable – earnings per share

This study controlled for earnings per share (EPS) because firms are more likely to engage in competitive strategy as the EPS increases. It has been documented that earnings per share can explain the competitive information of a firm's investment strategy (Griffin, 1976) and market-share strategy (Kama, 2009). Past studies have largely documented that earnings per share influence the profitability position (Taani, 2011; Purnamasari, 2015). This study finds earnings per share as suitable to determine competitive strategy and financial performance. EPS is measured as the ratio of net profit to number of outstanding shares.

Model specification

We estimated the impact of competitive strategies on the financial performance of non-financial firms in the tourism destination of Bangladesh. We conjecture that the impact of competitive strategies on financial performance is attributed to the heterogeneity impact of a firm's corporate financing decisions. In moderating for the role of capital structure, this indicates that the nature of firms' formulation and implementation of competitive strategies depends largely on firms' access and use of debt and equity financing. The empirical model connecting competitive strategies, capital structure and financial performance is presented as follows:

$$Differentiation_{it} = TDR_{it} + LTDR_{it} + STDR_{it} + EQR_{it} + EPS_{it} + \varepsilon_{it}$$
$$Cost\ Leadership_{it} = TDR_{it} + LTDR_{it} + STDR_{it} + EQR_{it} + EPS_{it} + \varepsilon_{it}$$
$$\begin{aligned}ROE_{it} = {}&\beta_1 Differentiation_{it} + \beta_2 CostLeadership_{it} + \beta_1 Differentiation * TDR_{it} \\ &+ \beta_2 CostLeadership * TDR_{it} + \beta_1 Differentiation * LTDR_{it} \\ &+ \beta_2 CostLeadership * LTDR_{it} + \beta_1 Differentiation * STDR_{it} \\ &+ \beta_2 CostLeadership * LTDR_{it} + \beta_1 Differentiation * EQR_{it} \\ &+ \beta_2 CostLeadership * EQR_{it} + \beta_3 TDR_{it} + \beta_3 LTDR_{it} + \beta_3 STDR_{it} \\ &+ \beta_4 EQR_{it} + EPS_{it} + \varepsilon_{it}\end{aligned}$$

The subscripts i and t represent each firm and time sampled for this study. This study examined both the cross-sectional and time effects of capital structure, competitive strategies, and financial performance. ROE is the return on equity. Cost leadership and differentiation are the two competitive strategies examined in this study. TDR is the total debt ratio. LTDR is the long-term debt ratio. STDR is the short-term debt ratio. EQR is the equity ratio and EPS is the earnings per share ratio. Error (ε) is the error term. We employed a static panel data estimator to analyse the empirical models.

Empirical results

In the previous section, we formulated the empirical models of competitive strategy and profitability, taking capital structure measures as regressors and moderating variables. This section presents the results of the empirical model.

The total debt ratio has the highest mean value of 48.32%, as shown in Table 14.3. Bangladesh firms use more short-term finance than long-term finance. The average equity ratio is 18.585. ROE in comparison to debt and

Table 14.3 Descriptive statistics

	SCAPEX	SCOGS	SPPE	RDS	TDR	LTDR	STDR	EQR	ROE	EPS
Mean	163.915	14.951	133.032	4.209	48.321	6.972	23.443	18.585	15.831	13.459
Median	96.155	7.336	106.839	2.605	47.081	6.434	22.059	17.958	15.920	13.746
Max.	422.823	46.173	248.416	11.561	94.400	46.117	74.485	61.924	19.114	17.217
Min.	1.904	0.108	2.116	0.005	13.209	-51.452	-43.085	-21.599	6.908	7.067
Std. Dev.	143.675	14.684	95.119	3.819	12.638	10.732	12.391	8.643	1.418	1.881
Skewness	1.241	1.666	0.7157	1.355	0.674	-0.010	0.600	0.521	-0.583	-0.777
Kurtosis	3.083	4.413	1.818	3.457	5.992	6.697	6.799	5.782	5.162	3.558

Note: TDR = total debt ratio, LTDR = long-term debt ratio, STDR = short-term debt ratio, EQR = equity ratio, ROE = return on equity, EPS = earnings per share, SCOGS = ratio of net sales to cost of goods sold, RDS = ratio of research and development expenses to net sales, SCAPEX = ratio of net sales to capital expenditures on items of plant, property and equipment, SPPE = ratio of net sales to net book value of plant, property and equipment.

Table 14.4 Correlation matrix

	1	2	3	4	5	6	7	8	9	10
1. TDR	1.000									
2. SCAPEX	0.131	1.000								
3. SCOGS	0.143	0.944	1.000							
4. SPPE	−0.134	0.723	0.807	1.000						
5. ROE	0.063	−0.561	−0.714	−0.700	1.000					
6. RDS	−0.261	−0.137	−0.133	−0.011	−0.200	1.000				
7. EQR	0.566	0.266	0.222	0.063	−0.078	−0.095	1.000			
8. EPS	−0.575	−0.408	−0.466	−0.408	0.439	−0.051	−0.574	1.000		
9. LTDR	0.658	0.285	0.273	−0.016	0.075	−0.288	0.829	−0.483	1.000	
10. STDR	0.673	0.243	0.320	0.053	−0.086	−0.270	0.640	−0.549	0.892	1.000

Note: TDR = total debt ratio, LTDR = long-term debt ratio, STDR = short-term debt ratio, EQR = equity ratio, ROE = return on equity, EPS = earnings per share, SCOGS = ratio of net sales to cost of goods sold, RDS = ratio of research and development expenses to net sales, SCAPEX = ratio of net sales to capital expenditures on items of plant, property and equipment, SPPE = ratio of net sales to net book value of plant, property and equipment.

equity finance is relatively good, at an average rate of 15.83%. All the debt finance variables are positively skewed except long-term debt, which is negatively skewed. The mean values of SCAPEX and SPPE are quite higher than the mean values of SCOGS and RDS. This is because SCAPEX and SPPE are differentiation strategies that are pursued by the firms using their items of plant, property and equipment. However, firms that pursue cost leadership strategy focus more on the cost of goods sold and expenses involving research and development. Firms try to reduce the cost of production and overheads to further pursue a cost leadership strategy.

Table 14.4 shows the correlation matrix. It shows that SCAPEX (cost leadership I) and SCOGS (differentiation strategy I) have weak positive correlation with total debt, while RDS (differentiation strategy II) and SPPE (cost leadership II) have negative weak correlation with total debt. In relation to multicollinearity test, the findings do not reveal serious multicollinearity. Four values are well above 0.80: 0.944, 0.807, 0.829 and 0.892. The values of 0.944 and 0.807 do not belong to the same models as they are different strategies predicted or influenced by capital structure decisions. The values of 0.829 (correlation between EQR and LTDR) and 0.892 (correlation between LTDR and STDR) depict slightly high values above the threshold but variables were not deleted from the model since the goal is to establish different measures of debt financing (total debt, long-term debt and short-term debt) and equity financing (EQR) in determining competitive strategies.

As shown in Table 14.5, the total debt ratio and long-term debt ratio have positive effects on differentiation strategy I (SCOGS), cost leadership I (SCAPEX) and cost leadership II (SPPE). Long-term debt ratio negatively impacts differentiation strategy II (RDS), while the total debt ratio has an insignificant impact on differentiation strategy II. Long-term debt ratio has more effects on differentiation

Table 14.5 Relationship between Capital Structure and Competitive Strategy

Variables	Differentiation Strategy I (SCOGS)		Differentiation Strategy II (RDS)		Cost Leadership I (SCAPEX)		Cost Leadership II (SPPE)	
Total debt ratio	0.0574	0.0678	-0.6098	-0.4374	0.0519	0.0584	0.0481	0.0642
	2.9016	4.3118	-1.3883	-1.1280	2.4219	3.3595	2.2486	3.7595
Long-term debt ratio	0.1172	0.1018	-2.1845	-2.5346	0.1363	0.1201	0.0955	0.0889
	4.4477	5.1657	-2.0639	-2.9371	4.7730	5.5125	3.3560	4.1535
Short-term debt ratio	-0.1560	-0.1422	0.2315	0.4671	-0.1624	-0.1442	-0.1429	-0.1396
	-5.2586	-5.3352	0.3155	0.7978	-5.0515	-4.8930	-4.4591	-4.8226
Equity ratio	-0.0829	-0.0664	1.6843	2.1428	-0.0836	-0.0647	-0.0697	-0.0605
	-3.4625	-3.8131	1.9750	3.2878	-3.2237	-3.3597	-2.6968	-3.1958
Earnings per share	-0.1299		-3.1163		-0.1333		-0.1374	
	-1.1024		-1.1857		-1.0439		-1.0798	
R-squared	0.9134	0.9132	0.1352	0.1214	0.9006	0.8981	0.9265	0.9242
Adjusted R-squared	0.8745	0.8772	0.0995	0.0941	0.8560	0.8559	0.8935	0.8927
F-statistic	29.2486	25.3649	4.8847	4.4551	20.2069	21.2508	28.1028	29.3831
Prob(F-statistic)	0.0000	0.0000	0.0032	0.0021	0.0000	0.0000	0.0000	0.0000
Durbin-Watson stat	1.5261	1.4796	1.2209	1.1810	1.4736	1.3810	1.8046	1.6982

Note: Coefficient values and t-values are presented in the first and second lines, respectively. SCOGS = ratio of net sales to cost of goods sold, RDS = ratio of research and development expenses to net sales, SCAPEX = ratio of net sales to capital expenditures on items of plant, property and equipment, SPPE = ratio of net sales to net book value of plant, property and equipment.

and cost leadership strategies than the total debt ratio when the coefficients of the two capital structure measures are compared. However, both short-term debt and equity ratio have negative effects on differentiation strategy I (SCOGS), cost leadership I (SCAPEX) and cost leadership II (SPPE). This finding suggests that total debt and long-term debt ratios enhance differentiation and cost leadership strategies, while short-term finance and equity finance reduce the competitive power of these strategies. Meanwhile, Bangladesh firms that focus on using R&D as a differentiation strategy can benefit more from the use of equity finance.

All the models are significant at the 1% level of significance with no serious problem of autocorrelation. The R square values show that capital structure has the most effect on cost leadership II ($R^2 = 0.9265$), more effect on differentiation strategy I ($R^2 = 9134$) and significant effect on cost leadership I ($R^2 = 0.9006$). The lowest effect of capital structure is seen when firms use R&D to pursue a differentiation strategy ($R^2 = 0.1332$). These results support the findings in Jordan et al. (1998), which found significant link between strategy and financial policy across UK firms.

Table 14.6 shows the moderating role of capital structure on the relationship between competitive strategy (SCOGS and SPPE) and financial performance. There was a significant increase in financial performance (ROE) after capital structure measures have been moderated for. In firms that consider only differentiation strategy, the moderating effect is that debt and equity finance on differentiation strategy increase financial performance while their moderating effects on cost leadership decrease financial performance. This is in line with the results of Frangouli (2002). However, in firms that pursue the two competitive strategies, long-term debt and short-term debt finance increase the effect of differentiation strategy on financial performance. Conversely, total debt and equity finance increase the effects of cost leadership on financial performance positively as against the negative effects of long-term debt and short-term debt finance on financial performance. Overall, the study finds that the moderating role of capital structure on the link between competitive strategy and financial performance varies with a combination of competitive strategies that a firm adopts.

Conclusion and policy implications

By employing the balanced panel data of 85 firms in Bangladesh, this chapter attempts to explore the heterogeneity effects of capital structure as a competitive power that can change the pattern of current competitive strategies such as differentiation and cost leadership. Generally, our empirical findings show that firms that engage in the use of high debt enhance their level of competition in the tourism industry. It implies that firms that engage more in the use of debt compete more strategically than their counterparts in a tourist destination. Furthermore, our results also show that capital structure increases cost leadership strategy while it reduces differentiation strategy. We find R square significantly increased through the competitive power of capital structure especially when debt finance is used. Theoretically, our findings support the trade-off theory over

Table 14.6 The moderating role of capital structure on the competitive strategy-finance link

Variable:	Model 1A	Model 1B	Model 2A	Model 2B	Model 2C	Model 2D	Model 3A	Model 3B	Model 3C	Model 3D
			Differentiation * Capital Structure – Finance Link				Cost Leadership * Capital Structure – Finance Link			
SCOGS	0.1302 / 3.0019	0.2197 / 3.5022					-0.6575 / -1.4392	0.2220 / 3.0347	0.1241 / 1.4051	-0.1746 / -1.3563
SPPE	-0.1331 / -3.4699	-0.2194 / -4.0042					0.5587 / 1.1814	-0.2201 / -3.3913	-0.1436 / -1.6842	0.1021 / 0.7942
SCOGS*Total debt ratio			0.0057 / 2.1741				0.0181 / 1.9986			
SCOGS*Long-term debt ratio				0.0057 / 1.6719				0.0009 / 0.2498		
SCOGS*Short-term debt ratio					0.0083 / 3.9872				0.0050 / 1.5474	
SCOGS*Equity ratio						0.0159 / 5.0399				0.0212 / 3.7976
SPPE*Total debt ratio			-0.0030 / -1.1555				-0.0136 / -1.4685			
SPPE*Long-term debt ratio				-0.0072 / -2.1156				-0.0013 / -0.3513		
SPPE*Short-term debt ratio					-0.0072 / -3.9403				-0.0031 / -0.9937	
SPPE*Equity ratio						-0.0123 / -4.2508				-0.0153 / -2.5584
Total debt ratio			0.0466 / 2.5998				0.0931 / 1.9562			
Long-term debt ratio				0.0255 / 0.9128				0.0256 / 0.9206		

The competitive power of capital structure 251

	(1)	(2)	(3)	(4)	(5)	(6)	(7)	(8)	(9)	(10)
Short-term debt ratio						0.0638 (3.7981)				0.0654 (2.5642)
Equity ratio								0.0507 (1.8353)		0.0884 (2.3769)
Earnings per share		0.1520 (7.1848)								
R-squared	0.9078	0.9180	0.9294	0.9167	0.9242	0.9195	0.9309	0.9197	0.9248	0.9203
Adjusted R-squared	0.8901	0.9001	0.9042	0.8979	0.9078	0.8990	0.9044	0.9010	0.9080	0.8993
F-statistic	51.1595	51.1509	36.8027	48.7559	56.4148	44.8729	35.1381	49.0220	55.0300	43.7959
Prob(F)	0.0000	0.0000	0.0000	0.0000	0.0000	0.0000	0.0000	0.0000	0.0000	0.0000

Note: Coefficient values and t-values are presented in the first and second lines, respectively. SCOGS = ratio of net sales to cost of goods sold, RDS = ratio of research and development expenses to net sales, SCAPEX = ratio of net sales to capital expenditures on items of plant, property and equipment, SPPE = ratio of net sales to net book value of plant, property and equipment.

Variable	Model 4A	Model 4B
SCOGS*Total debt ratio	−0.0191	−0.0183
	−3.1587	−3.1021
SCOGS*Long-term debt ratio	0.0173	0.0191
	2.2181	2.4200
SCOGS*Short-term debt ratio	0.0192	0.0128
	2.7549	1.7956
SCOGS*Equity ratio	−0.0205	−0.0179
	−2.9556	−2.5910
SPPE*Total debt ratio	0.0101	0.0109
	1.9202	2.1448
SPPE*Long-term debt ratio	−0.0201	−0.0246
	−2.7044	−3.1803
SPPE*Short-term debt ratio	−0.0165	−0.0104
	−2.4131	−1.4502
SPPE*Equity ratio	0.0231	0.0237
	3.2395	3.3805
SCOGS	0.6277	0.6476
	2.4832	2.5737
SPPE	−0.3301	−0.4416
	−1.5166	−1.9740
Total debt ratio	−0.1619	−0.1251
	−3.4386	−2.6109
Long-term debt ratio	0.1147	0.0759
	1.6214	1.0194
Short-term debt ratio	0.0969	0.0829
	1.7724	1.5388
Equity ratio	−0.1119	−0.0588
	−2.1480	−1.0082
Earnings per share		0.1063
		2.6913
R-squared	0.9930	0.9936
Adjusted R-squared	0.9889	0.9895
F-statistic	242.8940	242.5904
Prob(F-statistic)	0.0000	0.0000
Durbin-Watson stat	1.6596	1.6934

Note: Coefficient values and t-values are presented in the first and second lines, respectively. SCOGS = ratio of net sales to cost of goods sold, RDS = ratio of research and development expenses to net sales, SCAPEX = ratio of net sales to capital expenditures on items of plant, property and equipment, SPPE = ratio of net sales to net book value of plant, property and equipment.

the pecking order theory. Firms in tourist destinations that engage more in debt finance perform better and achieve more industry leadership than firms that live on ploughed back profitability, relying on internal financing. The demands for environmental sustainability within the tourism sector require huge investments that internal finance cannot address if the firm must grow, compete and attract more investors. We found that short-term debts and equity financing negatively affect profitability while both total debt and long-term debt significantly encourage firms to participate in differentiation and cost leadership strategies.

The main corporate policy implication of the results is that debt finance, i.e. total debt and long-term debt, play a vital role in determining competitive strategies in tourist-destination environments. To ensure that firms meet with the challenges of the tourism business and continuing demands for environmental sustainability, social innovation and corporate social responsibility, firms in a turbulent tourist environment should strive to use more debts to further enhance their competitive advantage. The use of internal financing and equity finance are not significant to enhance profitability and competitive strategies among firms in tourist environments. Additionally, investors in tourist environments will have less concern for agency problems as the use of additional debt will ensure the reduction of free cash flows into the hands of corporate managers.

The contributions of this study are two-fold. Theoretically, it extended the trade-off theory of capital structure to determine the competitive strategies of firms. While past studies have documented that differentiation strategy, focus strategy and cost leadership strategy determine corporate competitive strategies, this study found that the cost leadership strategy has more influence on a firm's profitability and competitive strategy than the differentiation strategy. In addition, the findings of the study have practical implications to control for agency problems. The use of additional debt reduces free cash flow, implying that cost leadership helps to manage agency problems while differentiation strategy may trigger increasing agency conflicts. Firms with cost leadership strategies will have high profitability which can influence their use of additional debt given the size of their profitability and their ability to secure creditors' claims on the firm's assets. Firms need additional debt to increase efficiency and productivity, improve quality and eliminate waste. Cost leadership firms need to invest heavily in technology in the production process and lean production methods to control costs.

However, this study has a few limitations. The study focuses more on profitability as a measure of financial performance. Future studies should consider other financial and non-financial measures of a firm's performance such a market share, customer service, customer satisfaction and employee retention. Future research should also focus on other types of competitive strategies besides differentiation and cost leadership. Areas such as Porter's six forces model can be examined in relation to financing strategy. Industry strategies are much more important than firms' level competitive strategies as the tourist industry faces continuous high competition and regulatory issues. Future studies should also examine the dynamic effect of capital structure on competitive strategies by using the panel generalized method of moments (GMM) estimator. By not focusing on these limitations, the findings of our study explain the practise of competitive strategies through a financing policy perspective. Financing policy hitherto best explains the heterogeneity in firms' competitive strategies.

References

Ajmera, P. (2017). Ranking the strategies for Indian medical tourism sector through the integration of SWOT analysis and TOPSIS method. *International Journal of Health Care Quality Assurance*, 30(8), pp. 668–679.

Bae, K. H., El Ghoul, S., Guedhami, O., Kwok, C. C. and Zheng, Y. (2019). Does corporate social responsibility reduce the costs of high leverage? Evidence from capital structure and product market interactions. *Journal of Banking are Finance*, 100, pp. 135–150.

Bakker, M. (2019). A conceptual framework for identifying the binding constraints to tourism-driven inclusive growth. *Tourism Planning are Development*, 16(5), pp. 575–590.

Balakrishnan, S. and Fox, I. (1993). Asset specificity, firm heterogeneity and capital structure. *Strategic Management Journal*, 14(1), pp. 3–16.

Banker, R. D., Mashruwala, R. and Tripathy, A. (2014). Does a differentiation strategy lead to more sustainable financial performance than a cost leadership strategy? *Management Decision*, 52(5), pp. 872–896.

Barton, S. L. and Gordon, P. J. (1988). Corporate strategy and capital structure. *Strategic Management Journal*, 9(6), pp. 623–632.

Birkinshaw, J., Morrison, A. and Hulland, J. (1995). Structural and competitive determinants of a global integration strategy. *Strategic Management Journal*, 16(8), pp. 637–655.

Borsekova, K., Vaňová, A. and Vitálišová, K. (2017). Smart specialization for smart spatial development: Innovative strategies for building competitive advantages in tourism in Slovakia. *Socio-Economic Planning Sciences*, 58, pp. 39–50.

Boskov, T., Fejzullai, L., Fejzullai, E. and Dimitrov, N. (2018). *Investment driven strategies in tourism sector engine for economic growth*. Saarbrücken Lambert Academic Publishing.

Camisón, C., Puig-Denia, A., Forés, B., Fabra, M. E., Muñoz, A. and Munoz Martinez, C. (2016). The importance of internal resources and capabilities and destination resources to explain firm competitive position in the Spanish tourism industry. *International Journal of Tourism Research*, 18(4), pp. 341–356.

Carrillo-Hidalgo, I. and Pulido-Fernández, J. I. (2019). Is the financing of tourism by international financial institutions inclusive? A proposal for measurement. *Current Issues in Tourism*, 22(3), pp. 330–356.

Chen, M. H. (2010). The economy, tourism growth and corporate performance in the Taiwanese hotel industry. *Tourism Management*, 31(5), pp. 665–675.

Chen, Y. J., Tsai, H. and Liu, Y. F. (2018). Supply chain finance risk management: Payment default in tourism channels. *Tourism Economics*, 24(5), pp. 593–614.

Chen, Z., Harford, J. and Kamara, A. (2019). Operating leverage, profitability, and capital structure. *Journal of Financial and Quantitative Analysis*, 54(1), pp. 369–392.

Damayanti, M., Scott, N. and Ruhanen, L. (2019). Coopetition for tourism destination policy and governance: The century of local power? In E. Fayos-Solà and C. Cooper (eds.), *The future of tourism*. Cham: Springer, pp. 285–299.

Dehning, B. and Stratopoulos, T. (2003). Determinants of a sustainable competitive advantage due to an IT-enabled strategy. *The Journal of Strategic Information Systems*, 12(1), pp. 7–28.

Della Corte, V. and Aria, M. (2016). Coopetition and sustainable competitive advantage. The case of tourist destinations. *Tourism Management*, 54, pp. 524–540.

Dwyer, L. (2018). Economics of tourism. *The SAGE Handbook of Tourism Management*, 173.

Eccles, R. (1991). The performance measurement manifesto. *Harvard Business Review*, 69(1), pp. 131–137.

Ezeuduji, I. O. (2015). Strategic event-based rural tourism development for sub-Saharan Africa. *Current Issues in Tourism*, 18(3), pp. 212–228.

Frangouli, Z. (2002). Capital structure, product differentiation and monopoly power: A panel method approach. *Managerial Finance,* 28(5), pp. 59–65.

Fuster, B., Lillo-Bañuls, A. and Martínez-Mora, C. (2018). Offshoring of services as a competitive strategy in the tourism industry. *Tourism Economics*, 24(8), pp. 963–979.

Griffin, P. A. (1976). Competitive information in the stock market: An empirical study of earnings, dividends and analysts' forecasts. *The Journal of Finance*, 31(2), pp. 631–650.

Guthrie, J. P., Spell, C. S. and Nyamori, R. O. (2002). Correlates and consequences of high involvement work practices: The role of competitive strategy. *International Journal of Human Resource Management,* 13(1), pp. 183–197.

Hambrick, D. C. (1983). An empirical typology of mature industrial-product environments. *Academy of Management Journal*, 26(2), pp. 213–230.

Hong, W. C. (2009). Global competitiveness measurement for the tourism sector. *Current Issues in Tourism*, 12(2), pp. 105–132.

International Finance Corporation (2017). *Doing Business 2017 Equal Opportunity for All.* Retrieved from: https://www.doingbusiness.org/content/dam/doing Business/media/Annual-Reports/English/DB17-Report.pdf (accessed: the 12th August 2019).Jordan, J., Lowe, J. and Taylor, P. (1998). Strategy and financial policy in UK small firms. *Journal of Business Finance are Accounting,* 25(1-2), pp. 1–27.

Kama, I. (2009). On the market reaction to revenue and earnings surprises. *Journal of Business Finance are Accounting,* 36(1-2), pp. 31–50.

Kotha, S. and Nair, A. (1995). Strategy and environment as determinants of performance: Evidence from the Japanese machine tool industry. *Strategic Management Journal*, 16(7), pp. 497–518.

Lee, C. F. and King, B. (2006). Assessing destination competitiveness: An application to the hot springs tourism sector. *Tourism and Hospitality Planning are Development*, 3(3), pp. 179–197.

Lee, J. W. and Manorungrueangrat, P. (2019). Regression analysis with dummy variables: Innovation and firm performance in the tourism industry. In S. Rezaei (ed.), *Quantitative tourism research in Asia: Current status and future directions.* Singapore: Springer, pp. 113–130.

Lowe, J., Naughton, T. and Taylor, P. (1994). The impact of corporate strategy on the capital structure of Australian companies. *Managerial and Decision Economics*, 15(3), pp. 245–257.

Martínez-Román, J. A., Tamayo, J. A., Gamero, J. and Romero, J. E. (2015). Innovativeness and business performances in tourism SMEs. *Annals of Tourism Research*, 54, pp. 118–135.

Mihalic, T. (2016). Sustainable-responsible tourism discourse – towards 'responsustable' tourism. *Journal of Cleaner Production*, 111, pp. 461–470.

Mistilis, N. and Daniele, R. (2005). Challenges for competitive strategy in public and private sector partnerships in electronic national tourist destination marketing systems. *Journal of Travel & Tourism Marketing*, 17(4), pp. 63–73.

Navickas, V. and Malakauskaitė, A. (2009). The possibilities for the identification and evaluation of tourism sector competitiveness factors. *Inžinerinė ekonomika*, 1, pp. 37–44.

Phillips, J., Faulkner, J. and International, S. (2017). *Tourism investment and finance: Accessing sustainable funding and social impact capital.* New York, NY: US Agency for International Development, USAID.

Porter, M. E. (1980). *Competitive strategy.* New York: Free Press.

Purnamasari, D. (2015). The effect of changes in return on assets, return on equity, and economic value added to the stock price changes and its impact on earnings per share. *Research Journal of Finance and Accounting,* 6(6), pp. 80–90.

Raymond, L., Bergeron, F., Croteau, A. M. and Uwizeyemungu, S. (2019). Determinants and outcomes of IT governance in manufacturing SMEs: A strategic IT management perspective. *International Journal of Accounting Information Systems,* 35, 100422.

Taani, K. (2011). The effect of financial ratios, firm size and cash flows from operating activities on earnings per share: (an applied study: on Jordanian industrial sector). *International Journal of Social Sciences and Humanity Studies,* 3(1), pp. 197–205.

Verreynne, M. L., Williams, A. M., Ritchie, B. W., Gronum, S. and Betts, K. S. (2019). Innovation diversity and uncertainty in small and medium sized tourism firms. *Tourism Management,* 72, pp. 257–269.

Wanzenried, G. (2003). Capital structure decisions and output market competition under demand uncertainty. *International Journal of Industrial Organization,* 21(2), pp. 171–200.

World Economic Forum (2017). *Travel and Tourism Competitiveness Report 2017.* Retrieved from: www.weforum.org/reports/the-travel-tourism-competitiveness-report-2017 (accessed: the 01st December 2019).

Yusuf, Y. Y., Gunasekaran, A., Adeleye, E. O. and Sivayoganathan, K. (2004). Agile supply chain capabilities: Determinants of competitive objectives. *European Journal of Operational Research,* 159(2), pp. 379–392.

Zhou, P. (2019). Development policy, prospective and investment outlook of China's tourism industry. In *The theory and practice of china's tourism economy (1978–2017).* Singapore: Springer, pp. 99–114.

15 Revenue management in the tourism and hospitality industry with special reference to Bangladesh

Md Yusuf Hossein Khan, Shahriar Tanjimul Islam and Azizul Hassan

Introduction

Currently tourism is emerging due to its ability to progress growth in a country, but also because it plays a huge role for social development. This sector is getting more and more integrated with production in national economies, employment, manufacturing and agriculture, hence it is getting further economic diversification and playing a huge role in strengthening developing countries' economies. According to Kostin (2018), this industry encompasses hotels, travel agencies, enterprises, food, construction and train hospitality professionals. He further emphasizes this field is one of the fastest growing fields in the modern economy. Following United Nations Conference on Trade and Development (2010), the industry is the top export earner in 60 countries. This sector also contributes to foreign exchange earnings for one-third of the developing countries and half of the least developed countries. This signifies the contribution of tourism sector. The main purpose of this chapter is however, observing the significance of revenue management in the tourism and hospitality sector in particular with hotels. Hotels always play a huge role for tourists to decide whether they will visit that particular place or not, and also play an important role in developing that local area, since many jobs are created with this industry (waiters, managers, guides, chefs/cooks etc.).

According to Business Dictionary (2019), revenue is defined as any income, which is generated for sale of goods/services, or from the use of capital or assets, associated with the main functions of an organization, before the deductions of any expenses. Hence it can be said that revenue is any direct income from customers, which the business receives, before deducting expenses. Therefore, revenue management (RM) maximizes the organization's raw income with the means of matching supply and demand of customers into different segments on their income and allocating capacities (El Haddad et al., 2008). Kimes and Wirtz (2003) define RM as the application of information systems and pricing strategies to provide precise amounts of services, as per the level of the customer, at right time. Since revenue is increased more with the level of customers arriving, marketing management has a major role to play to increase the demand of the customers, resulting in more revenue for the hotels (Cross et al., 2009) and managing consumer behaviour (Anderson and Xie, 2010).

Findings from this chapter will help us in understanding the nature and characteristics that influence the success of the RM implementation process. By doing so, it will help us to enrich our knowledge in successful dynamic strategies which will improve the implementation process. The chapter is organized to present the theoretical framework background of the implementation of the RM system and its practices in hotels. This is followed by the research, design and data collection methodology. Finally, a discussion on managerial implication and a conclusion will be drawn.

The term "reservation management" is rarely used in the literature. The key focus on this chapter will be yield management (YM) also known as revenue management (RM). The two terms however are not interchangeable. Sigala et al. (2001) explains that revenue management is about the process of how to do something; on the other hand, YM has an output focus on what to achieve. In strategic terms, reservation management focuses primarily on to the implementation stage, whilst the latter term, YM, mainly focuses on strategic goals.

RM or YM are common terms that are applied in the hotel industry to assist hotels as explained by El Haddad (2009) and Okumus (2004) in determining their room price and allocation to maximize revenue/income, simply by trying to sell the most rooms to the highest-paying customers. According to Cross et al. (2009) and Delain and O'Meara (2004), to practise RM, it is essential for an organization to assess the effectiveness of its various decisions. Kimes (1989) clearly defines RM/YM as a method that can help a firm sell the right inventory unit to the right customer at the right time and for the right price. This guides the decision of how to allocate undifferentiated units of limited capacity to available demand in a way that maximizes profit or revenue.

Loose forecasts may lead to suboptimal decisions and might prevent the organization from gaining maximum income. A good RM system is required in order to avoid an undesirable impact on hotel revenue and customers. This literature review will be more narrowly divided into three parts: the concepts of RM, implementation of RM and making improvements in Jones' model and lastly it will discuss the findings of the previous research regarding the impact of RM.

Concepts and models of revenue management

To start with the logical framework, we can look at the work of Noone and Mount (2008) which summarizes the key activities of hotel RM. It was adopted and extended into two main parts. Firstly, the core activities of hotel RM that were expanded to incorporate the RM process in other hospitality and tourism organizations. In particular, it is stated by business analysts that intelligence id needed for a successful implementation of RM, and along with this performance analysis and evaluations were added to their framework. Their work was further backed by the works of Kimes (1999) on restaurant RM and Kimes and McGuire (2001) on function space RM and hotel yield management. With so many works on these as extensions, the core activities of hospitality RM were then conceptualized as a cyclical process involving business analysis. Further analysis includes

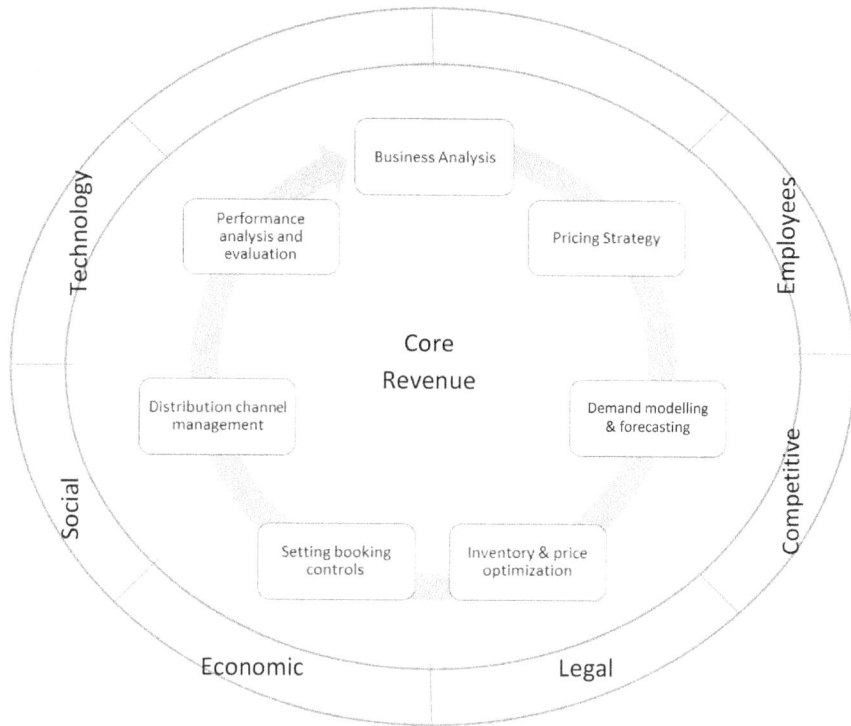

Figure 15.1 Core revenue management process

Source: Adapted and expanded from Noone et al., 2003

pricing strategy, demand modelling and forecast, inventory and price optimizations, distribution channel management, booking controls and performance analysis.

Figure 15.1 illustrates a concept map of the revenue management constellation of terms. In this chapter however we will focus how this whole RM system can be further improved with the introduction of loyalty cards to this method. The significance and benefits of this are further explained after the implementation stage. How the introduction to loyalty cards can improve the RM software and non-pricing techniques will be explained. The chapter will look further into how introductions to this can add more value in non-pricing techniques and better inventory management, as shown in Figure 15.2. In the end the chapter will try to give some significance to the pricing techniques to make the RM system more effective.

Implementation and improvements have been made in the RM model so far (Kimes, 1989; Yeoman and Watson, 1997) as these factors can be the greatest determinants of the success or failure of embracing and using RM (see

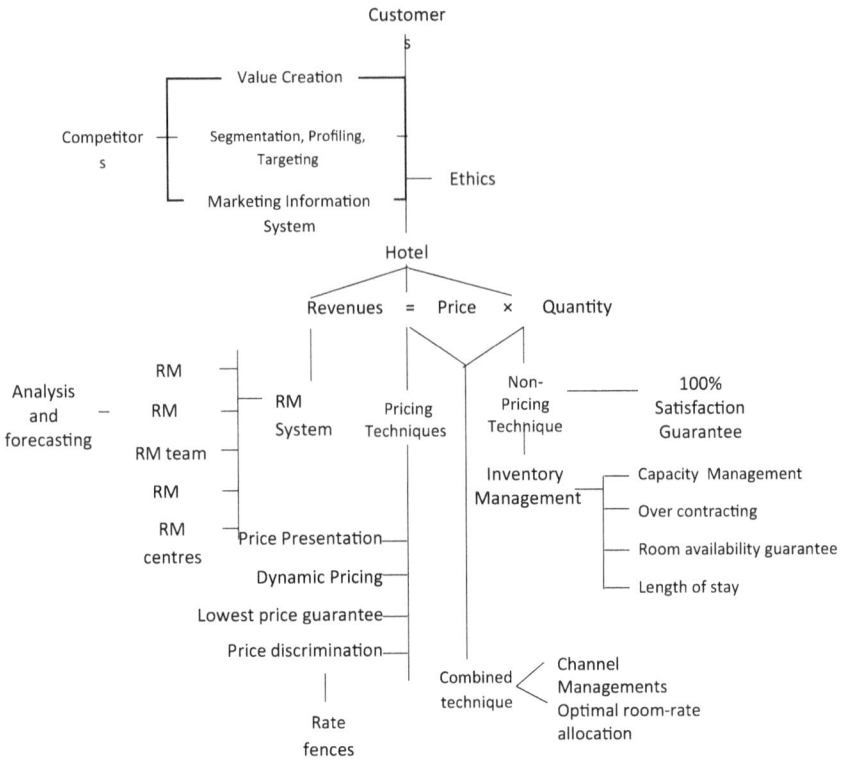

Figure 15.2 Revenue management constellation concept map
Source: Adapted from Ivanov (2014)

Figure 15.3). El Haddad (2015) suggests that any possible link should eliminate the existing association between reservation and strategic decision making units, as in reality, a short-term decision whether to accept, decline or reject a customer's booking request should not correlate with long-term tactics and strategies on RM.

Introduction of loyalty card programmes

Reward or loyalty programmes are becoming popular in the hotel industry for many reasons. According to studies conducted by Internet Marketing Inc. (2019), travellers are loyal to hotel brands they can trust, that demonstrate consistent value and that can influence customers both who are on property and between stays. The need for customer loyalty is important as this shows the strength of the relationship between an individual's relative attitudes and repeat patronage (Dick and Basu, 1994). Customer loyalty is the commitment of the customer towards a particular brand or company.

EXTERNAL ENVIRONMENT

CUSTOMERS

Feedback, Support & Praise

Hotel/Customer Interface

Decision System

Strategic Decision-making
- Data collection
- Optimum guest mix /
- Market segments
- Capacity levels
- Pricing
- Create strategy

- Management focus
- Support from management
- Decision to implement & objectives of RM

Operational Decision-Making (ODM)
- Daily operations
- Implemet RM techniques
- Advise on rates and restrictions

Evaluate RM system

Evaluate RM activites

Revise ODM

Communication Reports

Information System

Demand Analysis
- Analyze and track demand
- Analyze patterns
- Forecasts

Reservations
- Daily Operations
  - Accept
  - Decline
  - Deny

Technology
- Select software
- Information management technology

Human Support
- Incentive Scheme
- Yield culture
- Training
- Create RM committee/team

Decision Support System

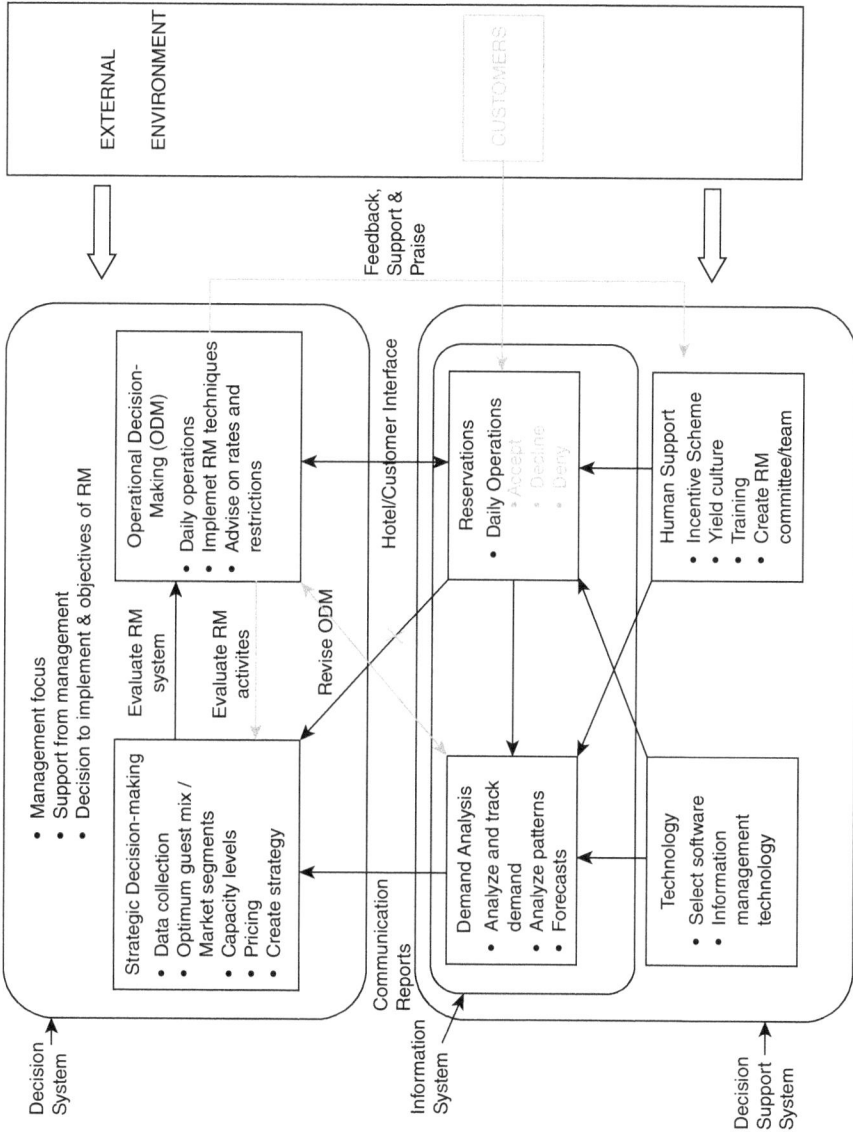

Figure 15.3 Revised hotel revenue management
Source: Adapted from Jones (1999, p. 1115)

In tourism, this loyalty has gained even more significance. Here looking at the attitudinal loyalty is also important, since the level of customer's intentions to revisit or repurchase at a business and their willingness to recommend a destination is measured (Li and Petrick, 2008; Yoon and Uysal, 2005). Loyalty researchers tend to observe attitudes to explain the behaviour and conceptualize loyalty from a casual perspective (Baldinger and Rubinson, 1996; Dick and Basu, 1994). Furthermore, Oliver (1999) found that positive attitudes towards a certain purchase experience also changed the attitudes toward the product or brand. It is very significant to understand that attitudinal loyalty towards a brand increases as the result of attitude development (Dick and Basu, 1994; Oliver, 1999).

From the literature just discussed we can understand the significance of customer loyalty. Hence for this to develop in the RM of the hospitality and hotel management sector, loyalty cards have been suggested. Card-based loyalty has gained a huge popularity especially during the last decade (Karolefski, 2003). Usage of these cards has become very common in consumers' day-to-day lives. One of the major benefits of giving loyalty cards to customers is the collection of useful information about consumer behaviour, which can be used for marketing decision making (Ergin et al., 2011). Laškarin (2013) also explains how these cards enable companies to customize their marketing mix variables. Through this approach, gathering useful information about the consumers and accumulated knowledge, the company can develop effective marketing strategies.

Improvements of RM through loyalty card programmes – suggested approach

Customer loyalty programmes and cards can be added as part of the programme in RM before accepting or declining a customer's booking request, influencing pricing and non-pricing techniques and inventory management systems. According to Sigala et al. (2001, p. 368), "Demand is not analysed by market segments; rather, it is analysed at a finer level of analysis concerned with the purchasing behaviour of different segments through different distribution channels against the demand conditions". Thus, loyalty cards should add a new feature into the RM system, creating a better customer database that helps build a long-term customer base, without the need for excessive promotional expenses. This database can help us in understanding the demand pattern of the customers more accurately, and more loyal customers will get more preference through this approach. Such an approach can make the RM system more effective and efficient, as the right promotional activities will be directed towards the right customers.

Implementation of this new approach

Sigala et al. (2001) proposed that in this network era, reservation management has become even more of a strategic implementation tool, with control at the head office. Through the usage of loyalty cards, more data about the customers can be stored for effective decision making and promotional activities. While doing so, the role of software is imperative. According to Guadix et al. (2010), the processing of a large database is not possible without appropriate RM software and the hotels that implement this gain strategic advantages over those that simply rely on intuition. Since so much customer data can be collected through loyalty cards and can be stored, it is important to have an effective customer relationship management (CRM). Generally, RM is closely linked with CRM. Hence there has been a great deal of study of these two functions amongst many researchers (Wang and Bowie, 2009). RM and CRM can have different objectives and time horizons. While RM is more short-term oriented, CRM intends to focus more on the long-term relationships between the company and its customers. On the contrary, Noone et al. (2003) have shown that CRM and RM should be viewed as complimentary business strategies, and RM tools can be effectively used in CRM practices. They found that RM plays a supportive role to CRM in the process of establishing and maintaining long-term profitable relationships between the hotel and its customers, whilst I will argue that CRM has to be part of RM, so that the most profitable and loyal customers will get the most from the hotels.

CRM systems allow more functionality to segment and analyse customer visits and spend, in the hopes of having a big customer database with information about their desire and ability to spend, and able to segment them by lifetime value (Go.duettocloud.com, 2019). There are many ways to segment customers, for instance on the basis of their length of stay, amount of expenditure to avail services and price sensitiveness, total revenue per room, total revenue per client and so on. It is one of the important components that is needed to be applied in RM (Xotels.com, 2019).

With a strong CRM system and RM practice, the next rational step would be getting connected and merged. With the help of CRM, we can calculate the profit at the customer level and segment by those with profitability statistics (i.e. REVPAR, room revenue per available room; TREVPEC, total revenue per client; and so on). Those segments then can be mapped to the Revenue Management System (RMS) to offer yieldable rates per profit tier or customer segment. Each tier then can receive independent yieldable prices, which would also correspond to rate codes in the RM to deal issues like reservations, availability, check in/out, guest profiles, report generation and so on. The key idea would be to have such a booking engine and call centre that would run RM software, allowing the customer to be recognized at the point of booking to decide the appropriate rate code.

This way, the most valuable customers will get maximum services and hotel chains will have more opportunities to retain them, plus the reservation team can work more efficiently. This approach will become very useful during the peak season or while deciding the correct amount of discounts the customers should receive.

Laškarin (2013) further explains that with proper reward schemes and introducing guest-friendly rules hotels can boost loyalty. Loyalty programmes are one of reasons why guests will select a specific hotel. The other three factors customers consider before choosing are location, price and past experience. Hence it can be said further that, with a proper rewarding system we can convert bargain-hunting guests to loyal guests, who are expected to come more frequently or

Level of profitability	Frequency of stay	
	Low	High
High	**POTENTIALLY LOYAL GUESTS** ☐ Members of a number of loyalty programs ☐ Need to be encouraged through relationship marketing.	**LOYAL GUESTS** ☐ Members of one loyalty program. ☐ CRM plays a vital role (maintaining relations with key clients).
Low	**BARGAIN HUNTERS** ☐ Price sensitive. ☐ Are on the lookout for bargains (group discounts). ☐ Very small chance they could become loyal guests.	**FREQUENT GUESTS** ☐ Price sensitive. ☐ Might be motivated if the advantages they could gain by becoming members of a loyalty program were pointed out to them.

Figure 15.4 Degrees of guest loyalty and profitability by level of profitability and frequency of stay

Source: Adapted from Laškarin, 2013

have longer stays, which will also have a big positive impact on the revenue source of the company.

Evaluation of this suggested approach

It is also important for management to understand that, in order to work effectively with RM, they will have to assess the potential risks of loyalty cards and consider the significant contributions of good pricing methods as well. From research of retail stores, Egin, Parilti and Ozsacmaci (2011) say that customers tend to utilize a range of store loyalty cards rather than just one. Hence from a business point of view, the company may not gain as much as it thought it has gained. O'Malley (1998) states that it is very important to consider that loyalty programmes with a strong operational plan can assist in developing a more cost-effective marketing strategies to increase lifetime value. A critical question lies in whether there is a way to give customers something valuable that cannot be duplicated by competitors. Such uniqueness mainly relies on the core benefit for customers; if done successfully, it will be less tempting for customers to switch to other brands/hotels. Such uniqueness can rely or brand value and pricing to maximize the profitability of the organization.

In this section, we have discussed the usefulness of loyalty cards and their implementation. By using loyalty/membership cards effectively, there is a good scope to create brand loyalty. Hence the significance of pricing methods is discussed next, so that the company can maximize its revenue and remain more unique at the same time.

Pricing techniques

Many scholars (i.e. Cross et al., 2009) have identified the importance of pricing and changes in price in accordance with the state of market in order to create a sustainable competitive advantage. Findings from Koenig and Meissner (2010) state that in the hotel industry the most commonly used pricing revenue management tools include dynamic pricing, price discrimination and lowest price guarantee.

Researchers like Palmer and McMahon-Beattie (2008) state that one of the fundamental concepts of pricing nowadays is dynamic pricing. It allows hotels to maximize revenue and yield by offering a price that will reflect the level of demand and availability of rooms and services. With this process, customers will pay different prices even when they have similar booking details such as length of stay, number and types of rooms, board basis and so on, depending on the moment of the reservation. However, such changes in price may lead to criticism by customers. Nevertheless, dynamic pricing can offer that extra bit of profitability which often may get sacrificed due to loyalty offers. Ivanov and Zhechev (2011) warned that charging at different prices should be applied with caution and should provide plenty of information to customers about booking terms and conditions.

Another important RM pricing tool, known as price discrimination, is considered to be the heart of pricing strategies (Hanks, 2002; Kimes and Wirtz, 2003). Price discrimination in practice means that a hotel charges their customers different prices for the same rooms or hotel facilities. The main reason behind this are the differences in price sensitiveness of hotels' market segments. For example, it can be seen that business travellers are less sensitive to price compared to leisure travellers, and are ready to pay higher prices if required. However, one major risk from this approach is that consumers may migrate from high-end services and products to low-end services and products. In order to avoid such migration, Zhang and Bell (2010) suggested introducing price fences, or conditions under which specific products are offered on the market. Many researchers (i.e. Hanks et al., 2002; Kimes and Singh, 2009) at different times suggested different types of price fences which include conditions like day of the week, duration of the stay, cancellation, amendment, payment, exclusive guests (i.e. president of a club, government officials, etc.), age and lead period.

The lowest price guarantee is another RM pricing tool. Carvell and Quan (2008) point out that hotels tend to provide their customers with the lowest price guarantee. This means that if the customer finds a lower price for the same or similar hotel within 24 hours after their booking, the hotel will match that lower price. This approach is also quite popular. According to Demirciftci et al. (2010), the lowest price claimed by several US hotel chains is a great source for advertisement on their websites.

Case study

A case study is summarized next, where little to some revenue management was practised. This case study takes place in Bangladesh, and involved exploratory research to analyse the data and their findings. Here the data were collected through a structured questionnaire and daily revenue was collected through data directly from field. Some pricing information was also collected from brochures or leaflets.

Tourism at Cox's Bazar, Bangladesh

According to Chowdhury and Chowdhury (2015), Cox's Bazar became the tourist capital in Bangladesh, amongst a lot of tourist spots. The researchers conducted and compared the results of their sample hotels between 2008 and 2014, where in 2008 they collected data of four private and four public hotels in Cox's Bazar, and during 2014 they collected those same eight hotels, with three more private hotels. To calculate guest occupancy and profit, some specific data was very important, and those data included variable cost per month, per bed daily revenue and occupancy. From their findings some interesting information were found in revenue in terms of private and public, as from 2008 to 2014 there were significant differences between the occupancy and profit of the public and private hotels. During 2014, private hotels had significant differences in occupancy and profit. Their findings signified that private hotels had better revenue management compared to public hotels. The researchers affirmed that the participation of the government was quite inadequate, and the contribution of investment from the public sector was very nominal. From the findings of Chowdhury and Chowdhury (2015), during 2008, public hotels enjoyed higher profits but in the course of seven years, the scenario reverses and it was the private hotels which were enjoying higher occupancy and profitability.

It has to be to be noted that during these seven years private hotels made significant investments, as backed by Islam (2013). He further states that the Bangladesh government also played a positive role in promoting domestic and international tourism from the private sector. Another factor that can affect tourism adversely was the political unrest during 2012, as elections took place. This was one of the many reasons why no further development took place in the public sector.

Suggestions

Tourism at Cox's Bazar, Bangladesh

Firstly, from an industry perspective, public hotels can provide seasonal discounts, better management of variable costs and have better understanding of the contribution margin per bed as suggested by Chowdhury and Chowdhury (2015). There is no doubt that the hotels in the public sector of Bangladesh need investment in areas like infrastructure and technology, in order to be more competitive and profitable like private hotels. Secondly, they need to have a more up-to-date RM system. From the findings of El Haddad (2015), implementation of RM has strongly impacted Unidom Hotels' sales and improved their profitability. This has helped the hotel to maximize most returns from their peak demand days. For public hotels in Cox's Bazar, they can start by implementing thorough pricing strategies, like introducing dynamic pricing or lowest price guarantees. With dynamic pricing a higher price can be effectively imposed during high season and a lower price charged during low season, and also if they book much earlier, then their booking price will be lesser and vice versa if they try and book at a very short notice. With lowest price guarantee strategies, they can attract more customers very easily, who are looking for budget hotels. Furthermore, loyal or profitable customers can be offered more facilities or services or kind gestures, like free drinks, breakfast, free mattresses and so on. These are some non-pricing strategies which can be implemented by public hotels.

Having a good RM system and a strong CRM system where a big customer database can be built will be very beneficial for business. The most profitable customers or services can easily be identified with a good RM system. This can help the company in sending them promotional rewards. The areas and scope for a good RM system are clearly shown in Figure 15.3. Customer response and feedback can be stored and can make the management decision and plans even more effective. The introduction of loyalty or membership cards will strengthen the CRM software, enabling them to provide better promotional offers to loyal customers. This will help customize the offers and may provide easier terms and conditions for booking (i.e. providing discounts in booking, etc.).

Conclusion

From the review of different academic literature and case studies in the field of hotel RM, it is evident there is more scope for research in this area. Hotel revenue practices have come a long way in the last two decades and developed a lot as they are used heavily in the hospitality industry. In the case study we have found that many public hotels in Cox's Bazar are underperforming compared to private hotels. Some of the basic problems that were noticed were due to lesser investment and no modern RM system. To enhance the performance of those hotels, some valuable suggestions are provided, from the perspective of revenue management. Ivanov and Zhechev (2011) state many big hotels have gone with advance forecasting models, I however believe that the whole hospitality industry has to evolve and cope with online distribution and more intelligent approaches through technology in order to thrive. Improving and depending on customer database and software technology is imperative in order to make effective loyalty programmes (through loyalty cards, etc.) or effective pricing. This way companies will greatly enhance the functions of revenue management systems, generating more revenue for the company (short and long term), and reduce operational costs and as a result gain more profits at the end.

In conclusion, revenue management in the old school sense cannot provide enough value for hotels. It has to become more of a strategic and tactical tool for revenue managers, other than simply becoming a tool for managing customers efficiently. This chapter has observed that the introduction of loyalty cards into RM, collecting more information from CRM and implementing the ideas in theory and case is possible. Thus, it was found that hoteliers will be able to provide their best services or available rooms to their most profitable customers. Companies will have more control over which customers to give discounts and not giving discounts and so on. Hence in the end of this literature paper a few RM tools for pricing were discussed.

References

Anderson, C. K. and Xie, X. (2010). Improving hospitality industry sales: Twenty five years of revenue management. *Cornell Hospitality Quarterly*, 51(1), pp. 53–67.

Baldinger, A. L. and Rubinson, J. (1996). Brand loyalty: The link between attitude and behaviour. *Journal of Advertising Research*, 36, pp. 22–36.

Business Dictionary. *Revenue*. Retrieved from: www.businessdictionary.com/definition/revenue.html (accessed: the 18th October 2019).

Carvell, S. and Quan, D. (2008). Exotic reservations – low-price guarantees. *International Journal of Hospitality Management*, 27(2), pp. 162–169.

Chowdhury, T. and Chowdhury, M. (2015). An analysis of guest occupancy and profit of private and public hotels in Cox's bazar. *Asian Business Review*, 4(3), p. 50.

Cross, R., Higbie, J. and Cross, D. (2009). Revenue management's Renaissance: A rebirth of the art and science of profitable revenue generation. *Cornwell Hospitality Quarterly*, 50(1), pp. 56–81.

DeLain, L. and O'Meara, E. (2004). Building a business case for revenue management. *Journal of Revenue and Pricing Management*, 2(4), pp. 368–377.

Demirciftci, T., Cobanoglu, C., Beldona, S. and Cummings, P. (2010). Room rate parity analysis across different hotel distribution channels in the U.S. *Journal of Hospitality Marketing and Management*, 19(4), pp. 295–308.

Dick, A. and Basu, K. (1994). Customer loyalty: Toward an integrated conceptual framework. *Journal of the Academy of Marketing Science*, 22(2), pp. 99–113.

El Haddad, R. (2009). The implementation of hotel revenue management practices and the implication on customers' behavioural intentions. *Unpublished PhD thesis*. Guildford: University of Surrey.

El Haddad, R. (2015). Exploration of revenue management practices- case of an upscale budget hotel chain. *International Journal of Contemporary Hospitality Management*, 27(8), pp. 1791–1813.

El Haddad, R., Roper, A. and Jones, P. (2008). The impact of revenue management decisions on customers' attitudes and behaviours: A case study of a leading UK budget hotel chain. *EuroCHRIE 2008 Congress*. Emirates Hotel School, Dubai: the 11th-14th October.

Ergin, E., Parıltı, N. and Özsaçmacı, B. (2011). Impact of loyalty cards on customers store loyalty. *International Business and Economics Research Journal (IBER)*, 6(2), pp. 77–82.

Go.duettocloud.com. (2019). *The ultimate guide to hotel revenue strategy – Duetto*. Retrieved from: http://go.duettocloud.com/revenue-strategy-guide (accessed: the 25th October 2019).

Guadix, J., Cortés, P., Onieva, L. and Muñuzuri, J. (2010). Technology revenue management system for customer groups in hotels. *Journal of Business Research*, 63(5), pp. 519–527.

Hanks, R. D., Cross, R. G. and Noland, R. P. (2002). Discounting in the hotel industry: A new approach. *The Cornell Hotel and Restaurant Administration Quarterly*, 43(4), pp. 94–103.

Internet Marketing Inc. (2019). *How loyalty programs benefit hotels*. Retrieved from: | Retrieved 2 November 2019, from www.internetmarketinginc.com/blog/loyalty-programs-benefit-hotels/ (accessed: the 25th October 2019).

Islam, S. M. N. (2013). Tourism Marketing in Developing Countries: A study of Bangladesh. Hospitality and Tourism Management. *Unpublished PhD Thesis.* Glasgow: University of Strathclyde.

Ivanov, S. (2014). Hotel revenue management: From theory to practice. *International Journal of Contemporary Hospitality Management*, 27(5), pp. 1048–1050.

Ivanov, S. and Zhechev, V. (2011). *Hotel revenue management – a critical literature review.* Retrieved from: www.academia.edu/1923325/Hotel_Revenue_Management_A_Critical_Literature_Review?auto=download (accessed: the 25th October 2019).

Jones, P. (1999). Yield management in UK hotels: A systems analysis. *The Journal of The Operational Research Society*, 50(11), pp. 1111–1119.

Karolefski, J. (2003). Sampling with sizzle; retailers are upgrading their promotions to engage shoppers and create a memorable experience. *Supermarket News:* the 20th January 2003.

Kimes, S. E. (1989). Yield management: A tool for capacity-considered service firms. *Journal of Operations Management*, 8(4), pp. 348–363.

Kimes, S. E. (1999). Implementing restaurant revenue management: A five-step approach. *Cornell Hotel and Restaurant Administration Quarterly*, 34(3), pp. 16–21.

Kimes, S. E. and McGuire, K. A. (2001). Function-space revenue management: A case study from Singapore. *Cornell Hotel and Restaurant Administration Quarterly*, 42(6), pp. 34–46.

Kimes, S. E. and Singh, S. (2009). Spa revenue management. *Cornell Hospitality Quarterly*, 50(1), pp. 82–95.

Kimes, S. E. and Wirtz, J. (2003). Has the revenue management become acceptable? Findings from an international study on the perceived fairness of rate fences. *Journal of Service Research*, 6(2), pp. 125–135.

Koenig, M. and Meissner, J. (2010). List pricing versus dynamic pricing: Impact on the revenue risk. *European Journal of Operational Research*, 204(3), pp. 505–512.

Kostin, B. K. (2018). Assessment of hospitality industry evolution and development in the Russian federation. *Journal of Eastern Europe Research in Business and Economics*, 2018, pp. 1–17.

Laškarin, M. (2013). Development of loyalty programmes in the hotel industry. *Tourism and Hospitality Management*, 19(1), pp. 109–123.

Li, X. and Petrick, J. (2008). Examining the antecedents of brand loyalty from an investment model perspective. *Journal of Travel Research*, 47(1), pp. 25–34.

Noone, B., Kimes, S. and Renaghan, L. (2003). Integrating customer relationship management and revenue management: A hotel perspective. *Journal of Revenue and Pricing Management*, 2(1), pp. 7–21.

Noone, B. and Mount, D. (2008). The effect of price on return intentions: Do satisfaction and reward programme membership matter? *Journal of Revenue and Pricing Management*, 7(4), pp. 357–369.

Okumus, F. (2004). Implementation of yield management practices in service organisations: Empirical findings from a major hotel group. *The Service Industries Journal*, 24(6), pp. 65–89.

Oliver, R. L. (1999). Whence consumer loyalty? *Journal of Marketing*, 63, pp. 33–44.

O'Malley, L. (1998). Can loyalty schemes really build loyalty? *Marketing Intelligence and Planning*, 16(1), pp. 47–55.

Palmer, A. and McMahon-Beattie, U. (2008). Variable pricing through revenue management: A critical evaluation of affective outcomes. *Management Research News*, 31(3), pp. 189–199.

Sigala, M., Lockwood, A. and Jones, P. (2001). Strategic implementation and IT: Gaining competitive advantage from the hotel reservations process. *International Journal of Contemporary Hospitality Management*, 13(7), pp. 364–371.

United Nations Conference on Trade and Development (2010). *United Nations conference on trade and development*. Retrieved from: www.unctad.org/en/docs/cid8_en.pdf (accessed: the 20th October 2019).

Wang, X. L. and Bowie, D. (2009). Revenue management: The impact on business-to-business relationships. *Journal of Services Marketing*, 23(1), pp. 31–41.

Xotels.com (2019). *Home*. Retrieved from: www.xotels.com/images/Revenue-Management-Manual-Xotels.pdf (accessed: the 24th October 2019).

Yeoman, I. and Watson, S. (1997). Yield management: A human activity system. *International Journal of Contemporary Hospitality Management*, 9(2), pp. 80–83.

Yoon, Y. and Uysal, M. (2005). An examination of the effects of motivation and satisfaction on destination loyalty: A structural model. *Tourism Management*, 26(1), pp. 45–56.

Zhang, M. and Bell, P. (2010). Fencing in the context of revenue management. *International Journal of Revenue Management*, 4(1), p. 42.

Part 9

Tourism marketing and green products

16 Environmental marketing

Tourists' purchase behaviour response on green products

Md. Nekmahmud

Introduction

Environmental concerns have been increasing steadily since the early 1970s (Dunlap, 1991; Caldwell, 1991) and increasing attention has been placed on the interface between society and the natural environment, including the importance of environmental consciousness as a factor that can influence human behaviour (Dunlap and Catton, 1979). Nowadays, people like to travel and spend their time with natural beauty. Thus, the tourism industry is rapidly growing and is the largest industry in the modern business world and one of the main international trade categories. Every country has focused on their tourism industry to attract both domestic and foreign tourists. Bangladesh as a developing country faces the challenges of globalization. It has beautiful sights and historical places which are able to get the attention of international tourists. Bangladesh has archaeological, natural, ecological, cultural and other tourism products to attract tourists. Therefore, Bangladesh has ample opportunity to become a tourist nation. Tourism, mostly a service industry, is more labor-oriented than other sectors of production in Bangladesh. Tourism development is seen as a way of improving a country's economy and social wellbeing (Meler and Ham, 2012).

While the travel and tourism sector accounted for 10.4% of global GDP and 9.9% of global employment in 2017, the contributions of this sector in Bangladesh are 4.3% of GDP and 3.8% of total employment (World Travel and Tourism Council [WTTC], 2018). Although this scenario is a positive development compared to the past, the global scenario suggests that the country is yet to realize its full potential. According to the World Tourism Organization (2017), globally, the tourism industry is ranked third in the world for export goods and services, contributing to 7% of the world's exports.

Environmental concern on sustainability and climate change has increased dramatically in the past decade and is affecting the way consumers behave. This change has led to a greater focus on green consumerism, and for the tourism industry, a greater interest in green tourism (Bergin-Seers and Mair, 2009). Green consumerism is one kind of behaviour that exists in different industries and markets. The introduction of green products and green consumerism influenced practitioners and researchers to study consumers' motives behind their green behaviour.

A supplementary strategy focused on making environment-friendly goods more affordable and more proficient and included more financial incentives (e.g. Van Vugt et al., 1995). In addition, motives that are more socially oriented are useful influencers of consumers' propensity to be green (Van Vugt, 2009) as these activate consumers' social, reputational and status-oriented perspectives.

Consumers are conscious of the huge effect that their purchasing behaviour has on the environment (Wahid et al., 2011). Consumers have started to develop an environmental consciousness in every market and industry for example in the restaurant industry (Laroche et al., 2001). In recent years, environmental friendly tourism products and services are the most important issues in the tourist industry in both developed and developing countries. Tourism is a potential industry in Bangladesh. Nowadays, Bangladeshi and foreign tourists are very conscious of purchasing green tourism products. Similarly, environmental friendly and green tourism products and services are very essential features for tourist behaviour response. Green tourism is the phenomenon of people away from their usual habitat in pursuit of leisure activities in the countryside.

There are different types of environmental friendly tourism services: (1) transportation services such as road, air and cruise ships and boats; (2) accommodation such as green hotels, motels; (3) restaurants, bars; (4) entertainment venues; and (5) other hospitality industry services such as resorts, spas and so on. Nevertheless, Bangladesh is facing the challenge to deliver environmental friendly tourist products, new experiences and services for potential local and foreign tourists.

Prior studies have focused on ecotourism attitude and interest, and their influence on ecotourism behaviours which has been empirically well established (Lai and Nepal, 2006; Oviedo-Garcı´a et al., 2016; Singh et al., 2007). However, ecotourism attitude and interest as determinants of behaviour remain essential (Lu et al., 2014), therefore, there is a need for research to focus on their antecedents of consumer pro-environmental behaviour and purchase behaviour of green tourism products and services. Research on green products and services is emerging to develop a green tourism marketing strategy, ecotourism and sustainable tourism to attract environmentalism tourists and foreign tourists for sustainable development.

The chapter highlights a comprehensive conceptual framework that helps to understand tourist's purchasing behaviour response in environment friendly tourism products and services. At the same time, however, it will focus on the main factors of environmental marketing that are influential to tourists' purchase intention of environmental friendly and green tourism services in Bangladesh. Moreover, the chapter addresses the overview and present scenario of environmental friendly and green tourism products and services in developing country of Bangladesh. This finding helps tourist agencies, the tourism industry and marketers to increase in value and be more considerate of tourists' current needs or demands for safer and better environmental friendly products and services.

In this present chapter, the literature review includes reviewing some important theories such as extended theory of planned behaviour (TPB), green tourism products and services, green perceived quality, environment concern, green purchase intention and behaviour response on green products and services. A previous study is presented to fill up the research gap. It attempts to differentiate the present study from past studies in green tourism industry.

Then the methodology deals with the selection of a sample of tourist respondents, methods of data collection, research questionnaire, measurement techniques and processing and analysis of data. The chapter proposes to develop the conceptual framework of environmental attitude, environment concern, green perceived quality, price awareness, purchase intention and behaviour by applying the theory of planned behaviour. Further the chapter attempts to prove in the construction of model and hypotheses development. Finally, the chapter concludes by proving hypothesizes and addressing a set of recommendations with a view to improving environmental friendly and green tourism products and services in Bangladesh.

Literature review

Green tourism products and services

Usually, the terms "green" and "environmental" products commonly mean those that are naturally non-toxic, made from recycled material or lightly packaged (Ottman, 2004). Green products are often seen as safer, healthier and gentler than other traditional products (Luchs et al., 2010). Green foods are those that are fine quality, safe to be consumed, nutritious, concerned with animal welfare and which are produced under the principle of sustainable development (Saleki and Seyedsaleki, 2012). Similarly, green products are locally grown, recycle/reusable, contain natural ingredients, contain recycled content, contain approved chemicals, do not pollute the environment and are not tested on animals (Mishra and Sharma, 2010). Repair, recycle, re-manufacture and reuse of the product (Charter, 1992; Prakash, 2002; Hasan et al., 2019) or part of it to reduce packaging and make products more durable, repairable, compostable, healthy and safe in shipment are the most common production strategies for green products (Mishra and Sharma, 2010).

In general, tourism products include food, accommodation, tours, transportation, recreational activities and historic sites. A green tourism product can be defined by the physical and psychological satisfaction it provides to tourists during their travelling to the destination that is good for health and has no bad impact on environment. Moreover, environmental friendly and green tourism products and services include healthy food, ozone-friendly products, eco-friendly accommodation and transportation and 3R products (refillable, reusable and recyclable). Green tourist products and services focus on facilities and services designed to meet tourist satisfaction with a good environmental impact. In Bangladesh, some of the popular green tourism products include accommodations in green hotels and environmental friendly transportation, villas and specific accommodations for families or small groups.

Theory of planned behaviour

The concept of the theory of planned behaviour (TPB) was proposed by Icek Ajzen (1985) to improve on the predictive power of the theory of reasoned action (Fishbein and Ajzen, 1975). The theory states (see Figure 16.1) that intention toward behaviour, subjective norms and perceived behavioural control together shape an individual's intentions and behaviours (Ajzen, 1991). In the context of tourism, attitude is the total evaluation of the tourism involvement response, which consists of two elements: belief about the likely consequence of tourism participation and values associated with the results. Subject norm is considered as the influence of others in particular on the decision to engage in the individual behaviour (Mohaidin et al., 2017). Perceived control significantly influences the intention to visit a tourist destination (Lam and Hsu, 2006). The TPB is extended by studying variables such as service quality, satisfaction, distraction image, word-of-mouth, knowledge and frequency of past behaviour to explain the green hotel consumers' revisit intentions (Mohaidin et al., 2017) and intention to select sustainable tourism destinations (Han and Kim, 2010). Therefore, to have a better explanation of the tourist's intention to purchase environmental friendly and green tourism products and services, this study attempts to add other variables to the TPB. More specifically, adding a separate environment concern, green perceived quality, awareness of price and safety and health concern components are derived in this chapter context. The TPB provided an alternative model that allows an in-depth understanding of the intention of tourists to purchase environmental friendly and green tourism products and services. Thus, the present model proposes that attitudinal factors (environmental attitude and environment concern, which are related to the attitude of TPB), awareness of

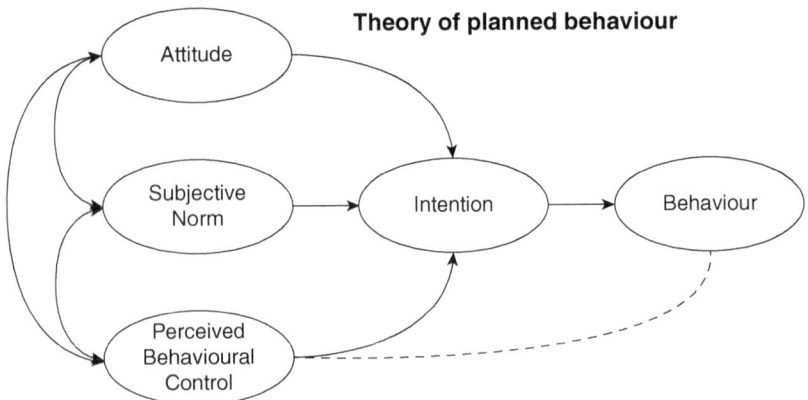

Figure 16.1 Theory of planned behaviour
Source: (Ajzen, 1991)

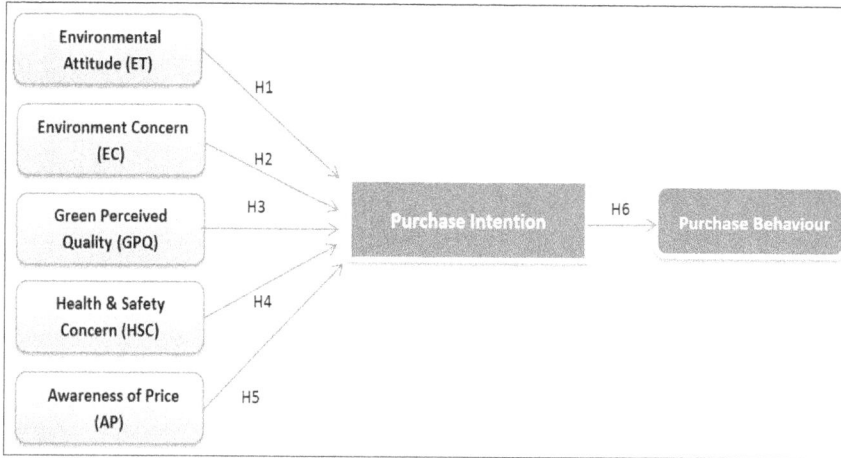

Figure 16.2 Conceptual framework of green purchase behaviour of environmental-
friendly and green tourism products and services

Source: Elaborated by the author

price, health and safety concern and green perceived quality can predict the intention behaviours of purchase of environmental friendly tourist products and services. Similarly, the model also considered the relationship between green purchase intention and tourist purchase behaviour to environmental friendly tourism products and services and it will present an opportunity to provide a greater explanation of the TPB model. The theoretical model for this research is presented in Figure 16.2.

Environmental attitude

Attitude is defined as a "person's degree of favourableness or unfavourableness on a psychological object" (Ajzen and Fishbein, 2000). It is considered as a key determinant that influences individuals to act with more environmentally responsible behaviour (Cottrell, 2003). During tours and travel, it has been verified that people who exhibit positive environmental attitudes will also portray a stronger desire to experience and indulge more with nature (Luo and Deng, 2008; Eagles and Higgins, 1998). Attitudes towards environmental issues are positively correlated to a willingness to purchase (Alwitt and Pitts, 1996; Chen and Chang, 2013). This indicates that stronger attitudes towards environmental issues can affect consumers' buying behaviour. As such, tourism studies showed that tourists positively influenced the behavioural intention of visitors in a different setting

(Chan and Lau, 2002; Cheung et al., 1999; Taylor and Todd, 1995). Hence, the following hypothesis is formed:

H1: *Environmental attitude has a positive influence on green tourism products and services and purchase intention.*

Environmental concern

Consumers who have a concern for the environment are optimistic about the green credentials of products and strongly inclined to purchase those products in order to gain a healthier lifestyle and to remain true to their principles (Agyeman, 2014; Magnier and Crie, 2015; Paul et al., 2016). Environment concern refers to people's awareness of the environmental issues and their willingness and support to resolve them(Hu et al., 2010). It is an important factor which influences purchase intention through its effect on attitude, subjective norms and purchase behaviour control (Chen and Tung, 2014). Hartmann and Apaolaza-Ibáñez (2012) stated that environment concern affects purchase intention directly as well as indirectly through the development of positive attitude toward green energy and environmental concern might be reflected by an increasing intention to purchase green products (Kalafatis et al., 1999).

Diamantopoulos, Schlegelmilch, Sinkovics and Bohlen (2003) mentioned environmental concern as a major factor in the consumer decision-making process. It uses three dimensions (knowledge about green issues, attitudes toward environmental quality and environmental sensitive behaviour).

In the Indian context, Paul et al. (2016) established significant direct and indirect effect of environmental concern on green purchase intentions through the mediation of TPB predictor variables. In a prior study (Arısal and Atalar, 2016), there were significant relations among environmental concern and ecological purchase intention. Environmental consumers are more internally controlled because they believe that an individual consumer can be useful in environmental protection (Gadenne et al., 2011; Hartmann and Apaolaza-Ibáñez, 2012). Therefore, we hypothesize the following:

H2: *Environmental concern has a positive influence on green purchase intention to environmental friendly and green tourism products or services.*

Green perceived quality

Perceived quality is the consumer's judgment about a product or service's overall excellence or superiority compared to alternatives (Zeithaml, 1988). Perceived quality is a significant factor that influences consumers for making purchase decisions (e.g. Zeithaml, 1988; Gutman and Reynolds, 1979) and measures customer satisfaction (Zeithaml et al., 1996; Kim et al., 2008). Green perceived quality is the customer's decision about a brand's overall environmental excellence (Chen and Chang, 2013).

Lee et al. (2007) defined service quality as a set of attributes comprising health and cleanliness, safety and security, facility quality, staff responsiveness and recreation settings. About the effect of perceived service quality on the behavioural intention of tourists, numerous studies on the relationship between service quality, satisfaction and behaviour of individuals have focused on behavioural intention (Baker and Crompton, 2000; Cronin et al., 2000; Yu et al., 2006; Nekmahmud and Rahman, 2018). Therefore, there is a relationship between quality service and tourist satisfaction, which in turn leads to destination loyalty (Hui et al., 2007). Also, quite a few studies in tourism industries have shown that service quality is a precursor factor of behavioural intentions (Li et al., 2011; Aliman and Mohamad, 2013; Ahmed and Azam, 2010). Thus, this chapter proposes that green perceived quality (services) influences tourists' intentions to purchase green tourism products or services. Most consumers believed that the green products had consistent quality, acceptable standard of quality and value for money (Mahesh, 2013). According to the previous studies and literature review, we hypothesize the following:

H3: *Green perceived quality of tourism products or services positively influences intention to purchase environmental products or services.*

Safety and health concerns

Safety and health concerns are defined as consumers' concerned regarding quality of life, health issues and the environment for humans and non-human species (Dunlap and Scarce, 1991; Qader and Zainuddin, 2011). The probability that individuals will be affected by one or more of these areas is an environmental issue (Dunlap and Van Liere, 1978; Mitchell, 1990). According to Wall (1995), safety and health concerns are considered to be the strongest predictor of attitude and behaviour and increasing concern with health and safety are becoming prominent factor in shaping people's attitudes towards the environment. However, Rundmo (1999) performed health behaviour, environmental behaviour as well as consumer behaviour related to purchasing green products. Consumers are very much concerned about food safety, particularly organic foods (Zhang, 2005). Educated and variety-seeking consumers are most likely to purchase organic food products in the future. Thus, safety and health are very vital variables of tourists when they are travelling. During travelling, tourists want to buy healthy environmental friendly products and expect green services, for example green hotels and restaurants. Therefore, we hypothesize the following:

H4: *Safety and health concerns positively influence green purchase intention of environmental friendly and green tourism products and services.*

Awareness of price

Price is the vital factor influencing consumers' purchase decisions. Green pricing can be defined as "setting prices for green products that offset consumers' sensitivity to price against their willingness to pay more for products' environmental performance" (Grove et al., 1996). Previous studies have identified price as a major barrier that keeps consumers from purchasing environmental friendly food (Jolly, 1991; Lockie et al., 2002). Nevertheless, consumers are willing to pay a premium price for green products, but perceived benefits and the product's category also influence the willingness to pay (e.g. Essoussi and Linton, 2010). Some research papers have analysed pricing decisions in the tourism service industry (e.g. Song et al., 2009; Njoya, 2019; He et al., 2019; Sharma and Nayak, 2020). Price is important when selecting hotel accommodation and identifying the price trigger points that would influence the purchasing behaviour of tourists (Lockyer, 2005). According to the previous studies and literature review, we hypothesize the following:

> H5: *Awareness of price positively influences green purchase intention of tourism products and services.*

Purchase intention (PI) and purchase behaviour (PB)

Attitude toward green purchase behaviour has been reported to relate positively to green purchase intention from different countries across a wide range of green products such as green hotels (Han and Yoon, 2015), beverages (Birgelen et al., 2009), organic food products (Zhou et al., 2013) and tourism (Barber et al., 2010).

Manaktola and Jauhari (2007) stated that attitude toward green practices in the lodging industry influences consumers' choice to stay in hotels adopting green practices. Prakash and Pathak (2017) reported positive association between attitude toward eco-friendly packing and intention to purchase products with such packaging. Similarly, Paul et al. (2016) and Yadav and Pathak (2017) demonstrated a positive linkage between attitude toward green products and green purchase intention. On the contrary, Ramayah et al. (2010) failed to find any significant association between attitude toward environmental consequences and green purchase intention. According to TPB, when the behaviour is voluntary in nature, its performance is the result of intention. In the context of green products, Yadav and Pathak (2017) found support for the positive association between behavioural intentions and green buying behaviour. Based on the theoretical framework of TPB and these arguments, we hypothesize the following:

> H6: *Green purchase intention has a positive influence on purchase behaviour of environmental friendly and green tourism products and services.*

Methodology

Participants and procedure

The chapter measures tourists' purchasing behaviour response to environment friendly tourism products and services. Moreover, it also a comprehensive conceptual framework that helps to understand tourists' purchasing behaviour response

Table 16.1 Summary of the socio-demographic profile of respondents

Variables	Frequency	Percent
Gender		
Male	130	64
Female	80	36
Age		
< 20 years	20	9
20–25 years	80	39
25–30 years	45	22
> 30 years	60	30
Level of Education		
Secondary education	15	8
Higher secondary	30	14
Undergraduate	80	39
Masters	75	36
PhD	5	3
Average Monthly Income		
0–10,000 BDT	60	30
10,000–20,000	55	27
20,000–30,000	65	31
> 40,000	25	12
Total Respondents	205	100

to environment friendly tourism products and services. This study is descriptive in nature, conducted based on mixed methods of primary and secondary data. The designated respondent for the chapter includes all tourists who regularly like to travel in new tourist destinations in Bangladesh. Table 16.1 shows the summary of the socio-demographic profile of tourists.

The random sampling method is used to select respondents from Bangladesh. To achieve the chapter goals, the primary data were collected over a 25-day period during January 2020 to February 2020. The online survey method was used to collect primary data by creating Google drive questionnaire interviews sent by email, social media (Facebook and Instagram) and some direct interviews. The secondary data are collected from different sources such as previous scientific articles, books, different related publications, news, reports and websites.

Questionnaire development and instrument

A questionnaire was designed as the major tool of the study. It had two sections: Section one identifies tourist social demographic criteria. Section two includes seven constructs (23 measurement questions) of independent variables (e.g. tourist environmental behaviour, environment concern, green perceived quality, health and safety concerns and awareness of price) and two dependent variables (e.g. purchase intention and purchase behaviour) (see Table 16.2 for the questionnaire constructs) through using of the theory of planned behaviour. Here a 5-point Likert scale was used to measure related questions for both independent variables and dependent variables, where 1 = strongly disagree and 5 = strongly agree.

Data were collected from direct and online interviews through the questionnaire. We distributed 250 questionnaires to tourists to measure purchasing behaviour response in environment friendly tourism products and services in Bangladesh. Among them, 210 respondents returned the completed questionnaires. Because of respondents' inability and excessive missing values, we had to drop 45 questionnaires. The size of the sample stands at 205.

The collected data were analysed by the partial least square–structural equation modelling (PLS-SEM) model which is a variance-based path modelling technique for analysing the structural equation modelling (SEM), measurement model and hypothesis testing by applying SmartPLS software, 3.2.8 version (Ringle et al., 2015). New developments in PLS-SEM are called a fully fledged SEM approach (Henseler et al., 2016; Valaei and Jiroudi, 2016). For the objectives of this study, PLS-SEM is most suitable for explaining how underlying key drivers predict purchase intention (Coelho and Henseler, 2012).

Results and discussion

Partial least squares (PLS)

PLS-SEM handles non-normal data, small sample sizes and uses formative indicators, the most prominent reasons for its application, and also allows the examination of more complex model structures or better handling of data inadequacies such as heterogeneity (Hair et al., 2014).

Measurement model

In the measurement model, the latent constructs in reliability and validity (e.g. internal consistency reliability, convergent validity and discriminant validity) of the construct measures were observed in this stage.

Reliability test

The chapter calculated the reliability which is generally measured via Cronbach's coefficient alpha and t composite reliability to check for internal consistency of the constructs. If the Cronbach's alpha values exceeded the criterion of 0.700, all constructs had no problems in reliabilities (Anderson et al., 2010). Table 16.2 demonstrates that the calculated value of the Cronbach's alpha values for all

Table 16.2 Standardized loadings, Cronbach's alpha, CR, AVE and VIF for the constructs

Items	Determinants	Loading	Cronbach's Alpha	rho_A	Composite Reliability[a]	(AVE)[b]	VIF
	Environmental Attitude		0.889	0.892	0.919	0.695	
ET1	I am concerned about wasting the resources of our planet	0.895					3.187
ET2	I believe that I should be personally involved in the preservation of wildlife and/or nature	0.803					1.967
ET3	All citizens have an obligation to protect and preserve wildlife and nature	0.803					2.009
ET4	I think more needs to be done to educate the general public about the importance of nature and wildlife to our planet	0.878					2.884
ET5	I think of myself as an environmentalist	0.782					1.783
	Environment Concern		0.640	0.674	0.802	0.577	
EC1	I am very concerned about the environment in the tourism sector	0.683					1.306
EC2	I would be willing to reduce my consumption to help protect the environment in the tourism industry	0.837					1.174
EC3	Anti-pollution laws should be enforced more strongly in tourism	0.750					1.377
	Green Perceived Quality		0.797	0.840	0.866	0.620	
GPQ1	I think green tourism products and services have an acceptable standard of quality	0.693					1.480
GPQ2	Green tourism products have consistent quality with respect to environmental concerns	0.719					1.601
GPQ4	Green tourism services in sustainable tourism destinations like nature are reliable	0.893					2.375
GPQ5	Green tourism products are recyclable, reusable and disposable	0.830					2.000
	Safety and Health Concerns		0.679	0.720	0.823	0.612	
SHC1	I think green tourism products and services are becoming safe and secure	0.867					1.618
SHC3	Green tourism products and services make me comfortable when I travelling	0.830					1.504
SHC4	I feel risk free when consuming green tourism products	0.629					1.179

(*Continued*)

Table 16.2 (Continued)

Code	Statement						
	Tourist Awareness of Price		0.734	0.750	0.848	0.651	
AP1	I think that the green tourism products are expensive	0.768					1.382
AP2	I would choose environmentally friendly tourism goods, services, campaigns or companies if the price were the same	0.811					1.600
AP3	If the price of green tourism products was less expensive I'd be willing to change my lifestyle by purchasing green products when I travel	0.840					1.452
	Purchase Intention		0.855	0.876	0.903	0.701	
PI1	I will consider buying green tourism products because they are less polluting	0.920					3.249
PI2	I plan to spend more on environmental friendly products rather than conventional products when I travel	0.822					2.007
PI3	I expect to purchase green tourism products in the future because of their positive environmental contribution	0.725					1.526
PI4	I plan to/ am willing to purchase green tourism products and services when travelling in near future	0.970					2.807
PB	**Purchase Behaviour**		1.000	1.000	1.000	1.000	
PB1	I have green purchasing behaviour for my travelling period	1.000					1.000

a Composite reliability = (square of the summation of the factor loadings) / {(square of the summation of the factor loadings) + (square of the summation of the error variances)}.

b AVE = (summation of the square of the factor loadings)/{(summation of the square of the factor loadings) + (summation of the error variances)}.

constructs surpassed the threshold value of 0.700 except health and safety concerns. This means the data is good and reliable and composite reliability ranges between 1.000 to 0.802 which all surpassed the boundary of 0.70 (Hair et al., 2014), signifying strong reliability among the measures and that the data is free from random error.

Convergent validity

The standardized loadings of all measurement items have been revealed by a bootstrapping analysis of 300 subsamples. In Table 16.2, the convergent validity was accomplished as the factor item loadings go beyond 0.60, the composite reliability exceeds 0.70 and the AVE is above 0.50 (Hair et al., 2014). All were significant (p<.001) with strong confirmation of convergent validity, and the measurement items were well loaded on their own constructs. Fornell and Larcker (1981) mentioned 0.50 is the minimum cut-off value for a reliable construct. After finalizing the analysis, five items (EC4, Major social changes are necessary to protect the natural environment in tourism; GPQ3, Quality of green services in this place is good value for money; SEC2, Green tourism products are good for my health because they are environmental friendly; PB1, I have been purchasing green tourism products; and PB3, I have green purchasing behaviour over the past travelling) were deleted for not meeting the criterion of loading value which is lower than 0.50.

On the other hand, the variance inflation factor (VIF) values of these analyses ranged from 1.000 (purchase behaviour) to 3.249 (purchase intention), which are less than the reference value of 5 (Hair et al., 2017). This indicates the structural mode result has no negative effect and no multicollinearity issues among the items or predictor constructs.

Structural model

Now we will explain the structural model of the PLS data analysis by observing the hypothesis. The endogenous variable of the R2 purchase intention was 0.871 and purchase behaviour was 0.048 which exceeded the minimum level of 10% suggested by Falk and Miller (1992), signifying that 84.8% and 48.0% respectively of the variance in tourist green purchasing intention and behaviour is explained by the independent variables that reflect strong power for the model. Even, the R2 value surpasses 20% is considered high for consumer behaviour studies stated by Rasoolimanesh et al. (2016).

Hair Jr. et al. (2017, p. 156) explained that "the path coefficient will be significant, if the confidence interval does not contain the value zero". Table 16.3 and Figure 16.3 demonstrate the results of the path coefficients and t-values which were itemized as outlined.

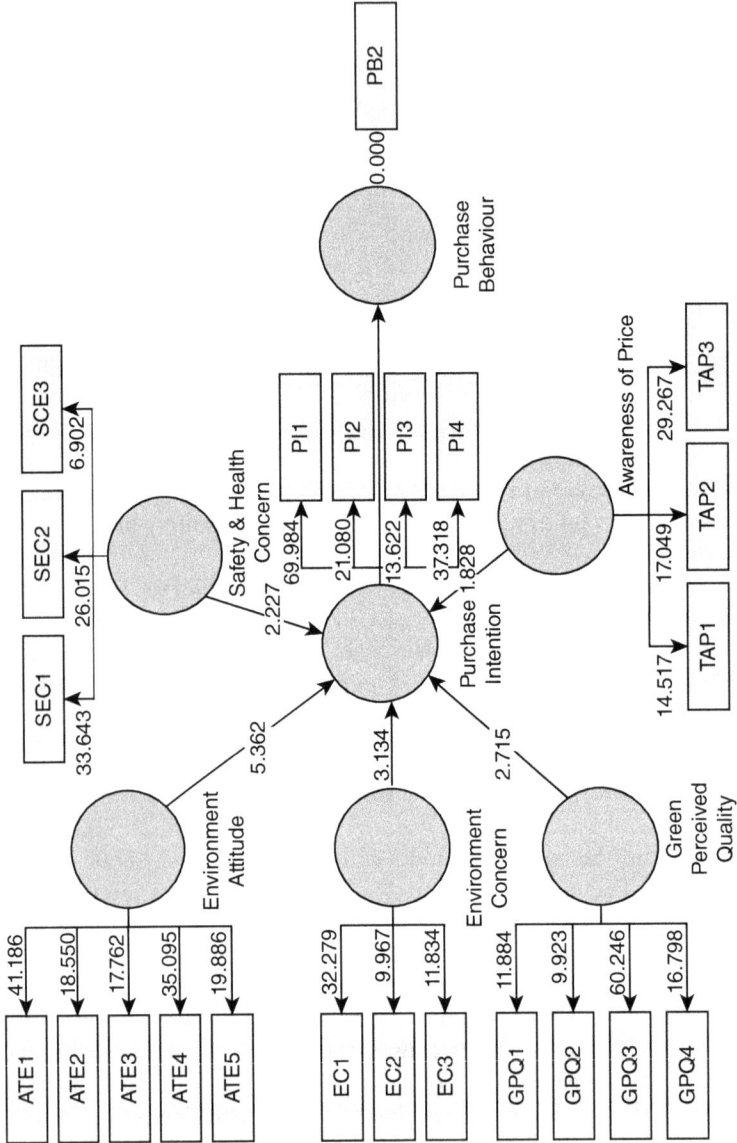

Figure 16.3 Structural model results (i.e. path coefficients with t-values, including level of significance; and R-square values)

The results of the hypotheses, interpretation and discussion

First hypothesis (H1)

The results of the structural equation and path coefficients analysis of the alternative hypothesis H1 as presented in Table 16.3 indicated that environment attitude is seen to have a significantly positive influence on green purchase intention, which is well within expectations ($\beta 1 = 0.376$, bootstrap t-value=5.436, p<0.05). Hence, H1 is therefore supported.

Second hypothesis (H2)

Tourists environment concern has a positive influence and significantly affects consumers' purchase intention on environmental friendly and green products and services in Bangladesh and disclosed a significant result ($\beta 1 = 1.014$; t-value = 3.172; p< 0.05), thus H2 is supported (see Table 16.3).

Third hypothesis (H3)

Tourists' green perceived quality (TPQ) has an insignificant influence on purchase decisions on environmental friendly and green tourism product and services in Bangladesh ($\beta 1 = -0.055$; t-value = 2.839 p > 0.05). Per the evidence, green perceived quality p-value is 0.005, which is lower than the value of 0.05 Thus, H3 is supported signifying that the significant positive relation between TPQ and PI (see Table 16.3).

Fourth hypothesis (H4)

In the alternative hypothesis, H4, tourist awareness of green price has a strongly negative significant influence on environmental-friendly and green tourism products and services in Bangladesh. The significance value for the hypothesis is 0.075, which is higher than the level of significance p ¼ 0.05. The path estimates noted that green awareness of price does indeed have a significant negative relationship with green purchase intention ($\beta 1 = 0.376$, bootstrap t-value = 1.786, p < 0.05), thus H4 is not supported (see Table 16.3).

Fifth hypothesis (H5)

Health and safety concerns have a positive significant influence on purchase intention on environmental-friendly and green tourism products and services and revealed a significant result ($\beta 1 = 1.452$; t-value = 2.204; p < 0.05), thus H5 is supported (see Table 16.3).

Table 16.3 Path coefficient and hypothesis testing

Hypothesized Paths	Mean (M)	Standard Deviation	Bootstrap t value	P Values (2-tailed)	Results/ Decisions
H1 Environment attitude -> Purchase intention	0.107	0.060	5.436	0.000	H1 supported
H2 Environment concern -> Purchase intention	0.206	0.066	3.172	0.003	H2 supported
H3 Green perceived quality -> Purchase intention	0.155	2.839	2.839	0.005	H3 supported
H4 Awareness of price -> Purchase intention	0.107	0.060	1.786	0.075	H4 not supported
H5 Safety and health concerns -> Purchase intention	0.192	0.086	2.204	0.028	H5 supported
H6 Purchase intention -> Purchase behaviour	0.216	0.093	2.347	0.019	H6 supported

Note: For two-tailed tests: * Statistically significant at p<0.05 (for t-value > 1.960).

Endogenous latent construct	Coefficient of determination (R2)	Adjusted R
Purchase intention	0.848	0.840
Purchase behaviour	0.048	0.038

Sixth hypothesis (H6)

The standardized beta coefficients reveal that green purchase intention was seen to significantly affect tourist purchasing behaviour in respect of environmental friendly and green tourism products and services ($\beta 1$ ¼ 0.174, bootstrap t-value =2.347, p < 0.05), thus H6 is supported (see Table 16.3).

Managerial Implications

The findings of this chapter have significant managerial implications. It will support generating new thoughts for researchers, academicians, consultants, policymakers, tourism agencies and marketing people about current tourist behaviour responses on green tourism products and services. Tourists are now more concerned about the environment and they have a positive attitude towards environmental issues. Visitors are now demanding an acceptable standard of eco-friendly products and services, so that they can travel conveniently, healthily and without risk. In general, people are travelling for refreshments, entertainment and wellness. So travellers are always expecting safety, healthy, comfortable tour. Travellers feel that green tourism products and services are good for their health and have a positive impact on the environment. Government and non-government organizations need to provide environmental-friendly tourism products and services from now on. Nevertheless, green tourism services in sustainable tourism destinations like nature are reliable. The consumer believes green tourism products are recyclable, reusable and disposable. Price is an important factor, where tourists are expecting a consistent price. The study also finds improvement areas for tourism investors and sellers for serving better than they did before and helps to achieve remarkable progress. People who are involved in tourism-related business and activities should pay attention to improving the quality and eco-friendly services, ensuring safety and to build a sustainable and environmental tourism industry.

Conclusions and recommendations

This chapter aims to understand tourist purchase intention and purchase behaviour of environmental friendly and green tourism products and services. In Bangladesh, the tourism industry has potential to contribute to GDP. Green tourism products and services could attract both foreign and domestic tourists and also contribute to environmental protection. In this chapter, the theory of planned behaviour (TPB) was explained to be a viable approach. The conceptual framework based on TPB theory improved the understanding of the intention to purchase green tourism products and services. The result of hypotheses about environmental attitude, environment concern, green perceived quality and health and safety concerns have a positive significant relationship with purchase intention to environmental friendly and green tourism products and services in Bangladesh. Nevertheless, Bangladeshi people are price sensitive. In general, the prices of green tourism products and services are higher than traditional services. Price also depends on income so an awareness of price has a negative relation with purchase intention. Finally, the chapter tried to prove the relation of purchase intention of environmental friendly tourist products and services with tourist purchase behaviour which has a negative relation.

Eco-friendly transportation facilities, easy online visas and immigration processes, easy money transfer process, safe and clean sightseeing opportunities, healthy eco-friendly food and accommodation facilities should be ensured by the public and private companies. In general, most consumers respond positively to green ads (e.g. print and television ads) (Huq et al., 2015; Huq et al., 2016) that predict tourists' purchase intention towards eco-friendly tourism products and services. Thus, the tourist agency and marketer should provide information to foreign tourists about environment-friendly tourism products and services by using green marketing tools, such as green branding, green advertising and ecolabels. The tourism policy of Bangladesh 2009 is not related with environmental issues. Environmental-friendly tourism policy is urgent for ensuring the eco-friendly tourism industry. Even, tourism policy and regulation should mention some guidelines to stakeholders on how they can ensure eco-friendly and healthy tourism products and services. The government should make green investment in the tourism sector to attract foreign tourists. Besides, tourist agency and marketing people can contribute to making a sustainable and eco-friendly tourism industry by applying environmental and sustainable marketing activities. To develop the tourism industry, Bangladesh should introduce green and environmental-friendly products and services. At present, a number of hotels and restaurants are now offering environmental-friendly service. High green investment, eco-friendly tourism policy and regulation, collaboration with several stakeholders, environmental marketing activities and green attitude could make the eco-friendly and sustainable tourism industry.

References

Agyeman, C. M. (2014). Consumers' buying behaviour towards green products: An exploratory study. *International Journal of Management Research and Business Strategy*, 3(1), pp. 194–195.

Ahmed, F. and Azam, M. S. (2010). Factors affecting the selection of tour destination in Bangladesh: An empirical analysis. *International Journal of Business and Management*, 5(3), pp. 142–157.

Ajzen, I. (1985). From intentions to actions: A theory of planned behaviour. In J. Kuhl and J. Beckman (eds.), *Action-control: From cognition to behaviour*. Heidelberg: Springer, pp. 11–39.

Ajzen, I. (1991). The theory of planned behaviour. *Organizational Behaviour and Human Decision Processes*, 50(2), pp. 179–211.

Ajzen, I. and Fishbein, M. (2000). Attitudes and the attitude-behaviour relation: Reasoned and automatic processes. *European Review of Social Psychology*, 11(1), pp. 1–33.

Aliman, N. K. and Mohamad, W. (2013). Perceptions of service quality and behavioural intentions: Mediation effect of patient satisfaction in the private health care in Malaysia. *International Journal of Marketing Studies*, 5(4), pp. 15–29.

Alwitt, L. and Pitts, R. (1996). Predicting purchase intensions for an environmentally sensitive product. *Journal of Consumer Psychology*, 5(1), pp. 49–64

Anderson, R. E., Hair, J. F., Black, J. W. C. and Babin, B. J. (2010). *Multivariate data analysis: A global perspective*. London: Pearson Education.

Arısal, İ. and Atalar, T. (2016). The exploring relationships between environmental concern, collectivism and ecological purchase intention. *Procedia-Social and Behavioural Sciences*, 235, pp. 514–521.

Baker, D. A. and Crompton, J. L. (2000). Quality, satisfaction and behavioural intentions. *Annals of Tourism Research*, 27(3), pp. 785–804.

Barber, N., Taylor, D. C. and Deale, C. S. (2010). Wine tourism, environmental concerns, and purchase intention. *Journal of Travel and Tourism Marketing*, 27(2), pp. 146–165.

Bergin-Seers, S. and Mair, J. (2009). Emerging green tourists in Australia: Their behaviours and attitudes. *Tourism and Hospitality Research*, 9(2), pp. 109–119.

Birgelen, M., Semeijn, J. and Keicher, M. (2009). Packaging and pro-environmental consumption behaviour: Investigating purchase and disposal decisions for beverages. *Environmental Behaviour*, 41(1), pp. 125–146.

Caldwell, B. J. (1991). Clarifying popper. *Journal of Economic Literature*, 29(1), pp. 1–33.

Chan, R. Y. and Lau, L. B. (2002). Explaining green purchasing behaviour: A cross-cultural study on American and Chinese consumers. *Journal of International Consumer Marketing*, 14(2/3), pp. 9–40.

Charter, M. (1992). *Greener marketing: A responsible approach to business*. Sheffield: Greenleaf Publishing.

Chen, M. F. and Tung, P. J. (2014). Developing an extended theory of planned behaviour model to predict consumers' intention to visit green hotels. *International Journal of Hospitality Management*, 36, pp. 221–230.

Chen, Y.-S. and Chang, C.-H. (2013). Towards green trust: The influences of green perceived quality, green perceived risk, and green satisfaction. *Management Decision*, 51, pp. 63–82.

Cheung, S. F., Chan, D. K. and Wong, Z. S. (1999). Reexamining the theory of planned behaviour in understanding wastepaper recycling. *Environment and Behaviour*, 31, pp. 587–612.

Coelho, P. S. and Henseler, J. (2012). Creating customer loyalty through service customization. *European Journal of Marketing*, 46(3/4), pp. 331–356.

Cottrell, S. P. (2003). Influence of social demographics and environmental attitudes on general responsible environmental behaviour among recreational boaters. *Environment and Behaviour*, 35(3), pp. 337–347.

Cronin, J., Brady, M. and Hult, G. (2000). Assessing the effects of quality, value, and customer satisfaction on consumer behavioural intentions in service environments. *Journal of Retailing*, 76(2), pp. 193–218.

Diamantopoulos, A. Schlegelmilch, B. B. Sinkovics, R. R. and Bohlen, G. M. (2003). Can socio demographics still play a role in profiling green consumers? A review of the evidence and an empirical investigation. *Journal of Business Research*, 56(6), pp. 465–480.

Dunlap, R. E. and Catton Jr, W. R. (1979). Environmental sociology. *Annual Review of Sociology*, 5(1), pp. 243–273.

Dunlap, R. E. and Scarce, R. (1991). Poll trends: Environmental problems and protection. *The Public Opinion Quarterly*, 55(4), pp. 651–672.

Dunlap, R. E. and Van Liere, K. (1978). The "new environmental paradigm": A proposed measuring instrument for environmental quality. *Social Science Quarterly*, 65, pp. 1013–1028.

Dunlap, T. R. (1991). *Saving America's wildlife*. Princeton, NJ: Princeton University Press.

Eagles, P. F. and Higgins, B. R. (1998). Ecotourism market and industry structure. In K. Lindberg, M. E. Wood, and D. Engldrum (eds.), *Ecotourism: A guide for planners and managers*. Washington, DC: The International Ecotourism Society, pp. 11–43.

Essoussi, L. H. and Linton, J. D. (2010). New or recycled products: How much are consumers willing to pay? *The Journal of Consumer Marketing*, 27(5), pp. 458–468.

Falk, R. F. and Miller, N. B. (1992). *A primer for soft modeling*. Akron, OH: University of Akron Press.

Fishbein, M. and Ajzen, I. (1975). *Belief, attitude, intention, and behaviour: An introduction to theory and research*. Reading, MA: Addison-Wesley.

Fornell, C. and Larcker, D. F. (1981). Structural equation models with unobservable variables and measurement error: Algebra and statistics. *Journal of Marketing Research*, 18(3), pp. 382–388.

Gadenne, D., Sharma, B., Kerr, D. and Smith, T. (2011). The influence of consumers' environmental beliefs and attitudes on energy saving behaviours. *Energy Policy*, 39(12), pp. 7684–7694.

Grove, S. J., Fisk, R. P., Pickett, G. M. and Kangun, N. (1996). Going green in the service sector. *European Journal of Marketing*, 30(5), pp. 56–66.

Gutman, J. and Reynolds, T. J. (1979). An investigation of the levels of cognitive abstraction utilized by consumers in product differentiation. In J. Eighmey (ed.), *Attitude research under the sun*. Chicago: IL: American Marketing Association, pp. 128–150.

Hair Jr, J. F., Sarstedt, M., Hopkins, L. and Kuppelwieser, V. G. (2014). Partial least squares structural equation modeling (PLS-SEM). *European Business Review*, 26, pp. 106–121.

Hair Jr, J. F., Sarstedt, M., Ringle, C. M. and Gudergan, S. P. (2017). *Advanced issues in partial least squares structural equation modeling*. London: SAGE publications.

Han, H. and Kim, Y. (2010). An investigation of green hotel customers' decision formation: Developing an extended model of the theory of planned behavior. *International Journal of Hospitality Management*, 29(4), pp. 659–668.

Han, H. and Yoon, H. J. (2015). Hotel customers' environmentally responsible behavioural intention: Impact of key constructs on decision in green consumerism. *International Journal of Hospitality Management*, 45(1), pp. 22–33.

Hartmann, P. and Apaolaza-Ibáñez, V. (2012). Consumer attitude and purchase intention toward green energy brands: The roles of psychological benefits and environmental concern. *Journal of Business Research*, 65(9), pp. 1254–1263.

Hasan, M. M., Nekmahmud, M., Yajuan, L. and Patwary, M. A. (2019). Green business value chain: A systematic review. *Sustainable Production and Consumption*, 20, pp. 326–339.

He, P., He, Y., Xu, H. and Zhou, L. (2019). Online selling mode choice and pricing in an O2O tourism supply chain considering corporate social responsibility. *Electronic Commerce Research and Applications*, 38, 100894.

Henseler, J., Hubona, G. and Ray, P. A. (2016). Using PLS path modeling in new technology research: Updated guidelines. *Industrial Management and Data Systems*, 116(1), pp. 2–20.

Hu, H. H., Parsa, H. G. and Self, J. (2010). The dynamics of green restaurant patronage. *Cornell Hospitality Quarterly*, 51(3), pp. 344–362.

Hui, T. K., Wan, D. and Ho, A. (2007). Tourists' satisfaction, recommendation and revisiting Singapore. *Tourism Management*, 28(4), pp. 965–975.

Huq, S. M., Alam, S. S., Nekmahmud, M., Aktar, M. S. and Alam, S. S. (2015). Customer's attitude towards mobile advertising in Bangladesh. *International Journal of Business and Economics Research*, 4(6), pp. 281–292.

Huq, S. M., Nekmahmud, M. and Aktar, M. S. (2016). Unethical practices of advertising in Bangladesh: A case study on some selective products. *International Journal of Economics, Finance and Management Sciences*, 4(1), pp. 10–19.

Jolly, D. A. (1991). Differences between buyers and nonbuyers of organic produce and willingness to pay organic price premiums. *Journal of Agribusiness*, 9(1), pp. 1–15.

Kalafatis, S. P., Pollard, M., East, R. and Tsogas, M. H. (1999). Green marketing and Ajzen's theory of planned behaviour: A cross-market examination. *Journal of Consumer Marketing*, 16(5), pp. 441–460.

Kim, C., Zhao, W. and Yang, K. H. (2008). An empirical study on the integrated framework of e-CRM in online shopping: Evaluating the relationships between perceived value, satisfaction, and trust based on customers' perspectives. *Journal of Electronic Commerce in Organizations*, 6(3), pp. 1–4.

Lai, P. H. and Nepal, S. K. (2006). Local perspectives of ecotourism development in Tawushan nature reserve, Taiwan. *Tourism Management*, 27(6), pp. 1117–1129.

Lam, T. and Hsu, C. H. (2006). Predicting behavioral intention of choosing a travel destination. *Tourism Management*, 27(4), pp. 589–599.

Laroche, M., Bergeron, J. and Barbaro-Forleo, G. (2001). Targeting consumers who are willing to pay more for environmentally friendly products. *Journal of Consumer Marketing*, 18(6), pp. 503–520.

Lee, S. Y., Petrick, J. F. and Crompton, J. L. (2007). The roles of quality and intermediary constructs in determining festival attendees' behavioural intention. *Journal of Travel Research*, 45(4), pp. 402–412.

Li, S. J., Huang, Y. Y. and Yang, M. M. (2011). How satisfaction modifies the strength of the influence of perceived service quality on behavioural intentions. *Leadership in Health Services*, 24(2), pp. 91–105.

Lockie, S., Lyons, K., Lawrence, G. and Mummery, K. (2002). Eating 'green': Motivations behind organic food consumption in Australia. *Sociologia ruralis*, 42(1), pp. 23–40.

Lockyer, T. (2005). The perceived importance of price as one hotel selection dimension. *Tourism Management*, 26(4), pp. 529–537.

Lu, A. C. C., Dogan, G. and Del Chiappa, G. (2014). Antecedents of ecotourism intentions and willingness to pay a premium for ecotourism products. *Journal of Travel Research*, 55(2), pp. 1–14.

Luchs, M. G., Naylor, R. W., Irwin, J. R. and Raghunathan, R. (2010). The sustainability liability: Potential negative effects of ethicality on product preference. *Journal of Marketing*, 74(5), pp. 18–31.

Luo, Y. and Deng, J. (2008). The new environmental paradigm and nature-based tourism motivation. *Journal of Travel Research*, 46, pp. 392–402.

Magnier, L. and Crié, D. (2015). Communicating packaging eco-friendliness. *International Journal of Retail & Distribution Management*, 43(4/5), pp. 350–366.

Mahesh, N. (2013). Consumer's perceived value, attitude, and purchase intention of green products. *Management Insight*, 9(1), pp. 37–43.

Manaktola, K. and Jauhari, V. (2007). Exploring consumer attitude and behaviour towards green practices in the lodging industry in India. *International Journal of Contemporary Hospitality Management*, 19(5), pp. 364–377.

Meler, M. and Ham, M. (2012). Green marketing for green tourism. *The 21th Biennial International Congress: Tourism & Hospitality Industry 2012: New Trends in Tourism and Hospitality Management.* Opatija, p. 130.

Mishra, P. and Sharma, P. (2010). Green marketing in India: Emerging opportunities and challenges. *Journal of Engineering, Science and Management Education*, 3(1), pp. 9–14.

Mitchell, V. W. (1990). Defining and measuring the quality of customer service. *Marketing Intelligence and Planning*, 8(6), pp. 11–18.

Mohaidin, Z., Wei, K. T. and Murshid, M. A. (2017). Factors influencing the tourists' intention to select sustainable tourism destination: A case study of Penang, Malaysia. *International Journal of Tourism Cities*, 3(4), pp. 442–465.

Nekmahmud, M. and Rahman, S. (2018). Measuring the competitiveness factors in telecommunication markets. In D. Khajeheian, M. Friedrichsen and W. Mödinger (eds.), *Competitiveness in Emerging Markets.* Cham: Springer, pp. 339–372.

Njoya, E. T. (2019). An analysis of the tourism and wider economic impacts of price-reducing reforms in air transport services in Egypt. *Research in Transportation Economics*, 79, 100795.

Ottman, J. (2004). *Green marketing: Opportunity for innovation.* New York, NY: Booksurge Publishing.

Oviedo-García, M. Á., Vega-Vázquez, M., Castellanos-Verdugo, M. and Reyes-Guizar, L. A. (2016). Tourist satisfaction and the souvenir shopping of domestic tourists: Extended weekends in Spain. *Current Issues in Tourism*, 19(8), pp. 845–860.

Paul, J., Modi, A. and Patel, J. (2016). Predicting green product consumption using theory of planned behaviour and reasoned action. *Journal of Retailing and Consumer Services*, 29(1), pp. 123–134.

Prakash, A. (2002). Green marketing, public policy and managerial strategies. *Business Strategy and the Environment*, 11(5), pp. 285–297.

Prakash, G. and Pathak, P. (2017). Intention to buy eco-friendly packaged products among young consumers of India: A study on developing nation. *Journal of Cleaner Production*, 141(1), pp. 385–393.

Qader, I. K. A. and Zainuddin, Y. B. (2011). The impact of media exposure on intention to purchase green electronic products amongst lecturers. *International Journal of Business and Management*, 6(3), pp. 240–348.

Ramayah, T., Lee, J. W. and Mohamad, O. (2010). Green product purchase intention: Some insights from a developing country. *Resources, Conservation and Recycling*, 54(12), pp. 1419–1427.

Rasoolimanesh, S. M., Dahalan, N. and Jaafar, M. (2016). Tourists' perceived value and satisfaction in a community-based homestay in the Lenggong Valley World Heritage Site. *Journal of Hospitality and Tourism Management*, 26, pp. 72–81.

Ringle, C. M., Wende, S. and Becker, J. M. (2015). *SmartPLS 3.* Boenningstedt: SmartPLS GmbH.

Rundmo, T. (1999). Perceived risk, health and consumer behaviour. *Journal of Risk Research*, 2(3), pp. 187–200.

Saleki, Z. S. and Seyedsaleki, S. M. (2012). The main factors influencing purchase behaviour of organic products in Malaysia. *Interdisciplinary Journal of Contemporary Research in Business*, 4(1), pp. 98–116.

Sharma, P. and Nayak, J. K. (2020). Understanding the determinants and outcomes of internal reference prices in pay-what–you-want (PWYW) pricing in tourism: An analytical approach. *Journal of Hospitality and Tourism Management*, 43, pp. 1–10.

Singh, T., Slotkin, M. H. and Vamosi, A. R. (2007). Attitude towards ecotourism and environmental advocacy: Profiling the dimensions of sustainability. *Journal of Vacation Marketing*, 13(2), pp. 119–134.

Song, H., Yang, S. and Huang, G. Q. (2009). Price interactions between theme park and tour operator. *Tourism Economics*, 15(4), pp. 813–824.

Taylor, S. and Todd, P. (1995). Understanding household garbage reduction behaviour: A test of an integrated model. *Journal of Public Policy and Marketing*, 14(2), pp. 192–204.

Valaei, N. and Jiroudi, S. (2016). Job satisfaction and job performance in the media industry. *Asia Pacific Journal of Marketing and Logistics*, 28(5), pp. 984–1014.

Van Vugt, M. (2009). Averting the tragedy of the commons: Using social psychological science to protect the environment. *Current Directions in Psychological Science*, 18(3), pp. 169–173.

Van Vugt, M., Meertens, R. M. and Van Lange, P. A. (1995). Car versus public transportation? The role of social value orientations in a real-Life social dilemma. *Journal of Applied Social Psychology*, 25(3), pp. 258–278.

Wall, G. (1995). Barriers to individual environmental action: The influence of attitudes and social experiences. *Canadian Review of Sociology/Revue canadienne de sociologie*, 32(4), pp. 465–489.

Wahid, N. A., Rahbar, E. and Shyan, T. S. (2011). Factors influencing the green purchase behaviour of Penang environmental volunteers. *International Business Management*, 5(1), pp. 38–49.

World Travel and Tourism Council (2017). *The economic impact of travel and tourism: Mauritius*. Retrieved from: www.wttc.org/-/media/files/reports/economic-impact-research/countries-2017/mauritius2017 (accessed: the 3rd October 2017).

World Travel and Tourism Council (2018). *Travel and tourism economic impact 2018: Bangladesh*. London: WTTC.

Yadav, R. and Pathak, G. S. (2017). Determinants of consumers green purchase behaviour in a developing nation: Applying and extending the theory of planned behaviour. *Ecological Economics*, 134(1), pp. 114–122.

Yu, C. H., Chang, H. C. and Huang, G. L. (2006). A study of service quality, customer satisfaction and loyalty in Taiwanese leisure industry. *The Journal of American Academy of Business*, 9(1), pp. 126–132.

Zeithaml, V. A. (1988). Consumer perceptions of price, quality, and value: A means-end model and synthesis of evidence. *Journal of Marketing*, 52(3), pp. 2–22.

Zeithaml, V. A., Berry, L. L. and Parasuraman, A. (1996). The behavioural consequences of service quality. *Journal of Marketing*, 60(2), pp. 31–46.

Zhang, X. (2005). Chinese consumers' concerns about food safety: case of Tianjin. *Journal of International Food & Agribusiness Marketing*, 17(1), pp. 57–69.

Zhou, Y., Thøgersen, J., Ruan, Y. and Huang, G. (2013). The moderating role of human values in planned behaviour: The case of Chinese consumers' intention to buy organic food. *Journal of Consumer Marketing*, 30(4), pp. 335–344.

Part 10

Tourism marketing and country image

17 The curious case of Bangladesh and Nepal

Tourism advertising to transform country image and empower developing countries

Imran Hasnat and Elanie Steyn

Introduction

Being categorized as a developing or developed country involves an intricate com-bination of factors "to better understand their social and economic outcomes" (Gbadamosi, 2020). The World Bank, for instance, uses World Development Indicators to "aggregate, group, and compare statistical data of interest" (The World Bank, 2020a). Similarly, the United Nations Development Programme (UNDP) uses the Human Development Index that combines "longetivity [*sic*], education and income" and considers "political freedom and personal security" (Gbadamosi, 2020).

Not only are many developing countries faced with historical and current reali-ties of complex economic, social and political systems but they also have to address issues of "crime, terrorism, political unrest, natural disasters, epidemics and acci-dents" (Avraham and Ketter, 2016). As Avraham and Ketter (2016) also point out, these realities and challenges leave developing countries having to function within systems with "limited effectiveness of public services, safety and security issues, public health issues, inadequate access to technology, poor public educa-tional services and low level of environmental sustainability" (also see Aliaskarov et al., 2017). These realities result in many developing countries having a poor international image, one often shaped around stereotypes (formed by media, the appeals of charity organizations and representation of "us" vs. "them") and the extent to which they deal with these issues (see Hallensleben, 2017).

How others see the world around them, including a country, its people, its government structures and its role in the international arena is the result of both informal and deliberate factors and efforts (Hasnat and Steyn, 2019). Tourism marketing and nation branding are two deliberate efforts governments and deci-sion makers in developing (and developed) countries use to counteract a negative country image or build upon a positive image in the eyes of the international world, especially the eyes of the international media (Avraham, 2017, p. 275).

This chapter focuses on the importance of tourism advertising for developing countries and highlights some benefits as well as challenges tourism advertis-ing brings to a country or destination. It also points out some strategies for developing countries as they implement tourism advertising as a tool to improve their country image. The authors finally apply these elements of tourism advertising to two case studies: The "Beautiful Bangladesh" campaign and the "Visit Nepal Year 2020: Lifetime Experiences" campaign.

The importance of tourism advertising for developing countries

Tourism advertising as a part of integrated destination marketing and branding has become almost an inevitable part of efforts to shape any country's international image (Ferreira, 2019). As Avraham (2017, p. 276) points out, however, these efforts are especially important for developing countries, as governments and other stakeholders realize the need to address an often negative international country image, for a variety of reasons. He indicates that leaders and decision makers in these countries know that "a negative image among world public opinion is an obstacle to the arrival of tourists, companies, and investors, and can lead to sanctions against a country." Similarly, research (e.g. Fullerton and Kendrick, 2013; Kendrick et al., 2015) has shown that positive messages about a country can change an audience's perception about people, government and a country's potential as a tourism destination (see Hasnat and Steyn, 2019).

Chibaya (2013) has found this to be true for Zimbabwe, as it changed its tourism advertising campaign from "Zimbabwe: Africa's Paradise" to "Zimbabwe: A World of Wonders" (p. 90). By taking into account the country's historical and political context, the effect of economic and other events on the country's tourism industry and by involving stakeholders from the government, private and public sectors, the tourism advertising campaign focused on the country's strengths to attract more tourists, investment and infrastructure development.

Other research (e.g. Fullerton et al., 2009; Hasnat and Steyn, 2019) indicates that mediated messages about a destination or a country can "spill over" or "bleed over" to positively affect more than the number of people visiting the country (Anholt, 2006; Rewtrakunphaiboon, 2007). It can help increase the sales and quality of a country's products (Erdem and Sun, 2002), increase and expand the number and scope of organizations and companies in a country and put it in a position to better compete in the international arena. This, in turn, helps developing countries to close the economic and development gap between themselves and other more powerful entities (Anholt, 2006). All in all, these spill-over effects contribute to growth in developing countries' GDP and the export of goods and services as a percentage of GDP, as well as increased direct and indirect employment opportunities for a country's workforce (see Salehi and Farahbakhsh, 2014; Tuhin and Majumder, 2010, p. 287).

All of this is the direct outcome of tourism advertising and a contribution to public diplomacy, a strategy Tuch (1990, p. 3) describes as "a government's process of communicating with foreign publics in an attempt to bring about understanding for its nation's ideas and ideals, its institutions and culture, as well as its national goals and policies." However, as Anholt (2007) indicates, governments and government agencies need to reach visitors and potential visitors with a tailor-made message to try and convince them to either visit a country or change their perception of that country. This is where tourism advertising comes in (Morgan et al., 2011; also see O'Guinn et al., 2012) as a "global communication between nations and travellers of all countries" (Salehi and Farahbakhsh, 2014, p. 124).

Specifically relevant for developing countries and their efforts to reach international audiences through tourism advertising campaigns is that messages take into account the opportunities and challenges of a "dynamic and multidimensional marketplace" (Pike, 2008, p. 268). As a result, these campaigns cannot ignore the political contexts and cultural realities of both the host country and the countries native to those they are trying to reach. As Ferreira (2019) points out, "different destinations have different challenges and opportunities. . . [and] they have different stakeholders too. . . . [Therefore,] it is very difficult to create a universal plan." However, tourism advertising plans should incorporate as many stakeholders as possible and combine a shared vision with specific timelines and outcomes.

In designing and implementing tourism advertising campaigns, role players, especially in developing countries, could take note of the following benefits and challenges related to the process and its outcomes.

Benefits of successful tourism advertising campaigns for developing countries

Al-Masud (2015, p. 14) points out that "developing countries are deprived of benefits derived from tourism." However, research has indicated that developing countries specifically can benefit most from the "socio-cultural, economic and environmental factors" that come with successful tourism advertising and marketing campaigns.

The most obvious benefit of successful tourism advertising campaigns for any country or community is an increase in tourism numbers (see Levine, 2015; Buckley, 2019). However, in addition, there is the potential of new joint ventures between different countries. As highlighted in the Model United Nations International School of the Hague Report (2014), visiting a country can lead to those with technologically advanced expertise or products collaborating with those in a developing country where raw materials are cheaper and labor wages lower. As these collaborations flourish, the international image of the developing country is likely to improve and tourism numbers likely to increase.

As tourism numbers increase, the local economy of the host country can benefit. Not only do visitors spend money on accommodation and food but they also visit local attractions, buy local products and in general boost the local economy. In addition, foreigners buying local products create international awareness of a product as physical products are taken back to the visitors' home countries. This does not only help the individual companies who manufacture and sell those products but it also creates opportunities for growth in employment, empowerment of local communities (e.g. women who earn an income, investment in education and infrastructure) and products leaving the physical borders of the country in which they are manufactured (see Robertson, 2013).

A benefit that comes from the above cycle is that poverty in developing communities could potentially be alleviated (Ali and Chowdhury, 2008, p. 8) through a better economy, more employment, investment of tourism money into infrastructure development and ultimately education. This, in turn, creates an environment of sustainable development and improvement if communities see a steady influx of tourists and tourism expenditure (see Croes and Rivera, 2014).

Many developing countries struggle with the reality of natural resources and the environment being exploited. However as Green (2018) points out, successful tourism advertising campaigns that focus on ecotourism can help developing countries and communities within those countries protect their natural resources and environment. In addition, ecotourism can help create jobs for local communities, empower local inhabitants with skills that they can transfer to other jobs or even assist them to start their own small businesses (Ali and Chowdhury, 2008, p. 13). Green (2018) states that travellers also benefit through these experiences, as they realize "the importance of conserving resources and avoiding waste, . . . [get educated to] live more sustainably at home, and . . . increase their understanding of and sensitivity toward other cultures."

Challenges of successful tourism advertising campaigns for developing countries

While the advantages of increased tourism and exposure through tourism advertising campaigns are plentiful, developing countries and communities should also be cognizant of the challenges increased numbers of tourists bring. The developers and implementers of these campaigns should collaborate and find ways to address the challenges so they can continue to reap the benefits these campaigns bring to their communities.

One of the most general challenges if tourism advertising campaigns are too successful is that, as Simpson (2008) points out, it has the potential of destroying cultures, "undermining social norms and economies, degrading social structures [and] stripping communities of individuality" (p. 1).

Related to this, Azarya (2004) points toward the dangers of marginalization as a result of an influx of tourists. While increased tourism and awareness has the potential to bring more income to a country and an indigenous community, for instance, it also has the potential of "freezing" them to the "peripheries of the world" (Azarya, 2004, p. 961). The uniqueness of these communities is what attracts tourists: their ceremonies, traditional food, art and customs. If not managed properly, these communities can become "tourism exhibits themselves".

In addition, mass tourism can have negative implications on societies and communities that the developers and implementers of tourism advertising campaigns do not always anticipate and that communities cannot always properly accommodate. In Lake Elsinore, California, local authorities responded to "Disneyland size crowds" "inundating" their city to take pictures of "hillsides blanketed in brightly coloured flowers" (Chiu, 2019). Similarly, local residents on Paris' Rue Crémieux "demanded that the city of Paris protect their privacy by closing the street to visitors on evenings and weekends" because of the influx of Instagrammers (O'Sullivan, 2019). Though both examples refer to communities in the developed world, the impact of these occurrences in the developing world might be even more significant. Many cities in developing countries are already feeling the strain of infrastructure and resources that cannot keep up with local demands. Huge influx of tourists to an area as a result of a successful tourism advertising campaigns or a major sporting event like the World Cup or the Olympic Games can make the situation even more dire (Model United Nations International School of The Hague, 2014).

In cases where mismanagement and corruption in developing countries are prevalent, communities stand to be affected by the negative impact of mass tourism even more. Nepali authorities have recently received significant criticism for mismanagement of permits to hikers applying to ascend Mount Everest, following "human traffic jams [and] . . . an aggressive, unruly atmosphere" (Sharma and Gettleman, 2019) that resulted in 11 deaths on Everest in 2019. Media organizations and insurance companies uncovered "a conspiracy by some guides, helicopter companies, teahouse owners and hospitals" to bill insurance companies after evacuating climbers who showed "minor signs of altitude sickness" (Sharma and Gettleman, 2019). This led authorities to announce that they will review the "old laws" and require climbers to provide better proof of their mountaineering experience and health conditions (Both, 2019). The impact of overcrowding Everest became even more apparent when several climbers died in early 2020 (Knowles, 2020).

Strategies developing countries could take into account as they develop tourism advertising campaigns

Baker and Cameron (2008) outline a number of strategies developing countries could keep in mind as they develop tourism advertising campaigns. They emphasize that these campaigns, and the impact thereof, should "satisfy the needs of all these stakeholders as well as target segments . . . it should occur not only on the demand side to increase visitor numbers, but also on the supply side to market the destination" (p. 82). In addition, they recommend that campaigns should "enhance the long-term prosperity of local people," satisfy the needs of visitors, strive to create maximum profitability for local businesses, optimize the multiplier effects that come from these relationships and ensure a "sustainable balance between economic benefits and socio-cultural and environmental costs" (Baker and Cameron, 2008, p. 82).

When discussing specific strategies and key points to keep in mind when developing tourism advertising campaigns, Ferreira (2019) highlights that countries should focus on including "physical items like local attractions, transportation and other facilities, and the infrastructure" in the planning mix. He also emphasizes that countries cannot overlook the role of people in effectively managing tourist activity and other outcomes that stem from implementing the campaigns. Ferreira (2019) states the best way to do this is by treating "residents of the local communities as industry stakeholders and make them aware of the different benefits of tourism." When the residents and communities themselves can benefit from the campaigns and their results, it is even more beneficial.

Loda et al. (2007) advise that when developing and implementing tourism advertising campaigns, stakeholders should keep the following in mind to provide potential visitors with more effective information. They highlight that publicity has a bigger impact than advertising and that the impact of the tourism advertising campaign can be increased by preceding it with a publicity campaign. The timing of publicity, followed by the advertising campaign, necessitates that campaigns should be planned well in advance. Planning and timing are especially important when being committed to involve larger groups of stakeholders in the campaigns and taking into consideration the physical realities of the destination the campaign focuses on (see Al-Masud, 2015, p. 13).

The impact of tourism advertising campaigns can also be increased if stakeholders in and implementers of these campaigns look for campaigns within campaigns or destinations within the bigger country destination. For instance, Al-Masud (2015, pp. 15–17) highlights how a country such as Bangladesh can, within the broader "Beautiful Bangladesh" campaign, benefit from site tourism, business tourism, archaeological tourism, educational tourism, religious tourism, medical tourism and adventure and recreational tourism (including nautical tourism or water tourism) (also see Sarker and Begum, 2013, p. 106). Rewtrakunphaiboon (2009) summarizes this approach as countries having to look for and invent "new destinations to be sold". As such, a country can focus on historical locations, locations with a specific political significance or even locations that are connected with the media and entertainment industries (locations for film shoots, hotels and restaurants featured in films or television programmes, locations included in travel or cooking shows etc.). Especially in cases where locations are featured in media and entertainment, the audience will have a positive connection with them. Tourism advertising campaigns can use this positive destination image to remind potential visitors of this connection, drive more visitors to the area or to focus on the elements that make these locations unique (Rewtrakunphaiboon, 2009).

To summarize, the most optimal strategy that developing countries can take to make their tourism advertising campaigns as effective as possible is to take an integrated approach between government actors, private and public sector role players, local communities and those impacted by the campaigns the most (see Ahmmed, 2013, p. 38). Through this integrated approach and proper planning for designing, implementing and evaluating the campaigns, developing countries can ensure that all benefit from it, not just those in higher positions in government and elsewhere (see Al-Masud, 2015, p. 15; Tuhin and Majumder, 2010, p. 295; Ali and Chowdhury, 2008, p. 1).

The next section of the chapter highlights ways in which two countries in South Asia, Bangladesh and Nepal, have taken advantage of tourism advertising to change their country image and reap the additional benefits just highlighted. It also focuses on some of the previously mentioned strategies these countries have implemented in their "Beautiful Bangladesh" and "Visit Nepal Year 2020: Lifetime Experiences" campaigns.

"Beautiful Bangladesh" tourism advertising campaign

Bangladesh is a country located in a delta of the Bay of Bengal. The roots for Bangladesh's poor international image go back several decades, right after its independence from Pakistan. On the one hand, the newly independent country faced the aftermath of the independence struggle but on the other hand it had to confront and resolve realities of poverty, natural disasters and uncontrolled population growth. These struggles caused then–U.S. Secretary of State Henry Kissinger to refer to the country as the "bottomless basket" of the world (Bari, 2008). The international media reinforced this image from the 1980s onward and only a few years into the 21st century is the country seeing its international image change into a somewhat more positive one (see Abi-Habib, 2018; Tinne, 2013; Islam, 2009).

Changing Bangladesh's image from a struggling, developing country with little to offer (see Sarker and Begum, 2013, p. 104; Tuhin and Majumder, 2010, p. 288) into one that is attracting tourists, showing signs of significant economic growth and technological advancement (The World Bank, 2020c; 2020d), eradicating poverty and making improvements in healthcare (Scholte, 2014; Tinne, 2013) has taken deliberate efforts from various role players (Leung, 2012). One such initiative is launching the country's first national branding campaign through the Bangladesh Tourism Board in 2011. The "Beautiful Bangladesh" campaign promoted the natural beauty, friendly people and tourism attractions of the nation as well as the message that Bangladesh is open for the world, hinting at its renewed secular and liberal foreign policies. The initial campaign, with the slogan "Beautiful Bangladesh: School of Life," ran concurrently with Bangladesh co-hosting the Cricket World Cup with India and Sri Lanka, giving it a higher visibility among international audiences. It continued in 2014, when the Tourism Board commissioned "Beautiful Bangladesh: Land of Stories." The "School of Life" campaign television commercial was widely broadcast during the 2011 International Cricket Council (ICC) Championship, with the original telecast being during the opening ceremony of the tournament on February 17. An international audience in more than 180 countries saw this telecast. It continued to air on different local and international television channels throughout the cricket tournament. It was also featured at several tourism film festivals and won the award for the third best television commercial at the Zagreb Film Festival (Bangladesh Tourism Board, n.d.).

Research shows that actors in the Bangladeshi government and tourism industry seem to have successfully tapped into the benefits of the "Beautiful Bangladesh" tourism advertising campaign and the strategies developing countries can implement as they develop these campaigns (Al-Masud, 2015). As Howlader

(2016) points out, "the present government is paying much importance to tourism. As part of various initiatives for the development of the tourism industry, the prime minister declared 2016 as a Visit Bangladesh Year. The Visit Bangladesh Year will be observed . . . until 2018." The Visit Bangladesh initiative includes members of various communities, including a focus on youth and involvement by the Bangladesh missions abroad. However, critics mention a "lack of pragmatic initiatives" and a lack of focus on international tourists as elements that might hamper its success (see UNB, 2016).

Indicators also show that Bangladesh is already seeing the positive impact of the "Beautiful Bangladesh" campaign as its effects "bleed over" into other areas in the country. This can be seen, for instance, in the country's booming (yet sometimes controversial) ready-made garments industry and the "Made in Bangladesh" brand (see Sherman, 2014; Kamlani, 2013; Haider, 2007). Similarly, Bangladesh sees a growing entrepreneurial ecosystem and organizations such as the Grameen Foundation play a significant role in economically empowering women, poorer communities and disenfranchised groups through micro loans and other support. Tourism numbers and expenditure in Bangladesh are also increasing. The latter has grown from $87 million in 2010 to $357 million in 2018 (The World Bank, 2020c) and tourism's contribution to GDP has increased by one-third between 2011 and 2017 (World Travel & Tourism Council, 2017), contributing to 4.4% of the country's GDP in 2018 (Knoema, 2020). Similarly, the export of goods and services as a percentage of GDP has increased from 16% in 2010 to almost 18% in 2016 (Trading Economics, 2018). However, this percentage did decline to just below 15% in 2018 (The Global Economy, 2020; The World Bank, 2020d).

The impact of the "Beautiful Bangladesh" campaign seems to continue, as the World Economic Forum's Travel and Tourism Competitiveness Report 2019 (pp. 31–33) stated:

> Bangladesh had the world's greatest percentage improvement on its overall TTCI (Travel & Tourism Competitiveness Index) score, helping it move up five spots to rank 120th globally. The country enhanced its safety and security (123rd to 105th), ICT readiness (116th to 111th), T&T prioritization (127th to 121st), (and) price competitiveness (89th to 85th) . . . [indicating] the nation's high potential for upward mobility. . . . [These improvements] are likely to make Bangladesh more conducive for travel.

"Visit Nepal Year 2020: Lifetime Experiences" tourism advertising campaign

Similar to its South Asian neighbour, Bangladesh, Nepal is also in the process of rebuilding its international image. Situated between China and India, Nepal, with an area slightly larger than New York State, is home to just over 30 million people (The World Factbook, 2020). In 2006, the Comprehensive Peace Agreement saw the end of a decade-long conflict in the country that cost it significantly in terms of loss of lives, political instability and subsequent economic impact (The World Bank, 2020b). This started the road toward transformation in the country, beginning with a new Constitution in 2015 and successful elections in 2017, GDP growth of more than 7% in 2019 (following three consecutive years of GDP growth of above 6%), growth in the tourism industry positively impacting the retail, hotel and restaurant sector and an influx in remittance money from abroad (The World Bank, 2020b).

However, this process suffered a significant setback in April 2015 when an earthquake measuring 7.8 on the Richter scale left "9,000 people dead, 22,000 injured and several million people without homes" (Carswell, 2017). Not only did Nepal face food, energy and infrastructure crises following the earthquake (Lorch, 2015), but one of its major contributors to GDP, the tourism industry, suffered a severe blow. International media images showed the reality of people left dead, injured and panicked following the quakes (Taylor, 2015). While the Nepali government tried to assure international travellers that it is safe to travel to the country, more earthquakes struck. This saw the number of tourists to Nepal fall from close to 800,000 in 2014 to just over 500,000 in 2015 (Carswell, 2017).

In an effort to counteract the realities left by the earthquakes, socialtours. com, a responsible tourism company in Nepal (www.socialtours.com/about/) launched the "I Am in Nepal now" social media campaign. This campaign encouraged travellers to Nepal to hold up a placard stating that they are in Nepal now, take a picture and post it to social media (The Kathmandu Post, 2015) and featured both dignitaries and everyday travellers (Lorch, 2015). The Nepal Tourism Board joined forces with socialtours.com (and other tour operators) and set up the Nepal Now website (https://nepalnow.org), featuring facts, figures, news, events, stories and a blog related to tourism happenings in Nepal. This initiative received support from the Centre for the Promotion of Imports (CBI) in the form of training programmes and expert advice (CBI, 2017). From a tourism advertising strategy perspective, this approach speaks to the idea of collaboration, involving different role players and benefiting a variety of entities in a country or community. Not only did this collaboration empower local tour operators through training but it also provided expertise to the Nepal Tourism Board to "independently manage and update the campaign website and their own destination website" (CBI, 2017). As an additional element to this campaign, the Nepal Tourism Board set up the WelcomeNepal website (welcomenepal.com) as a "permanent source of information about Nepal". Following these initiatives, tourism numbers for 2016–2017 were restored to the same levels as before the earthquake (Cuskelly, 2017), reaching close to one million. These numbers increased by 24% between 2017 and 2018 (Thapa, 2020a).

This prompted the Nepali government to set a goal of bringing two million international visitors to Nepal by the end of 2020, resulting in the "Visit Nepal Year 2020: Lifetime Experiences" tourism advertising campaign (Thapa, 2020a; 2020b). In August 2019, the Minister of Culture, Tourism and Civil

Aviation, Hon. Yogesh Bhattarai summarized the focus of the campaign (Visit Nepal, 2020):

> The need at this hour is to improve our products, to design the campaign carefully, to refine our marketing strategies, to develop human resource and to upgrade use of technology. Our success however is incumbent on the synergy of our efforts. We must lay emphasis on sustainable and responsible tourism practices. The campaign initiatives must reflect that we are working towards the highest standards in promoting Nepal as a preferred travel destination, and a country that values and safe guards its environment and heritage. Tourism is a powerful way to advance inclusive economic growth. It creates jobs and provides livelihoods in the remotest corners of our country. It promotes social mobility, develops critical infrastructure, connects economies to global value chains, increases trade and investment, and when done carefully, protects the environment and preserves our cultural heritage. Its transformative impact can drive our country on the path of shared prosperity.

Though the minister's words ring true to some of the benefits of successful tourism advertising highlighted earlier, as well as some of the strategies developing countries could take into account when designing and implementing successful tourism advertising campaigns, commentators warn of the challenges the country will face. As such, Thapa (2020a) points out that "lack of budget, resources and security and rampant corruption . . . [lack of] good air connectivity with major tourist source markets, . . . [and] poor services" (at the Tribhuvan International Airport in Kathmandu) are among some of the hurdles the government needs to address to deal with the potential benefits of a successful tourism advertising campaign. On the flipside, it is estimated that, in preparation for the results of a successful VNY2020 campaign, "about four thousand new hotel rooms of star level will be added in Kathmandu, Chitwan and Pokhara" in 30 new hotels, including five international hotel chains (Thapa, 2020b).

It is too soon to tell whether the campaign will be successful. However, Nepal's previous attempts at implementing tourism advertising campaigns (from the first "Visit Nepal 1998" to "Nepal Tourism Year 2011") have seemingly been successful in increasing tourism numbers, adding to local employment opportunities, growing the local economy and highlighting the need to improve infrastructure (Prasain, 2020).

Conclusion

Given the consequences of planned and unplanned circumstances, many developing countries face the reality of having a poor international image, struggling economies and unstable political conditions. Add to this international media coverage that often does not portray these countries in a positive light, many have to counter a negative country image and a lack of interest from international audiences. Tourism is the fastest-growing industry in the world, not only generating revenue and "cultural wealth" for a country but also driving economic growth and development as it is estimated to provide "20% of total world employment since 2013" (Loss, 2019). Successful tourism advertising campaigns are therefore key initiatives through which a country can transform its country image and empower communities, individuals and organizations.

This chapter highlighted the benefits developing countries especially can gain from tourism advertising campaigns. At the same time, we outlined some of the challenges many of these countries have to overcome in their attempts to reap the benefits. We highlighted some strategies tourism organizations, governments and community stakeholders can implement as they think about developing tourism advertising campaigns that would most benefit their constituencies. Finally, we applied these to two case studies: the "Beautiful Bangladesh" campaign that was launched hand-in-hand with the country hosting a major sports event in 2011 and the "Visit Nepal Year 2020: Lifetime Experiences" campaign being launched at the beginning of this year. We illustrated how the former campaign (in conjunction with other developments and initiatives in the country) has set Bangladesh on a path to promote a more positive country image internationally, strengthen its economy and make major strides in its competitiveness as an international tourist destination. While the "Visit Nepal Year 2020" campaign is still in its infancy and it faces many challenges (the latest being the death of several hikers following an avalanche the Annapurna region in January 2020 and the outbreak of the coronavirus in the same month), previous tourism advertising campaigns in Nepal have contributed to the development of Nepal in positive ways.

While other countries can learn from the challenges Bangladesh and Nepal are facing as they implement their tourism advertising campaigns, they can most certainly also capitalize on the benefits these campaigns have brought to the two small South Asian countries that strive to improve their economies, become more politically stable and promote to the world the depth of their culture and traditions.

References

Abi-Habib, M. (2018). *Violence intensifies as student protests spread in Bangladesh.* Retrieved from: www.nytimes.com/2018/08/06/world/asia/bangladesh-student-protests.html (accessed: the 1st February 2020).

Ahmmed, M. (2013). An analysis on tourism marketing in Bangladesh. *International Proceedings of Economics Development and Research*, 67(8), pp. 35–39.

Ali, M. M. and Chowdhury, S. (2008). Different aspects of tourism marketing strategies with special reference to Bangladesh: An analysis. *Business Review*, 6(1&2), pp. 1–18.

Aliaskarov, D. T., Beisenova, A. S., Irkitbaev, S. N., Atasoy, E. and Wiskulski, T. (2017). Modern changes in Zhezkazgan City: Positive and negative factors of tourism development (Kazakhstan). *GeoJournal of Tourism and Geosites*, 20(2), pp. 243–253.

Al-Masud, T. M. (2015). Tourism marketing in Bangladesh: What, why and how. *Asian Business Review*, 5(1), pp. 13–19.

Anholt, S. (2006). *Brand new justice: How branding places and products can help the developing world.* Amsterdam: Elsevier.

Anholt, S. (2007). *Competitive identity: The new brand management for nations, cities and regions.* London: Palgrave Macmillan.

Avraham, E. (2017). Changing the conversation: How developing countries handle the international media during disasters, conflicts, and tourism crises. *Journal of Information Policy*, 7, pp. 275–297.

Avraham, E. and Ketter, E. (2016). *Tourism marketing for developing countries. Battling stereotypes and crises in Asia, Africa and the Middle East.* London: Palgrave Macmillan.

Azarya, V. (2004). Globalization and international tourism in developing countries: Marginality as a commercial commodity. *Current Sociology*, 52(6), pp. 949–967.

Baker, M. J. and Cameron, E. (2008). Critical success factors in destination marketing. *Tourism and Hospitality Research*, 8(2), pp. 79–97.

Bangladesh Tourism Board. (n.d.) *Promotion and marketing.* Retrieved from: http://tourismboard.gov.bd/activities/promotion-marketing/ (accessed: the 5th February 2020).

Bari, M. R. (2008). The basket case. *Forum*, 3(3). Retrieved from: archive.thedailystar.net/forum/2008/march/basket.htm (accessed: the 5th February 2020).

Both, A. (2019). *Nepal's proposed changes to climbing permits for Mount Everest.* Retrieved from: www.reuters.com/article/us-nepal-everest-factbox/nepals-proposed-changes-to-climbing-permits-for-mount-everest-idUSKBN1YM0DG (accessed: the 1st March 2020).

Buckley, J. (2019). *The X-rated marketing campaign increasing tourism to this little visited country.* Retrieved from: www.cnn.com/travel/article/vilnius-g-spot-europe/index.html (accessed: the 20th February 2020).

Carswell, H. (2017). *How Nepal's tourist industry is bouncing back two years on from devastating earthquake.* 21 April. Retrieved from: www.independent.co.uk/travel/asia/nepal-earthquake-tourist-industry-bouncing-back-a7688611.html (accessed: the 1st March 2020).

Centre for the Promotion of Imports (CBI). (2017). *NepalNOW campaign – Nepal's tourism recovery efforts.* Retrieved from: www.cbi.eu/news/nepalnow/ (accessed: the 20th February 2020).

Chibaya, T. (2013). From "Zimbabwe Africa's paradise to Zimbabwe a world of wonders": Benefits and challenges of rebranding Zimbabwe as a tourist destination. *Developing Country Studies*, 3(5), pp. 84–91.

Chiu, A. (2019). *"Poppy Apocalypse": Small California city overrun by thousands of tourists declares "public safety crisis".* Retrieved from: www.washingtonpost.com/nation/2019/03/18/poppy-apocalypse-small-california-city-overrun-by-thousands-tourists-declares-public-safety-crisis/?utm_term=.1a687ef1b10a (accessed: the 1st February 2020).

Croes, R. and Rivera, M. (2014). *Poverty alleviation through tourism development. A comprehensive and integrated approach.* Palm Bay, FL: Apple Academic Press.

Cuskelly, C. (2017). *Nepal tourism soars to higher levels than before the devastating earthquake.* Retrieved from: www.express.co.uk/travel/articles/798036/Nepal-earthquake-tourism-recovery (accessed: the 1st March 2020).

Erdem, T. and Sun, B. (2002). An empirical investigation of the spillover effects of advertising and sales promotions in umbrella branding. *Journal of Marketing Research*, 39(4), pp. 408–420.

Ferreira, M. (2019). *Role of destination management and destination marketing in tourism*. Retrieved from: https://pragueeventery.com/role-of-destination-management/ (accessed: the 15th February 2020).

Fullerton, J. A. and Kendrick, A. (2013). Strategic uses of mediated public diplomacy: International reaction to U.S. tourism advertising. *American Behavioural Scientist*, X(XX), pp. 1–18.

Fullerton, J. A., Kendrick, A. and Kerr, G. (2009). Australian student reactions to U.S. tourism advertising: A test of advertising as public diplomacy. *Place Branding and Public Diplomacy*, 5(2), pp. 141–150.

Gbadamosi, A. (2020). *Understanding the developed/developing country taxonomy*. Retrieved from: www.a4id.org/policy/understanding-the-developeddeveloping-country-taxonomy/ (accessed: the 1st February 2020).

Green, J. (2018). *Advantages of ecotourism*. Retrieved from: https://traveltips.usatoday.com/advantages-ecotourism-61576.html (accessed: the 2nd March 2020).

Haider, M. Z. (2007). Competitiveness of the Bangladesh ready-made garment industry in major international markets. *Asia-Pacific Trade and Investment Review*, 3(1), pp. 3–27.

Hallensleben, L. (2017). *How the media influences our view of developing countries*. Retrieved from: https://frontier.ac.uk/blog/2017/07/27/how-the-media-influences-our-view-of-developing-countries (accessed: the 1st February 2020).

Hasnat, I. and Steyn, E. (2019). Toward a "Beautiful Bangladesh": The bleed-over effect of tourism advertising. *Place Branding and Public Diplomacy*. Retrieved from: https://doi.org/10.1057/s41254-019-00142-6 (accessed: the 1st February 2020).

Howlader, Z. H. (2016). *Parjatan Corporation takes campaigns to boost tourism*. Retrieve from www.thedailystar.net/business/parjatan-corporation-takes-campaigns-boost-tourism-1289269 (accessed: the 1st February 2020).

Islam, S. (2009). *Tourism marketing in developing countries: A study of Bangladesh*. Retrieved from: http://home.wmin.ac.uk (accessed: the 15th January 2020).

Kamlani, T. (2013, October 11). *Made in Bangladesh*. Retrieved from: www.cbc.ca/fifth/episodes/2013-2014/made-in-bangladesh (accessed: the 15th January2020).

Kendrick, A., Fullerton, J. A. and Broyles, S. J. (2015). Would I go? U.S. citizens react to a Cuban tourism campaign. *Place Branding and Public Diplomacy*, 11(4), pp. 249–262.

Knoema. (2020). *Bangladesh – contribution of travel and tourism to GDP as a share of GDP*. Retrieved from: https://knoema.com/atlas/Bangladesh/topics/Tourism/Travel-and-Tourism-Total-Contribution-to-GDP/Contribution-of-travel-and-tourism-to-GDP-percent-of-GDP (accessed: the 1st March 2020).

Knowles, H. (2020). *Seven still missing, 200 rescued after avalanche hits Himalaya trekkers*. Retrieved from: www.washingtonpost.com/gdpr-consent/?next_url=https%3a%2f%2fwww.washingtonpost.com%2fworld%2f2020%2f01%2f20%2fannapurna-avalanche-nepal%2f (accessed: the 1st March 2020).

Leung, M. (2012). *Seeing Bangladesh in a positive light: Mikey Leung at TEDxDhaka*. Retrieved from: www.youtube.com/watch?v=SvgPxOoLdgU (accessed: the 15th January 2020).

Levine, A. (2015). *Why tourism advertising is more powerful than you think*. Retrieved from: www.forbes.com/sites/andrewlevine2/2015/03/19/why-tourism-advertising-is-more-powerful-than-you-think/#6e72e4a645de (accessed: the 1st February 2020).

Loda, M. D., Norman, W. and Backman, K. F. (2007). Advertising and publicity: Suggested new applications for tourism marketers. *Journal of Travel Research*, 45, pp. 259–265.

Lorch, D. (2015). *"I am in Nepal now" says new tourism campaign, but Nepal is in chaos.* Retrieved from: www.npr.org/sections/goatsandsoda/2015/09/10/439195188/ i-am-in-nepal-now-says-new-tourist-campaign-only-nepal-is-in-chaos (accessed: the 1st March 2020).

Loss, L. (2019). *Tourism has generated 20% of total world employment since 2013.* Retrieved from: www.tourism-review.com/tourism-industry-is-the-pillar-of-economy- news11210 (accessed: the 1st March 2020).

Model United Nations International School of The Hague. (2014). *Special confer- ence I: Security and globalization. Reducing the impact of mass tourism in developing countries.* Retrieved from: www.munish.nl/pages/downloader?code=spc101&com code=spc1&year=2014 (accessed: the 1st February 2020).

Morgan, N., Pritchard, A. and Pride, R. (2011). *Destination brands: Managing place reputation.* New York, NY: Routledge.

O'Guinn, T. C., Allen, C. T. and Semenik, R. J. (2012). *Advertising and integrated brand promotion.* Mason, OH: South-Western Cengage Learning.

O'Sullivan, F. (2019). *Their street is famous on Instagram, and they can't take it anymore.* Retrieved from: www.citylab.com/life/2019/03/rue-cremieux-paris-instagram- tourists-where-to-take-pictures/584164/ (accessed: the 1st February 2020).

Pike, S. (2008). *Destination marketing. An integrated marketing communication approach.* Oxford: Butterworth-Heinemann.

Prasain, S. (2020). *Why Nepal's tourism campaigns have – and haven't – worked.* Retrieved from: https://kathmandupost.com/money/2020/01/02/why-nepal- s-tourism-campaigns-have-and-haven-t-worked (accessed: the 1st February 2020).

Rewtrakunphaiboon, W. (2009). *Film-induced tourism: Inventing a vacation to a location.* Retrieved from: https://go.aws/2Il0zRO (accessed: the 1st February 2020).

Robertson, L. (2013). *How tourism can alleviate poverty.* Retrieved from: www.bbc. com/travel/story/20130320-how-tourism-can-alleviate-poverty (accessed: the 1st March 2020).

Salehi, H. and Farahbakhsh, M. (2014). Tourism advertisement management and effective tools in tourism industry. *International Journal of Geography and Geology*, 3(10), pp. 124–134.

Sarker, M. A. H. and Begum, S. (2013). Marketing strategies for tourism industry in Bangladesh: Emphasize on niche market strategy for attracting foreign tourists. *International Refereed Research Journal*, IV(1), pp. 103–107.

Scholte, M. (2014). *No "basket case".* Retrieved from: www.dandc.eu/en/article/ bangladesh-development-success-not-basket-case (accessed: the 15th January 2020).

Sharma, B. and Gettleman, J. (2019). *Nepals says Everest rules might change after traffic jams and deaths.* Retrieved from: www.nytimes.com/2019/05/29/world/ asia/mount-everest.html (accessed: the 1st March 2020).

Sherman, L. (2014, March 13). *American apparel's creative director explains the "Made in Bangladesh" campaign.* Retrieved from: http://fashionista.com/2014/ 03/american-apparel-made-in-bangladesh-campaign (accessed: the 15th January 2020).

Simpson, M. C. (2008). Community benefit tourism initiatives – a conceptual oxy- moron? *Tourism Management*, 29(1), pp. 1–18.

Taylor, A. (2015). *Nepal after the earthquake.* Retrieved from: www.theatlantic.com/photo/2015/04/nepal-after-the-earthquake/391481/ (accessed: the 1st March 2020).

Thapa, L. B. (2020a). *VNY 2020: Opportunities and challenges.* Retrieved from: https://therisingnepal.org.np/news/34393 (accessed: the 1st March 2020).

Thapa, L. B. (2020b). *Visit Nepal 2020. Opportunities and challenges.* Retrieved from: http://fnwonline.com/visit-nepal-2020-opportunities-and-challenges/ (accessed: the 1st March 2020).

The Global Economy. (2020). *Bangladesh: Exports, percent of GDP.* Retrieved from: www.theglobaleconomy.com/Bangladesh/exports/ (accessed: the 1st March 2020).

The Kathmandu Post. (2015). *"I am in Nepal now" campaign picks up in social media.* Retrieved from: https://kathmandupost.com/miscellaneous/2015/07/07/i-am-in-nepal-now-campaign-picks-up-in-social-media (accessed: the 1st March 2020).

The World Bank. (2020a). *Data.* Retrieved from: https://datahelpdesk.worldbank.org/knowledgebase/articles/378834-how-does-the-world-bank-classify-countries (accessed: the 15th February 2020).

The World Bank. (2020b). *The World Bank in Nepal. Overview.* Retrieved from: www.worldbank.org/en/country/nepal/overview (accessed: the 15th February 2020).

The World Bank. (2020c). *International tourism, receipts (current US$).* Retrieved from: https://data.worldbank.org/indicator/ST.INT.RCPT.CD?locations=BD (accessed: the 1st March 2020).

The World Bank. (2020d). *Export of goods and services (% of GDP).* Retrieved from: https://data.worldbank.org/indicator/NE.EXP.GNFS.ZS (accessed: the 1st March 2020).

The World Factbook. (2020). *South Asia: Nepal.* Retrieved from: www.cia.gov/library/publications/the-world-factbook/geos/np.html (accessed: the 1st March 2020).

Tinne, W. S. (2013). Nation branding: Beautiful Bangladesh. *Asian Business Review*, 2(3), pp. 31–36.

Trading Economics. (2018). *Bangladesh – exports of goods and services (% of GDP).* Retrieved from: www.wttc.org/-/media/files/reports/economic-impact-research/countries-2017/bangladesh2017.pdf (accessed: the 1st February 2020).

Tuch, H. N. (1990). *Communicating with the world: U.S. public diplomacy overseas.* New York, NY: St. Martin's Press.

Tuhin, K. W. and Majumder, T. H. (2010). An appraisal of tourism industry development in Bangladesh. *European Journal of Business and Management*, 3(3), pp. 287–297.

UNB. (2016). *"Solid campaign" needed to attract foreign tourists.* Retrieved from: www.theindependentbd.com/printversion/details/36122 (accessed: the 1st February 2020).

Visit Nepal 2020. (2020). *Nepal.* Retrieved from: https://visitnepal2020.com/wp-content/themes/visitnepaltwenty/assets/pdf/Tabloid_VNY_Aug2019.pdf (accessed: the 1st March 2020).

World Economic Forum. (2019). *The travel & tourism competitiveness report 2019. Travel and tourism at a tipping point.* Retrieved from: http://www3.weforum.org/docs/WEF_TTCR_2019.pdf (accessed: the 1st March 2020).

World Travel & Tourism Council. (2017). *Travel & tourism. Economic impact 2017. Bangladesh.* Retrieved from: www.wttc.org/-/media/files/reports/economic-impact-research/countries-2017/bangladesh2017.pdf (accessed: the 1st February 2020).

Part 11

Future trends, implications and challenges

18 Potentials of tourism products and services in Bangladesh

Azizul Hassan and Haywantee Ramkissoon

Introduction

"Potentials" is a buzzword that relies on many diverse aspects. Potentials must be spotted, well-planned and executed. The successful outcome of potentials carries weight based on valid and justified policy planning and measures to implement. Thus, the theoretical concept of potentials comes to reality in terms of successful implementation for the well-being of society and mankind. Tourism in Bangladesh has undoubted potentials. Currently, the potentials of tourism in Bangladesh are recognized when development efforts are reflected in the development programmes and policies. Bangladesh, as a land of natural beauty and diversified cultures, possesses huge potentials for the development of its tourism industry. Ancient tourists argued that this land has always been able to attract a large number of tourists, priests, traders and wanderers from many different parts of the world. Based on the ancient relics, natural beauty and indomitable hospitality, Bangladesh offers enormous potentials to appear as an amazing tourist destination. This research finds that the country will experience a sharp growth of domestic tourists who will benefit from disposable income and the availability of leisure time (The Financial Express, 2018). There may be great potentials, however, these will require that tourist demands are met. This chapter analytically explains diverse aspects related to tapping the potentials of tourism products and services in Bangladesh.

Tourism in Bangladesh: potentials at the baseline

Every development design needs to have a set goal to reach. In principle, each of the initiatives is oriented towards reaching these goals. The tourism industry of Bangladesh is not the exception. The government of Bangladesh has declared 2021 as the "Year of Tourism" that deserves attention from the relevant policy planners, stakeholders and beneficiaries. The tourism industry of Bangladesh in this way is left to reach the set goals, which are believed to be attainable. The government of Bangladesh has set a clear vision for developing tourism in the country and has been working to attain this vision. The tourism industry of Bangladesh can possibly contribute to achieving the country's vision for 2021 in many different ways.

Lonely Planet, the world's most popular travel guide, placed Bangladesh in its top ten "Best Value" destinations list for 2019 and says, "Bangladesh creates astonishingly few ripples given everything it has to offer" (LonelyPlanet, 2019). Bangladesh as a tourist destination is also advertised in famous tourism and travel outlets. However, these are inadequate. The promotional and advertising activities of Bangladesh in the global media have never been satisfactory leaving a considerable gap in receiving the benefits of the global tourism industry. For decades, Bangladesh has been characterized as a natural calamity stricken and poor nation in the global media. It has not been branded as a country with tourism offerings. This is a significant drawback when the potentials of tourism in Bangladesh are already recognized but the publicity and media coverage remain poor and inadequate. The image of tourism in Bangladesh needs to be appealing in both the local and global media. Also, Dhaka, the capital of Bangladesh, is declared as "Dhaka, the OIC City for Tourism 2019" (Arab News, 2019).

Beyond diverse adversities, the tourism industry in Bangladesh has been developing rapidly. In the last few years, domestic tourism has expanded with good economic benefits. In many countries in Africa and Asia, expanding domestic tourism is often used a poverty alleviation tool (The Economist, 2019). Bangladesh is no exception. Domestic tourism can possibly contribute to overall tourism benefits including poverty alleviation and employment generation. However, there are not as many foreign tourists. The prime reason for the facilitation of domestic tourism is the rise of an affluent middle class in Bangladesh. This social class possesses both leisure time and disposable income. Tour operating businesses in the country also continue to expand mainly due to the involvement of this social class's financial capacities. However, although the country witnesses a sharp growth of domestic tourism, there has always been inadequacy in terms of facility offerings for domestic tourists. Many of the existing and popular tourist attractions in the country still lack adequate tourist facilities and infrastructures, security and safety arrangements and useful communication codes.

In terms of nature-based tourist attractions, Bangladesh has a lot to offer. The most popular tourist attraction in Bangladesh is the Cox's Bazar sea beach. The Sundarbans is the world's largest mangrove forest (Islam et al., 2017). The Chattogram Hill Tracts are next with three districts: Rangamati, Bandarban and Khagrachari. St. Martin Island and Kuakata are also popular. Sylhet comes next with the tea gardens, Hazrat Shahjalal and Shahporan Mazars and many other resources, most of which are still unexplored and less known. Gazipur, nearby the city of Dhaka, is becoming a popular resort destination. Apart from all of these, several destinations in the country, like Netrokona, Sitakunda, Sherpur, Bhola and Barisal, are becoming popular in terms of domestic tourism. The trend of establishing attractive tourist facilities becomes visible in many parts of the country. For example, Paharpur in Naogaon or Sylhet's Bichanakandi have no accommodation facilities for tourists, leaving these tourist attractions' potential quite blurred and questionable.

There have not been any beneficial and tourism-friendly development policies and strategies aimed to develop the tourism industry of the country. The government framed the new Tourism Policy in 2010 (Hassan and Burns, 2014).

It emphasizes the tourism potentials underlining the development of community tourism, eco-tourism, pilgrimage tourism, rural tourism, archaeological tourism and riverine tourism within the cultural and traditional perspectives of Bangladesh. The National Tourism Policy prioritized the tourism industry, being led by the private sector. A tourism board was formed primarily for publicity and marketing abroad. A specialized law known as the Exclusive Tourist Zone and Tourism Protected Area was enacted for attracting foreign investment and sustainable development of the sector (Hassan and Kokkranikal, 2018). Tourist facilities have also been moderately developed. Still, more standard facilities are required for keeping up with the changing tastes and trends.

Bangladesh, a South Asian country, is a democratic republic that has political stability in recent times. This stability actually has resulted in a consistent growth of its GDP, which was 7.86% and 7.28% in financial years 2018–2019 and 2017–2018, respectively (World Bank, 2019). This growth is followed by the socio-economic development and investment in its tourism industry. Travel and tourism accounts for 3.8% of employment and 4.3% of the GDP (World Travel and Tourism Council, 2018).

As shown in Table 18.1, Bangladesh is in the 9th position; this decade, the economy of Bangladesh will have a significant GDP contribution from travel and tourism. In reality, domestic tourists in Bangladesh have disposable income along with sufficient time for tourism and leisure (Honeck and Akhtar, 2014). Also, the government has a commitment to support tourism as a major economic industry.

Bangladesh and neighbouring countries

Bangladesh is a relatively new destination in the global tourism arena with enormous tourism resources and unlimited potentials. Tourism in Bangladesh witnessed slower growth in recent decades. The other neighbouring countries of

Table 18.1 Countries where the contribution of travel and tourism to GDP will grow the fastest from 2019–2029

Fastest Growth	*10-Year Real Growth*
Travel & Tourism GDP	Annualized %
1 Qatar	7.8
2 Myanmar	6.9
3 India	6.8
4 China	6.6
5 Azerbaijan	6.4
6 Anguilla	6.3
7 Uzbekistan	6.2
8 Benin	6.1
9 Bangladesh	6.1
10 Kyrgyzstan	6.1

Source: World Travel and Tourism Council (2019a)

Bangladesh (i.e. India, Nepal and Maldives) are said to have vastly developed their tourism industry infrastructures and turned it to a major source of foreign currency earning. In reality, Bangladesh is far behind (Hassan et al., 2020). However, Bangladesh has a tourism market that definitely has promise for expansion and further development. In terms of tourism service capacities and tourism offerings, Bangladesh always held a solid position. Thus, the availability and abundance of resources for tourism activities are well-versed with the intervention of effective policy planning and implementation. Resources for tourism products and services need to be readily accessible for the rapid and sustained development of the tourism industry. Most of the tourism natural resources of Bangladesh, tangible and non-tangible, could contribute to promoting the destination to neighbouring countries. Sri Lanka, Nepal, Bhutan and Maldives have plans for the rapid and quick formulation and implementation of long-term policies for future tourism development. The government of Bangladesh is keeping pace in this regard to catch up the success of these countries (Dhaka Tribune, 2019).

Tourism resource development and sustainability concerns

One of the most important and crucial aspects of tourism is sustainability, that is, the way to keep intact tourism resources for the next generations (Pulido-Fernández et al., 2019; Ramkissoon and Sowamber, 2018). Examples include ecological destruction in the Sundarbans, St. Martin's Island and many other natural tourist attractions. In the last decade, Cox's Bazar has witnessed a sharp increase in its tourism infrastructure development. However, most of these developments came in an unplanned way. This happened because of the urgency to accommodate the pressure of extra tourists. A similar consequence happened in St. Martin's Island when the influx of mass tourism started leaving drastic effects on the island's fragile environment. Negative effects of mass tourism also took place in other major tourist attractions of the country such as Sylhet's Tamabil, Jaflong and Bichanakandi and the Sundarbans. Future generations might not be able to enjoy the beauties of these tourist attractions. Involvement of local indigenous knowledge and resource adequacy are important for sustainable tourism development in Bangladesh. Sustainability needs to be ensured in almost every initiative geared towards tourism development. Significant challenges need to be dealt with, including environmental sustainability for protecting biodiversity and keeping it unharmed from tourism practices and initiatives (Islam and Shamsuddoha, 2019).

In order to align with its sustainable development goals (SDGs), Bangladesh needs a consistent and reliable flow of foreign currencies. At present, the main sources of foreign currencies are the export of ready-made garments (RMG), the remittances sent by the Bangladeshi diaspora working abroad and the export of a few other commodities (Shahzalal and Hassan, 2019). However, these have never been adequate for meeting the development needs of the country. Thus, the country needs to rely on loans and grants from other countries. The urgency to explore more foreign currency earning avenues is important for halting domestic and external loan increases. In this regard, tourism can be a solid and reliable source of ensuring foreign currency inflow to Bangladesh. The country is adorned with panoramic beauty, there is also the rich cultural Buddhist heritage and natural and cultural heritage (Shabnam et al., 2019). These tourist spots can be an addition to the foreign currency earning platform.

Tourism resources development

Active role of the government

The government of Bangladesh has taken initiatives for the general development of the tourism industry. However, some of them have left positive impacts and some others have not. There is further need for more work on the development initiatives so they can generate more support for tourism development. The government needs to actively move forward with timely and effective plans to capitalize the potentials of the tourism industry. There is a pressing need for governmental agencies to attract new entrepreneurs to invest in the tourism industry of Bangladesh, and perhaps governmental financial organizations need to provide more incentives. New tourism entrepreneurs need to be welcomed to initiate their plans and investments. The government also needs to continue research and monitoring for identifying the drawbacks of the tourism industry and implement policies for stakeholder engagement (Hassan and Burns, 2014; Dewnarain et al., 2019).

Involving the private sector

With the involvement of private sector in the tourism industry, the country has witnessed some encouraging outcomes. Bangladesh has now many five-star hotels and other accommodation, good transportation and other relevant tourist facilities. The country will badly need larger private sector investments to accommodate future tourist demands. Private sector partnerships are being encouraged to promote Bangladesh's tourism industry. This has resulted in the building of tourism resources including motels, hotels and restaurants. This also resulted in a range of benefits for locals in terms of employment and livelihood support. A good number of local youths work as tour guides across Bangladesh and in the Sundarbans and Lawachhara, in particular. With the growth of domestic tourism demand, there is a need for more tourist facilities. Local people in the community will also get the opportunity to sell locally made tourist products and services. World Travel and Tourism Council (2019b) listed Bangladesh in the "Top 20 Countries: Fastest Growing in Terms of T&T GDP", stating that in 2018, travel and tourism GDP growth was 11.6% and by 2029, a total of 741,000 jobs will be created. So far, there has not been a good inflow of direct foreign investments. A strong and effective private-public partnership is yet to be established to encourage and promote tourism enterprises in the country.

Using indigenous knowledge

The locals should be involved in sustainable tourism development (Islam and Carlsen, 2016; Ramkissoon and Sowamber, in press; Nunkoo and Ramkissoon, 2016). Locals can offer accommodation for tourists, creating the chances to interact with the tourists as well as earn a livelihood (Situmorang et al., 2019). This initiative also gives rise to community-based tourism enterprises in terms of tourism promotion and attracting foreign tourists. All of these initiatives can contribute to a better quality of life for the locals (Ramkissoon et al., 2018) and contribute further to positive economic impacts in Bangladesh. Involvement of the locals in tourism promotion promotes domestic tourism, and remote tourist attractions with natural and cultural assets are shown on the map. It is very important that in the process policy guidelines are followed, and public support for tourism development is essential (Megeihi et al., 2020). Following gradual development in the tourism industry, the local people may enjoy better lifestyles. Local people from many remote parts of Bangladesh can showcase their tourism products and services and make a livelihood from tourism (Sowamber and Ramkissoon, 2019).

Considering stakeholders' opinions

Apart from the popular ones, there are a number of new tourist spots in Bangladesh having enormous potentials and largely unexplored (Dhaka Tribune, 2018). One of the reasons for this is the lack of proper development and promotion planning for the tourism industry (Zarei and Ramkissoon, in press). Some of the tourist spots include relatively less-known hills, beaches, forest areas and other places that are becoming more popular mainly due to the rising influence of social media. Still, most of these tourist spots remain unexplored by both foreign and local tourists. There is a need for the government and destination marketers to sustainably promote these unexplored cultural and natural assets (Ramkissoon, 2016; Ramkissoon and Uysal, 2011, 2014). Diverse local tourist groups including adventure-seeking youths are becoming attracted to new tourist spots rather than the conventional ones such as the Sundarbans, tea estates in Sylhet and Cox's Bazar. Proper initiatives need to be taken to develop tourist spots in relatively remote areas including the Nafakhum and Bandarban districts, Remakri, Paddamukh, Thanchi and Amiakhum and Sylhet's Bichhnakandi and Ratargul, along with some other recently popular areas.

The government needs to offer the necessary logistical facilities along with other initiatives to further assist the growth of domestic tourism, as well as promote Bangladesh's tourist offerings to attract international tourists. There are social media groups and private tourist agencies working closely both to market and develop these new and uncommon places and redefining them as vibrant tourism attractions. The coordination between the relevant ministries and departments attached to the tourism sector is essential to represent the government's willingness to work with the private sector for sustainable tourism development (Ramkissoon et al., 2013).

One of the popular tourist magazines in Bangladesh, the "Bangladesh Monitor" believes that the tourism industry in Bangladesh is a rather neglected one. The government has not taken any initiative to prepare a master plan aimed to develop this industry, no effective tourist guidelines have been deployed and there is no concrete plan to target and encourage tourists (The Daily Star, 2013, 2018). No master plan has been developed, which means the authorities are less likely to work coherently to improve the new tourist attractions with immense potentials. Bangladesh Parjatan Corporation (BPC), Bangladesh Tourism Board (BTB) and the Ministry Civil Aviation and Tourism (MOCAT) should work together to identify new and unexplored tourist spots and develop and promote these spots with logistical support. These collective efforts may be able to attract both domestic and overseas tourists as well as enrich the tourism industry. The government encouraged these initiatives in an attempt to make the tourism industry economically viable and attractive. The government appears keen to offer relevant and necessary policy support for this industry to develop (Hassan et al., 2020; Hassan and Kokkranikal, 2018).

Understanding geo-politics: Bay of Bengal Initiative for Multi-Sectoral Technical and Economic Cooperation

According to The Independent (2019) and The New Nation (2019), Bangladesh expects to realize potentials of the tourism industry through the Bay of Bengal Initiative for Multi-Sectoral Technical and Economic Cooperation (BIMSTEC) tourism cooperation. From 30 to 31 August 2018, the 4th Summit of the BIMSTEC Member States was held in Kathmandu, Nepal, where members agreed to take solid initiatives for promoting intra-BIMSTEC tourism. These Member States tasked the relevant authorities to devise strategies to create opportunities based on earlier initiatives. This includes the plan that was adopted earlier, the Plan of Action for Tourism Development and Promotion for the BIMSTEC Region. The BIMSTEC Member States agreed to take concrete measures to facilitate tourism by ensuring the security and safety of tourists supported by smooth transport connectivity. These states insisted in their commitment to develop and promote the Temple Tourist Circuit, the Buddhist Tourist Circuit, eco-tourism, the Ancient Cities Trail and medical tourism.

Understanding geo-politics: focusing on the Asian regional tourism market

Again according to The Independent (2018), Chinese President Xi Jinping visited Bangladesh on 14 October 2016. Bangladesh and China came to an agreement for expanding cooperation and exchanges in education, culture, tourism and other relevant areas. These two countries also agreed to promote interactions between the youths, think tanks, media, non-governmental groups, women's organizations and local authorities. Xi Jinping proposed that China offers huge importance on cross-border tourism under the Belt and Road Initiative (BTI). At present, China generates the largest number of outbound tourists in the world, numbering 144 million per year (China Daily, 2018). Bangladeshi tourism industry insiders believe that the country with its tourism products and resources has the capacity to attract many foreign tourists mostly originating from countries like India, Myanmar and China. These tourists, known as special interest tourists, can enjoy destinations and resources attached to Buddhism. India has been promoting its Buddhist Circuit to attract Chinese tourists. In Myanmar, tourists from China are the second largest group in a few popular tourist destinations like Mandalay and Yangon. These Chinese tourists also visit religious sites in Myanmar including Hpa-An, Bago, Mon State, Pindaya and the ancient cities like Mrauk-U, Bagan and Nabule Ngapal.

Unlocking the potentials of the Buddhist tourism circuit

Bangladesh, as a country with a majority of Muslims, has several traditional sites for religious tourism. The country also attracts pilgrims to its Islamic Holy sites. In line with this current trend, the country has many different sites with importance attached to Buddhism that are largely untapped and niche (Bhandari, 2019). Millions of Buddhists mostly live in the South East Asia, East Asia and Far East countries. The Bangladesh Ministry of Civil Aviation and Tourism, as the responsible ministry, aims to promote sustainable tourism in the country (MOCAT, 2019). To facilitate potential Buddhist tourist flow to the country, the government of Bangladesh has stressed regional cooperation to develop the strategic tourism resources of the country. Examples can be found in the ancient Buddhist heritage sites that scatter across the country from Bogra's Paharpur to Cumilla's Shalban Bihar, and others for which China has initiated development support (China Daily, 2018). The panoramic natural beauty of Bangladesh coupled with the world's largest sea beach, Cox's Bazar, remain the conventional attractions for the tourists. Bangladesh must adequately develop the potential of these tourist attractions, followed by establishing relevant tourist facilities.

In Bangladesh, a good number of people can possibly be engaged both in direct and indirect Buddhist tourism activities. Even though there is also a serious shortage of skilled manpower and human resources in the tourism and hospitality sector, the tourism industry in Bangladesh can contribute to training and development, contribute to poverty alleviation and address unemployment issues in the country. The generation of employment and the reduction of poverty can be made possible by reaching a specific level of tourism development in a country. Tourism in Bangladesh can be a useful and important means to achieve its sustainable development goals (SDGs), and Bangladesh's tourism sector will benefit from both the policy support and effective implementation.

Tourist facilities development

All forms of transportation including road, rail, water and air must be developed to allow the Buddhist tourists to reach their destinations without unexpected hassles (Hasnat and Hasan, 2018). Separate, special training facilities are on the way to connect Bangladesh, India, Myanmar and China. The inland waterways of Bangladesh have immense potential for the development of Buddhist tourism and the services need to be developed adequately for this purpose. Bangladesh as an extended tourism destination possesses enormous potentials for attracting tourists originating from China, Japan, Myanmar and India. These potentials need to be unlocked by both the private sector and the government to allow the tourism industry of Bangladesh to grow. Some major obstacles may need to be removed perhaps by providing on-arrival visa ease through both land and sea ports. A positive role of the government is essential to remove obstacles so that these foreign tourists can consider Bangladesh an extended destination. Private sector tourism enterprises also need to establish wider engagements with India's and Myanmar's tour operators. Also, ocean cruise tourists have created new opportunities that Bangladesh can explore.

Conclusion

The tourism industry in Bangladesh has experienced ongoing development relying on different factors. The country has mostly been able to oversee its tourism industry in a great way, but priority tourism niches require attention. The tourism industry in Bangladesh will thrive if it can capitalize on its potentials. To date, the country is said to have been partly successful in reaping such potentials. The injection of both private and public investments, increasing attention of expatriate and non-resident Bangladeshis to visit their friends and family members, disposable income of the locals and adequate time for leisure activities all support the tourism industry of the country. Bangladesh already has its National Tourism Policy and supporting legislative frameworks aiming to make positive contributions to the tourism industry. These help the tourism industry of Bangladesh to develop. However, the national tourism policy has not been updated on the basis of local and global changing perspectives. The National Tourism Policy needs to be able to spot and address the barriers and challenges for achieving a well-deserved tourism industry. Thus, the policy and complete set of legislative frameworks are required to be both reorganized and modernized. The global tourism industry is highly competitive and demands a stakeholder approach for sustainable tourism development in the country to address not only the country's but also broader societal goals (Ramkissoon and Sowamber, 2012). Bangladesh still requires several short, mid- and long-term projects supported with a considerable amount of budgetary allocation. The tourism industry also needs to be prioritized in the national development agendas, plans and programmes. Drawing on limited existing literature on tourism development in Bangladesh, this chapter aims to encourage future researchers to further explore the country's potential in developing its tourism sector aligning with its sustainable development goals.

References

Arab News (2019). *Bangladesh looks to boost tourism from OIC states.* Retrieved from: www.arabnews.com/node/1524806/world (accessed: the 5th September 2019).

Bhandari, K. (2019). *Tourism* and the geopolitics of *Buddhist* heritage in Nepal. *Annals of Tourism Research*, 75, pp. 58–69.

China Daily (2018). *Ruins of Buddhist site in Bangladesh set to draw tourists.* Retrieved from: https://bit.ly/2mucxRi (accessed: the 5th September 2019).

Dewnarain, S., Ramkissoon, H. and Mavondo, F. (2019). Social customer relationship management: An integrated conceptual framework. *Journal of Hospitality Marketing & Management*, 28(2), pp. 172–188.

Dhaka Tribune (2018). *Experts: Emerging tourist spots remain uncared for, unexplored.* Retrieved from: https://bit.ly/2mlTrN0 (accessed: the 5th September 2019).

Dhaka Tribune (2019). *Does Bangladesh need foreign tourists?* Retrieved from: www.dhakatribune.com/opinion/op-ed/2019/07/09/does-bangladesh-need-foreign-tourists (accessed: the 5th September 2019).

Hasnat, M. M. and Hasan, S. (2018). Identifying tourists and analyzing spatial patterns of their destinations from location-based social media data. *Transportation Research Part C: Emerging Technologies*, 96, pp. 38–54.

Hassan, A. and Burns, P. (2014). Tourism policies of Bangladesh – a contextual analysis. *Tourism Planning & Development*, 11(4), pp. 463–466.

Hassan, A., Kennell, J. and Chaperon, S. (2020). Rhetoric and reality in Bangladesh: Elite stakeholder perceptions of the implementation of tourism policy. *Tourism Recreation Research*.

Hassan, A. and Kokkranikal, J. (2018). Tourism policy planning in Bangladesh: Background and some steps forward. *e-Review of Tourism Research (eRTR)*, 15(1), pp. 79–87.

Honeck, D. and Akhtar, M. S. (2014). *Achieving Bangladesh's tourism potential: Linkages to export diversification, employment generation and the 'Green Economy'.* Retrieved from: www.econstor.eu/bitstream/10419/104756/1/798048549.pdf (accessed: the 5th September 2019).

Islam, F. and Carlsen, J. (2016). Indigenous communities, tourism development and extreme poverty alleviation in rural Bangladesh. *Tourism Economics*, 22(3), pp. 645–654.

Islam, M. M. and Shamsuddoha, M. (2019). Coastal and marine conservation strategy for Bangladesh in the context of achieving blue growth and sustainable development goals (SDGs). *Environmental Science & Policy*, 87, pp. 45–54.

Islam, S., Hossain, M. K. and Noor, M. (2017). Determining drivers of destination attractiveness: The case of nature-Based tourism of Bangladesh. *International Journal of Marketing Studies*, 9(3), pp. 10–23.

LonelyPlanet (2019). *Best in travel: 2019 best value.* Retrieved from: www.lonelyplanet.com/best-in-travel/value (accessed: the 5th September 2019).

Megeihi, H., Woosnam, K., Rebeiro, A., Ramkissoon, H. and Denley, T. (2020). Employing a value-belief-norm framework to gauge Carthage residents' intentions to support sustainable cultural heritage tourism. *Journal of Sustainable Tourism* https://doi.org/10.1080/09669582.2020.1738444

Ministry of Civil Aviation and Tourism (2019). *Home.* Retrieved from: https://bit.ly/2muSnXv (accessed: the 31st December 2019).

Nunkoo, R. and Ramkissoon, H. (2016). Stakeholders' views of enclave tourism: A grounded theory approach. *Journal of Hospitality & Tourism Research*, 40(5), pp. 557–558.

Pulido-Fernández, J. I., Cárdenas-García, P. J. and Espinosa-Pulido, J. A. (2019). Does environmental sustainability contribute to tourism growth? An analysis at the country level. *Journal of Cleaner Production*, 213, pp. 309–319.

Ramkissoon, H. (2016). Place satisfaction, place attachment and quality of life: Development of a conceptual framework for island destinations. In P. Modica and M. Uysal (eds.), *Sustainable island tourism: Competitiveness and quality of life*. Oxfordshire: CABI, pp. 106–116.

Ramkissoon, H., Mavondo, F. and Uysal, M. (2018). Social involvement and park citizenship as moderators for quality-of-life in a national park. *Journal of Sustainable Tourism*, 26, pp. 341–361.

Ramkissoon, H., Smith, L. and Weiler, B. (2013). Relationship between place attachment, place satisfaction, and pro-environmental behaviour in an Australian National Park. *Journal of Sustainable Tourism*, 21(3), pp. 434–457.

Ramkissoon, H. and Sowamber, V. (2018). Environmentally and financially sustainable tourism, ICHRIE Research report, pp. 1–4, ICHRIE Research reports, Translating Research Implications: Industry's commentary.

Ramkissoon, H. and Sowamber, V. (in press). Local community support in tourism in Mauritius. Ray of light by LUX*. In M. Novelli, E. Adu-Ampong and A. Ribeiro (eds.), *Routledge handbook of tourism in Africa*.

Ramkissoon, H. and Uysal, M. S. (2011). The effects of perceived authenticity, information search behaviour, motivation and destination imagery on cultural behavioural intentions of tourists. *Current Issues in Tourism*, 14(6), pp. 537–562.

Ramkissoon, H. and Uysal, M. S. (2014). Authenticity as a value co-creator of tourism experiences. In N. K. Prebensen, J. S. Chen and M. Uysal (eds.), *Creating experience value in tourism*. Wallingford: CABI, pp. 113–124.

Shabnam, S., Ramkissoon, H. and Choudhury, A. (2019). Role of ethnic cultural events to build an authentic destination image: A case of 'Pohela Boishakh' in Bangladesh. In A. Hassan and A. Sharma (eds.), *Tourism events in Asia: Marketing and development*. Oxon: Routledge, pp. 47–63.

Shahzalal, M. and Hassan, A. (2019). Communicating sustainability: Using community media to influence rural people's intention to adopt sustainable behaviour. *Sustainability*, 11(3), pp. 1–28.

Situmorang, R., Trilaksono, T. and Japutra, A. (2019). Friend or Foe? The complex relationship between indigenous people and policymakers regarding rural tourism in Indonesia. *Journal of Hospitality and Tourism Management*, 39, pp. 20–29.

Sowamber, V. and Ramkissoon, H. R. (2019). Sustainable tourism as a catalyst for positive environmental change: The case of LUX* Resorts & Hotels. In D. Gursoy and R. Nunkoo (eds.), *The Routledge handbook of tourism impacts: Theoretical and applied perspectives*. Oxon: Routledge, pp. 338–349.

The Daily Star. (2013). *The great potential of tourism*. Retrieved from: www.thedaily star.net/news/the-great-potential-of-tourism (accessed: the 5th September 2019).

The Daily Star. (2018). *Tourism: Potential remains untapped*. Retrieved from: https://bit.ly/2muSnXv (accessed: the 5th September 2019).

The Economist. (2019). *Who needs foreigners? Domestic travellers have revived Bangladesh's tourism industry*. Retrieved from: https://econ.st/2kM6Vl2 (accessed: the 5th September 2019).

The Financial Express. (2018). *Prospects for domestic tourism*. Retrieved from: https://bit.ly/35JExT0 (accessed: the 5th September 2019).

The Independent. (2018). *Bangladesh can become 'extended destination for global tourists'*. Retrieved from: https://bit.ly/2m1AbnY (accessed: the 5th September 2019).

The Independent. (2019). *Bangladesh has potentials to be a destination for global tourists*. Retrieved from: www.theindependentbd.com/post/193275 (accessed: the 5th September 2019).

The New Nation. (2019). *Huge potential for tourism in Bangladesh: India, Nepal, Bhutan, Sri Lanka and Maldives implementing long-term policy.* Retrieved from: https://bit.ly/2m0a3tM (accessed: the 5th September 2019).

World Bank (2019). *GDP growth.* Retrieved from: https://data. worldbank.org/indicator/ny.gdp.mktp.kd.zg?locations= bdandyear_high_desc=false (accessed: the 5th September 2019).

World Travel and Tourism Council. (2018). *Economic impact 2018: Bangladesh.* London: World Travel and Tourism Council.

World Travel and Tourism Council. (2019a). *Travel and tourism global impact 2019.* Retrieved from: file:///C:/Users/USER/Downloads/Global%20Economic%20Impact%20Trends%202019.pdf (accessed: the 31st December 2019).

World Travel and Tourism Council. (2019b). *Economic impact-2019.* Retrieved from: https://bit.ly/2m7L3k1 (accessed: the 5th September 2019).

Zarei, A. and Ramkissoon, H. (in press). Sport tourists' preferred event attributes and motives: A case of Sepak Takraw, Malaysia. *Journal of Tourism & Hospitality Research.*

Index

For Product Safety Concerns and Information please contact our EU
representative GPSR@taylorandfrancis.com
Taylor & Francis Verlag GmbH, Kaufingerstraße 24, 80331 München, Germany

www.ingramcontent.com/pod-product-compliance
Lightning Source LLC
Chambersburg PA
CBHW060803220326
41598CB00022B/2524